Regenerative Medicine, Artificial Cells and Nanomedicine – Vol. 1

ARTIFICIAL CELLS

Biotechnology,

Nanomedicine,

Regenerative Medicine,

Blood Substitutes,

Bioencapsulation,

and Cell/Stem Cell Therapy

Regenerative Medicine, Artificial Cells and Nanomedicine – Vol. 1

ARTIFICIAL CELLS

Biotechnology,
Nanomedicine,
Regenerative Medicine,
Blood Substitutes,
Bioencapsulation,
and Cell/Stem Cell Therapy

Thomas Ming Swi Chang
McGill University, Canada

World Scientific

NEW JERSEY · LONDON · SINGAPORE · BEIJING · SHANGHAI · HONG KONG · TAIPEI · CHENNAI

Published by

World Scientific Publishing Co. Pte. Ltd.

5 Toh Tuck Link, Singapore 596224

USA office: 27 Warren Street, Suite 401-402, Hackensack, NJ 07601

UK office: 57 Shelton Street, Covent Garden, London WC2H 9HE

Library of Congress Cataloging-in-Publication Data
Chang, Thomas Ming Swi.
    Artificial cells : biotechnology, nanomedicine, regenerative medicine, blood substitutes,
bioencapsulation, cell/stem cell therapy / by Thomas Ming Swi Chang.
      p. ; cm. -- (Regenerative medicine, artificial cells and nanomedicine ; v. 1)
    Includes bibliographical references and index.
    ISBN-13: 978-981-270-576-1 (hardcover) -- ISBN-10: 981-270-576-7 (hardcover)
    ISBN-13: 978-981-270-778-9 (pbk.) -- ISBN-10: 981-270-778-6 (pbk.)
    1. Artificial cells. 2. Blood substitutes. 3. Nanotechnology. I. Title. II. Series.
    [DNLM: 1. Blood Substitutes. 2. Biotechnology. 3. Cells. 4. Nanomedicine.
5. Regenerative Medicine. WH 450 C456a 2007]
    RM171.7.C478 2007
    615'.39—dc22

                                         2007013738

British Library Cataloguing-in-Publication Data
A catalogue record for this book is available from the British Library.

Typeset by Stallion Press
Email: enquiries@stallionpress.com

Printed by Fulsland Offset Printing (S) Pte Ltd, Singapore

Dedication

My wife, Lancy, for her selfless dedication, moral support and encouragement throughout my 50 years of research on artificial cells.

My children, Harvey, Victor, Christine and Sandra.

My grandchildren, Josh, Jared, Micaela, Matthew, Emma and Katya.

Acknowledgements

This research would not have been possible without access to the vast amount of the scientific knowledge gathered by numerous investigators throughout the centuries around the world.

More immediately, there are many people who have at one time or another given invaluable encouragement, support or advice. It is not possible to acknowledge all, except to mention a few early examples when this work was at its very beginning. Sir Arnold Burgen, F.R.S., while at McGill, has through his teaching and discussions initiated my deep interest in the area of membrane and cell physiology. In 1956, he supported my request to Professor F. C. MacIntosh, F.R.S., then Chairman of the Department of Physiology, to move my research from my dormitory room to a corner of the teaching laboratory in the department. Professor MacIntosh has, in addition, given me much encouragement and support at the beginning. In 1962, Professor S. G. Mason, was interested in using my 1957 artificial cells for some of his rheology experiments and invited me to his laboratory for six months. I am thankful to the Canadian biomedical scientists for their open enthusiasm and standing ovation when I first presented this study at the Annual Conference of the Canadian Federation of Biological Societies in 1963. This and the acceptance of my paper as the sole author by *Science* in 1964, suggested to me that this research might have some future potential for use in medicine. This played an important role in my decision to continue with the research rather than to go into clinical practice. Professor D. V. Bates, then Chairman of the Department of Physiology, for his support and help in greatly expanding my research facilities. Professor J. Beck, Physician-in-Chief of the Royal Victoria Hospital, for his support in my first clinical trials on hemoperfusion and the nephrologists at the McGill Teaching hospitals for referring their patients to me to carry out the hemoperfusion procedures — especially Dr. A. Gonda, Dr. Paul Barre, Dr. M. Levy, and Dr. J Dirks. My uncle, Professor Robert SM Chang, especially for his advice on academic "protocols." Professor S. Freedman, while dean of medicine, recognized this research effort as the Artificial Cells and Organs Research Centre at McGill University. My sincere appreciation also goes to Professor W. Kolff, Professor A. Burton, Professor O. Denstedt, and many others for their interests, encouragement, and support especially in the early phase of this research. This included professors, chairmen of departments, deans of faculties, vice-principals and principals at McGill University and the many colleagues around the world.

This research would not have been possible without my wife, Lancy. Through her cheerful acceptance she has made it possible for me to combine this research with the heavy load of a medical student in the early years. Since 1958, in addition to her moral support, she has looked after everything for me at home and even in my office so as to allow me to concentrate completely and fully on my research and other academic activities. She has also contributed extensively to this monograph by carrying out the very time consuming and laborious task of inputting, typing, arranging and editing the very large reference section.

It has been a pleasure to work in my laboratory with the many past and present graduate students, research fellows, research assistants, research associates and others, especially (alphabetically): Barre, P, Bourget, L, Bruni, S, Cameron, D, Campbell J, Cattaneo, M, Chang, H, Chang, V, Chawla, AS, Chirito, E, Chow, KM, Chuang, S, Coffey, JF, Cole, C, Coromili, V, Cousineau, J, D'Agnillo, F, Daka, J, Dixit, V, Ergan, F, Espinosa-Melendez, E, Francoeur, TE, Fustier, C, Garofalo, F, Grunwald, J, Grunwald, J, Gu, JS, Gu, KF, Hertzman, CM, Hewish, M, Holeczek, K, Ilan, E, Ito, Y, Johnson, L, Kashani, SA, Keipert, PE, Khanna, R, Kuntarian, N, Kuruvilla, S, Lee Burns, T, Lesser, B, Lister, C, Liu, ZC, Lloyd-George, I, Lo, KS, Malave, N, Malouf, C, Migchelsen, M, Mobed, M, Mohsini, K, Morley, D, Morton, P, Nalchesky, P, Nasielski, P, Ning, J, Nishiya, T, O'Keefe, P, Piskin, E, Pont, A, Powanda, D,

Poznansky, M, Prakash, S, Prichard, S, Quebec, EA, Ransome, O, Razack, S, Reiter, B, Resurrección, E, Rodrigue, A, Rong, J, Rosenthal, AM, Safos, S, Sarkissian, CN, Shi, ZQ, Shu, CD, Sipehia, R, Siu-Chong, E, Sorrini, P, Stark, A, Stefanescu, A, Tabata, Y, Tani, T, Taroy, E, Varma, R, Verzosa, A, Wahl, HP, Walter, S, Watson W, Wong, H, Wong, N, Yu, BL, Yu, WP, Yu, YT, Yuan, ZY, Zheng, WH, Zhou, MX, Zolotareva, E and others.

I gratefully acknowledge the financial support as follows: My parents in the first summer of 1957. The Faculty of Medicine at McGill for support throughout the next three summers. When this research was carried out on a full time basis from 1962, the Medical Research Council of Canada for the research support and for the awards, respectively, as Medical Research Fellow (1962–1965), Medical Research Council Scholar (1965–1968), and MRC Career Investigator from 1968 until the end of the career investigator program in 1999. After the MRC became the new Canadian Institutes of Health Research, it continued with ongoing research supports. When the Quebec Ministry of Education, Science and Technology started the new "Virage Centre of Excellence in High Technology" program, they awarded me one of the first centers. The Bayer/Canadian Red Cross Society/Medical Research Council of Canada Joint Funds program had supported my research on blood substitutes even before the H.I.V. crisis in donor blood. The more recent Quebec Ministry of Health's Hemovigillance and Transfusion Medicine Program has awarded me its first Research Group (d'equipe) to work on "Biomedical Research on Blood Substitutes in Transfusion Medicine." In addition to the above major long-term research supports, there were also smaller and shorter term research supports from other agencies and organizations over the last many years.

It has been a pleasure to accept the invitation of World Scientific Publishing to prepare this monograph. Their scientific editor, Ms. SC Lim, has been most helpful throughout the preparation of this monograph.

Contents

Dedication   v

Acknowledgements   vii

Chapter 1. 50th Anniversary of Artificial Cells   1

1.1.   Background   1
1.2.   Starting with Artificial Red Blood Cells   2
1.3.   Further Research and the First Routine Clinical Use of Artificial
       Cells   5
1.4.   Importance of Progress in Parallel Areas of Biotechnology,
       Molecular Biology, and Regenerative Medicine   7
1.5.   Historical Milestones   7

Chapter 2. Basic Principles   12

2.1.   Basic Features of Artificial Cells   12
       Basic features of early artificial cells   12
       Present status of the basic features of artificial cells of macro,
       micron, nano and molecular dimensions   14
2.2.   Nanotechnology and Nanobiotechnology   15
       Nanobiotechnology and artificial cells   15
       Nanobiotechnology and the assembling of hemoglobin
       with enzymes that remove oxygen radicals   17
       Nanobiotechnology for the assembling of hemoglobin
       with other enzymes   17

      Conjugation of polymer with proteins 18
2.3.   Cell Homogenate, Organelle, Enzymes and
      Multienzyme Systems 18
      Cell homogenate 18
      Enzymes and enzyme therapy 19
      Multienzyme systems with cofactor recycling 20
2.4.   Artificial Cells Containing Intracelluar Compartments 21
2.5.   Artificial Cells Containing Biologics and Magnetic Material . . 22
2.6.   Cells, Islets, Stem Cells, Genetically-engineered Cells and
      Microorganisms 22
      Artificial cells containing cells 22
      Stem cells 24
      Genetically-engineered cells 24
2.7.   Artificial Cells Containing Bioadsorbents 25
2.8.   Research on Membrane Model Systems 27
2.9.   Cell Physiology 28
2.10. Drug Delivery 28
      Polymeric semipermeable microcapsules 28
      Biodegradable polymeric artificial cells, nanoparticles,
      nanocapsules 29
      Liposomes — lipid membrane artificial cells 29
2.11. Other Systems 30

Chapter 3.   Oxygen Carriers Based on Nanobiotechnology   **31**

3.1.   Introduction 31
3.2.   Hemoglobin as Oxygen Carrier 32
3.3.   Modified Hb 32
      PolyHb based on nanobiotechnology (Fig. 3.1) 33
      Conjugated Hb (Fig. 3.1) 34
      Intramolecularly crosslinked single tetrameric Hb (Fig. 3.1) . . 35
      Recombinant human Hb (Fig. 3.1) 35
      Present status 35
3.4.   PolyHb in Clinical Trials 36
3.5.   Nanobiotechnology-based PolyHb Compared with other
      Modified Hb 39
      Vasopressor effects 39

Theories why vasopressor effects are observed only in some
types of modified Hb? 39
Effects of tetrameric Hb on vasoactivity and ECG 40
3.6. PolyHb with <2% Tetrameric Hb 43
3.7. Nanobiotechnology and Immunogenicity 45
Antigenicity measured as antibody titers 45
Effects of infusion 46
3.8. Bridging the Gap between Safety Studies in
Animal and Humans 48
In vitro screening test using human plasma before use in
humans 48
Basic principles of screening test 49
Use in research 50
Use in industrial production 50
Correlation of in vitro complement activation to clinical
symptoms 51
Clinical trials and use in humans 52
3.9. Design of Animal Study for Hemorrhagic Shock 52
Hemorrhagic shock model 52
Experimental designs 53
3.10. General Discussions 55
3.11. Methods and Procedures 56
Special method for preparation of PolyHb containing low levels
of tetrameric Hb (0.2 to 2%) 56
In vitro screening test using human plasma before use
in humans 58
Animal study on hemorrhagic shock and exchange transfusion 59

Chapter 4. A Nanobiotechnologic Therapeutic that Transports
Oxygen and Remove Oxygen Radicals: For Stroke,
Hemorrhagic Shock and Related Conditions **62**

4.1. Introduction 62
4.2. Ischemia-Reperfusion 62
4.3. PolyHb-CAT-SOD 63
4.4. Enzyme Activities after Intravenous Injections 64
Plasma Hb concentration 64

SOD activity in plasma 65
CAT activity in plasma 66
Discussions 66
4.5. Tests in Animal Models for Stroke and Hemorrhagic Shock . . . 67
4.6. Ischemia-Reperfusion of the Intestine 67
Severe hemorrhagic shock and the intestine 67
Test on effects of reperfusion on ischemic intestine 68
4.7. Test in Combined Hemorrhagic Shock and Stroke Rat Model . 69
Prolonged hemorrhagic shock and stroke model 69
Rat model of combined hemorrhagic shock and stroke 71
Duration of stroke and reperfusion injuries 71
Effects of different solutions on reperfusion injuries 72
Disruptions of blood-brain barrier 72
Brain edema 73
4.8. Discussion 74
4.9. Preparation and Characterization 75
Method of preparation 75
Molecular characterization of PolyHb-CAT-SOD 75
Oxygen dissociation curve 77
Effects of the crosslinking procedure on metHb formation . . . 77
4.10. In Vitro Studies of the Antioxidant Properties of Crosslinked HB-
SOD-CAT 78
Scavenging of superoxide ($O_2^{\bullet-}$) 79
Scavenging of hydrogen peroxide (H_2O_2) 80
Effects of oxidative challenge on the degradation of Hb and iron
release in PolyHb and PolyHb-CAT-SOD 81
Absorbance spectra and Hb degradation 81
Release of iron 83
Lipid peroxidation and salicylate hydroxylation 84
4.11. Methods and Procedures 85
Preparation of PolyHb-SOD-CAT 85
Measurement of SOD activity 86
Measurement of CAT activity 87
Hydrogen peroxide scavenging 87
Ratio of SOD/CAT and Hb oxidation 87
Measurement of FerrylHb 87

Iron release study   88
Circulation time studies   88
Rat model of intestinal ischemia-reperfusion   89
Measurement of oxygen radical   89
Combined hemorrhagic shock and transient global cerebral
  ischemia-reperfusion   90

Chapter 5.   Nanotechnology-based Artificial Red Blood Cells (RBCs)   93

5.1.   Introduction   93
5.2.   Micron Dimension Artificial RBCs (ARBCs) with Ultrathin
       Polymeric Membrane   95
       Hemoglobin   95
       Carbonic anhydrase   95
       Catalase   97
       Immunological studies   97
5.3.   Intravenous Injection of ARBCs   98
       Polymeric membrane artificial
         RBCs 1 micron and larger   98
       Neuraminic acid and circulation time of red blood cells   99
       Effects of surface charge of polymeric particles on
         circulation time 101
       Artificial rbc's with sulfonated nylon membrane 101
       Polysaccharide incorporation 102
       Summary 103
5.4.   Submicron Lipid Membrane Artificial RBCs (LEH) 104
       Composition and preparation 105
       Methemoglobin 106
5.5.   LEH in Animal Studies 107
       Circulation time after infusion 107
       Safety of lipid membrane artificial rbc's 108
       Efficacy in exchange transfusion, hemorrhagic shock
         and hemodilution 109
       Summary 110
5.6.   Nanodimension Biodegradable Polymeric Membrane RBCs . . 110
       Introduction 110
       Nanodimension biodegradable polymeric membrane
         artificial rbc's 111

Safety and reasons for selection of polylesters as the
    biodegradable polymer membrane 112
5.7. Characterization of Nano Artificial RBCs 113
    Electromicroscopic appearance 113
    Size distribution of polylactide membrane artificial rbc's 113
    Other properties 114
    Steady shear viscosity determination 115
5.8. Safety and Efficacy of Nano Artificial RBCs 115
    Amount and fate of PLA membrane 115
    Efficiency in nanoencapsulating hemoglobin 117
    Oxygen affinity, Hill coefficient and Bohr effect 117
    Enzymes and multienzymes 118
    Nano artificial rbc's containing metHb reductase system 119
    Nano artificial rbc's permeable to reducing factors 120
    Enzymes and multienzymes in PLA nano artificial rbc's 121
5.9. Circulation Time of Nano Artificial RBCs 121
    Circulation half-life of nano PLA artificial rbc's 121
    Different types of PEG-PLA copolymers for nano
        artificial rbc's 121
    Analysis of results 126
    Relevance for clinical application in humans 127
5.10. General 127
5.11. Method of Preparation of Biodegradable Polymeric Nano
    Artificial RBCs 128

**Chapter 6. Use of Enzyme Artificial Cells for Genetic Enzyme
        Defects that Increase Systemic Substrates to Toxic Levels 130**

6.1. Introduction 130
    General 130
    Enzyme replacement therapy 132
6.2. Acatalasemia: Congenital Defect in the Enzyme Catalase . . . 132
6.3. Preliminary Study on the Use of Catalase Artificial Cells to
    Replace Defective Catalase in Acatalasemia 133
    Implantation of catalase artificial cells 133
    Catalase artificial cells retained in a chamber for body fluid
        perfusion 135

6.4.   *In Vivo* Kinetics of Catalase Artificial Cells for
        Acatalasemic Mice 135
        Summary 135
        Methods of preparation of catalase artificial cells and
            experimental procedures 136
        Characteristics of acatalasemic mice 136
        Kinetics of recovery of injected perborate 136
6.5.   Immunological Studies on Catalase Artificial Cells
        in Acatalasemic Mice 137
        Effects of immunizing doses of catalase solution and catalase
            artificial cells on antibody production 137
        Permeability of membrane of collodion artificial cells
            to catalase antibody 138
        Response of immunized acatalasemic mice to injections of
            catalase artificial cells or catalase solution 139
        Appearance of catalase activity in the blood after
            injection of catalase solution or catalase artificial cells . . 140
6.6.   Conclusion from Results of Basic Study Using the Acatalasemic
        Mice Model 141
6.7.   Oral Xanthine Oxidase Artificial Cells in a Patient with Lesch-
        Nyhan Disease 142
        Volume ratio of artificial cells/substrate solution 143
        Effects of pH on enzyme activity 144
        Patient with Lesch-Nyhan disease 144
6.8.   Conclusion Based on Study 146
6.9.   Phenylketonuria: Genetic Defect in Enzyme Phenylalanine
        Hydroxylase 147
        Phenylketonuria 147
        Enzyme replacement therapy 147
        Oral PAL artificial cells for PKU based on novel theory
            of enterorecirculation of amino acids 148
6.10.   Research Leading to Proposal of Enterorecirculation of
        Amino Acids 149
        Classical theory of the main source of amino acids in
            the intestinal lumen 149
        Research to test the classical theory 149

Animals on 24 h of protein-free diet 149
Results of study 149
Proposed theory of extensive enterorecirculation of amino
    acids 150
6.11. Oral Enzyme Artificial Cells for Genetic Enzyme Defects that
    Result in Elevated Systemic Amino Acid Levels 152
General design for study of use of oral PAL artificial cells
    in PKU rats 152
Methods of preparation of catalase artificial cells and
    experiment procedures 154
Phenylketonuria rat model 154
Intestinal phenylalanine 154
Plasma PHE levels in PKU rats 155
CSF cerebrospinal fluid in PKU rats 155
Growth as shown by body weight changes 155
Congenital PKU mice model 157
Summary discussion 158
Oral enzyme artificial cells to deplete other amino acids 158

**Chapter 7. Enzyme Artificial Cells in Substrate-dependent Tumors
and Activation of Prodrug   160**

7.1. Introduction 160
7.2. Asparaginase in Lymphosarcoma and Lymphoblastic Leukemia 161
7.3. Asparaginase Artificial Cells for Lymphosarcoma 162
Asparaginase artificial cells on 6C3HED lymphosarcoma
    in mice (Fig. 7.2A) 162
Effects on plasma asparagine level 164
Asparaginase activities in the body 165
Asparaginase does not leak out of artificial cells
    after injection 166
Discussions 167
7.4. Crosslinking Asparaginase to Hemoglobin
    (PolyHb-Asparaginase) 168
7.5. Crosslinking of Asparaginase to Polymer
    (Polyethylene Glycol — PEG) 170
7.6. PolyHb-Tyrosinase Artificial Cells for Melanoma 172

Introduction 172
Polyhemoglobin-tyrosinase for melanoma 172
Studies on optimization, in vitro and in vivo
    characterization 176
Summary 178
7.7.   Tyrosinase Artificial Cells by Oral Route 179
Introduction 179
Effects of Long-term Oral Administration of Polymeric
    Microcapsules Containing Tyrosinase on Maintaining
    Decreased Systemic Tyrosine Levels in Rats 180
Enzyme kinetics 182
7.8.   Discussions on PolyHb-Tyrosinase and Tyrosinase Artificial
    Cells 185
7.9.   Enzyme Artificial Cells for Activating Prodrugs 186
Introduction 186
Clinical trial results 187
7.10. Materials and Methods 188
Asparaginase artificial cells 188
PolyHb-tyrosinase in melanoma 189
Preparation of PolyHb-tyrosinase 189
Determination of tyrosinase activity 190
Effects of intravenous injection of PolyHb-tyrosinase in rats . . 190
B16-F10 melanoma cells and culture conditions 191
Intravenous injection of PolyHb-tyrosinase in normal mice . . 191
B16F10 melanoma bearing mice model 191
Oral tyrosinase artificial cells 192
Materials 192
Control artificial cells 192
Tyrosinase artificial cells 192
In vitro studies using rat intestinal juice 193
Animal studies 193

Chapter 8.   Artificial Cells for Cell Encapsulation   **195**

8.1.   Introduction 195
8.2.   Artificial Cells Containing Islets, Liver Cells, Endocrine Cells and
    Others 198
Artificial cells containing islets 198

      Artificial cells containing hepatocytes 202
      Other areas 203
8.3.   Challenges of Artificial Cell Encapsulated Cells in Therapy . . . 203
      Availability of cells 203
      Mass transfer, biocompatibility and stability 204
      Cell protrusion or entrapment in artificial cell membrane . . . 205
      Method to prevent cell protrusion or entrapment in artificial cell
          membrane 207
      Other improvements to microencapsulation method 207
      Capillary fiber cell encapsulation 208
      Oral administration 208
8.4.   Artificial Cells Containing Genetically-engineered Cells 209
      General 209
      Growth hormone, monoclonal antibodies, factor XI,
          erythropoietin 210
      Neurological disorders 210
      Tumors 211
      Discussions 212
8.5.   Artificial Cells Containing Stem Cells 212
8.6.   Artificial Cells Containing Microorganisms 213
8.7.   Concluding Remarks 216
8.8   Method I: Standard Alginate-Polylysine-Alginate Method 217
      Materials 217
      Droplet generators 217
8.9.   Method II: 2-Step Method 219
8.10. Method III: Preparation of Cells for Encapsulation 221
      Preparation of rat hepatocytes 221
      Bone marrow stem cells from rats 222
      Genetically-engineered E. coli DH5 cells and microorganism . 222
8.11. Method IV: Artificial Cells with Macromolecular Cutoffs 223
      Microorganism 223
      Procedure 224

Chapter 9.   Artificial Cells Containing Hepatocytes and/or
             Stem Cells in Regenerative Medicine   **225**

9.1.   Introduction 225

Artificial cells and regenerative medicine 225
Design of study 227
9.2.   Artificial Cells Containing Hepatocytes 228
Galactosamine-induced fulminant hepatic failure in rats 228
Gunn rats with severe elevation of bilirubin,
    hyperbilirubinemia 229
Summary 231
9.3.   Immunoisolation 231
Can artificial cells protect hepatocytes from immunorejection? 231
Why do rat hepatocytes inside artificial cells recover their
    viability after implantation into mice? 233
Aggregation of artificial cells after implantation 233
Artificial cells prepared by standard method: cell viability after
    implantation 234
Artificial cells prepared by two-step method: cell viability after
    implantation 236
9.4.   Artificial Cells Containing Hepatocytes or Hepatocytes Plus
    Stem Cells 236
Introduction 236
Viability of free cells in culture 237
In vitro viability of encapsulated cells 237
Discussion of above results 239
Study using Gunn rat model 240
9.5.   Artificial Cells Containing Hepatocytes in Rats with 90% of Liver
    Surgically Removed 241
Introduction 241
Artificial cells prepared using 2-step method on survival of rats
    with 90% of liver surgically removed 241
Factors to be considered 242
9.6.   Artificial Cells Containing Bone Marrow Stem Cells 243
Viability of isolated cells 244
Survival rates of rats with 90% of liver surgically removed . . . 244
Remnant liver weight 246
Blood chemistry 246
Plasma hepatic growth factor (HGF) levels 246
Laparotomy and histology 247

Immunocytochemistry 249
PAS glycogen stain 249
*Possible mechanisms responsible for recovery of 90%
     hepatectomized rat model* 249
9.7.   Artificial Cells Containing Stem Cells in Regeneration Medicine 251

**Chapter 10.   Hemoperfusion in Poisoning, Kidney Failure,
                Liver Failure, and Immunology   252**

10.1. Introduction 252
Artificial kidney machine 252
Hemoperfusion based on artificial cells 252
10.2. Development and Clinical Trials 253
*Treatment of patients with severe accidental or suicidal
     poisoning* 255
*Hemoperfusion for removal of unwanted or toxic substances
     from blood under other conditions* 256
Treatment of patients with terminal kidney failure 256
Treatment of patients with hepatic coma 257
Protein-coated artificial cells in immunoadsorption 258
10.3. Basic Principle of Artificial Cell in Hemoperfusion 258
Permeability and transport characteristics 258
Experimental analysis 260
10.4. Artificial Cells Containing Activated Charcoal in
       Hemoperfusion 261
Why use activated charcoal for artificial cell hemoperfusion? . 261
Effects on embolism 262
Effects of hemoperfusion on platelets 262
Clearance of the ACAC artificial cell artificial kidney 263
Improvements in hydrodynamics 265
10.5. Hemoperfusion in Acute Suicidal or Accidental Poisoning . . . 266
Preclinical studies 266
First clinical trials in acute poisoning 266
*Result of clinical trials on 11 adult patients with suicidal or
     accidental drug poisoning* 267
Pediatric patient with accidental theophylline overdose 273
Routine clinical uses in patients around the world 275

10.6. Hemoperfusion in Terminal Renal Failure Patients 276
10.7. Hemoperfusion Alone Supplemented by Hemodialysis 279
   General clinical results 279
   First patient on long-term hemoperfusion 279
   Hemoperfusion and removal of uremic metabolites 283
10.8. Conjoint Hemoperfusion-Hemodialysis 283
10.9. Miniaturized Artificial Kidney Based on Hemoperfusion-
   Ultrafiltration 286
10.10. Hemoperfusion in Liver Failure 288
   First observation of recovery of consciousness in hepatic coma 288
   Results around the world on effect of hemoperfusion on hepatic
   coma 290
   Control studies in galactosamine-induced hepatic failure rats . 291
10.11. Immunoadsorption 294
   Protein coated artificial cells in immunoadsorption 294
10.12. Detailed Procedures for the Laboratory Preparation of ACAC
   Hemoperfusion Device 295
   The extracorporeal shunt chamber 295
   Preparation of activated charcoal 295
   Preparation of artificial cells containing activated charcoal . . . 296
   Procedure for hemoperfusion 297

Chapter 11. Perspectives on the Future of Artificial Cells as
            Suggested by Past Research   299

11.1. Introduction 299
11.2. Membrane Material 300
11.3. Polymeric Membrane 300
   Porosity of polymeric membrane of artificial cells 300
11.4. Lipid Membrane Artificial Cells 304
11.5. Artificial Cells with Lipid-Polymer Membrane and Incorporation
   of Macrocyclic Carrier, NA-K-ATPase and Other Carriers . . . 305
   Artificial cells with lipid-polymer membrane 305
   Incorporation of channels into lipid-polymer membrane of
   artificial cells 307
   Incorporation of Na-K-ATPase in the membrane of artificial
   cells 310
   Nanobiosensors 312

11.6. Surface Properties of Artificial Cell Membranes 312
11.7. Drug Delivery 314
 General 314
 Polymeric semipermeable microcapsules 316
 Biodegradable polymeric artificial cells, nanoparticles,
  nanocapsules 316
 Liposomes evolved into lipid vesicles that are lipid membrane
  artificial cells 316
 Polymer(PEG)-lipid membrane artificial cells or PEG-lipid
  vesicles 318
 Polymersomes: polymeric membrane artificial cells 319
11.8. Artificial Cells Containing Multienzyme Systems with Recycling
 of ATP and NADH 320
 Artificial cells containing multienzyme system for recycling of
  ATP 320
 Artificial cells containing multienzyme system for recycling of
  NAD(P)H 320
 Artificial cells containing urease, glutamate dehydrogenase and
  glucose-6-phosphate dehydrogenase 321
 Dextran-NADH retained and recyled inside artificial cells
  (Fig. 11.4) 322
 Multienzyme with NAD-dextran for conversion of waste, urea,
  into useful essential amino acids 324
 Recycling of free NADH retained within the lipid-polymer
  membrane artificial cells for conversion of urea into amino
  acid 324
 Summary 327
11.9. Artificial Cells Containing Microsomes, Cytosol Ribosomes and
 Polymerases 327
 Introduction 327
 Artificial cells containing liver microsomes and cytosol 328
 Towards a "living" artificial cell containing polymerases,
  ribosomes and transcription/translation system 328
11.10. New Generations of Computer System and Nanoscale Robotics
 Based on Artificial Cells 331
11.11. The Future of Artificial Cells 332

Appendix I. 1957 Report on "Method for Preparing Artificial Hemoglobin Corpuscles" 335

ERYTHROCYTES 335
   1. Structure 335
   2. Function 336
   3. Enzymes and inclusion bodies 336
HEMOGLOBIN 337
   1. Molecular structure 337
   2. Hemoglobin content of the blood 337
   3. Methemoglobin 337
   4. Derivatives of hemoglobin 338
MEMBRANE 338
   1. Reason for using collodion membrane 338
   2. Studies made on collodion membrane 339
   3. Final constituents of collodion solution used in
       this experiment 339
METHOD 340
   Method I 340
   Method II 343
   Experimentation with collodion 343
   Method III 343
   Method IV — successful method 344
   Procedure 345
   Proof that collodion membrane is formed 347
RESULTS 348
   i. Size of corpuscles 348
   ii. Content of corpuscles 348
   iii. Oxygen content 348
DISCUSSIONS 350
   Transportation of O_2 and CO_2 351
   Enzyme systems in the erythrocyte 353
   Buffering action of erythrocyte 353
SUMMARY 354
BIBLIOGRAPHY 354

Appendix II.   Methods not Described in Detail in the 11 Chapters   355

A.   EMULSION METHODS FOR ARTIFICIAL CELLS 355
*A.1. Cellulose nitrate membrane artificial cells of micro
       dimensions* 355
A.2. Polyamide membrane artificial cells of micro dimensions . 357
*A.3. Lipid-polymer membrane artificial cells of micro
       dimensions that retain ATP and NAD(P)H* 360
A.4. Double emulsion methods 361
B.   DROP METHODS FOR LARGER ARTIFICIAL CELLS 362
B.1. Polymer membrane artificial cells 362
B.2. Lipid-polymer membrane artificial cells 364
*B.3. Lipid-polymer membrane artificial cells with macrocyclic
       carrier* 365
*B.4. Incorporation of Na-K-ATPase to membrane of
       artificial cells* 365
*B.5. Standard alginate-polylysine-alginate artificial cells
       (tissues, cells, microorganisms)* 366
*B.6. Two-step method for alginate-polylysine-alginate artificial
       cells (tissues, cells, microorganisms)* 372
B.7. Macroporous agar membrane artificial cells 374

References   377

Index   441

50th Anniversary of Artificial Cells

This first chapter is not a scientific chapter. It is a personal "story" on the occasion of the 50th Anniversary of the first report on artificial cells, aimed at answering the following questions. Why artificial cells and what are they? Where and how were they first prepared? What is the time line of the new ideas related to artificial cells since they were first reported 50 years ago?

1.1. Background

Back in the 1950s, Professor F.C. MacIntosh, chairman of Physiology, and Sir Arnold Burgen started a special "honors physiology" program in the faculty of medicine at McGill University. The farsighted and challenging program consisted of advance cell physiology combined with advance courses in polymer chemistry, physical chemistry and radiation chemistry. I was one of the four chosen to be the "guinea pigs" for this new program and it started my interest in applying basic research to medical treatment. The problem was where to start.

We all feel humble in the face of the ingenuity and complexity of nature. Yet, it need not prevent us from studying, examining and even attempting to prepare clinically useful systems having a few of the simpler properties of their natural counterparts. Indeed, working on a molecular level, researchers at that time had already synthesized a number of biological molecules.

However, on the cellular level, despite the basic importance of cells, no one seemed to be interested in "artificial cells." Perhaps this was premature, since our basic knowledge of biological cells was still incomplete. Yet, organ substitutes like Kolff's artificial kidneys were hardly an exact replicas of their biological counterparts. Despite this, these substitutes were already an accepted method of treatment for kidney failure patients at that time. Thus, I thought that to prepare clinically useful artificial cells, one might not have to prepare replicas of biological cells.

1.2. Starting with Artificial Red Blood Cells

Being an uninitiated beginner, I thought that one can easily prepare artificial red blood cells for use in patients. After all, red blood cells are one of the simplest biological cells. Furthermore, there are practical reasons for doing this. Red blood cells are the best material for use in transfusion, but as shown in Fig. 1.1, there are a number of problems involved. Artificial red blood cells would solve many of these problems (Fig. 1.1).

When I asked around for a method to do this, I was politely told that I was asking for the impossible task of preparing a water emulsion suspended in water. Most people were also taken back by this "far-fetched" idea. Fortunately, my parents have instilled in me the sense that hard work and determination can turn impossible ideas into possibilities. They cited my grandfather as an example of someone who was able to build a manufacturing empire in Swatow (Shantou), starting with nothing as a poor preacher's son. Thus, I quietly and stubbornly started some simple experiments in my dormitory room at McGill's Douglas Hall of residence. Beginner's luck plus the excellent teaching in advance cell physiology and advance chemistry had allowed me to prepare some very crude artificial red blood cells.

I showed this preliminary result to Sir Arnold Burgen and he was most enthusiastic. He helped me persuade the chairman, Professor MacIntosh, to let me use this research for the required honors physiology research project. The department assigned me a corner of the teaching laboratory. By working out a drop method to first prepare larger artificial red blood cells (Fig. 1.2), I was able to use this principle, but using emulsification to prepare microscopic artificial red blood

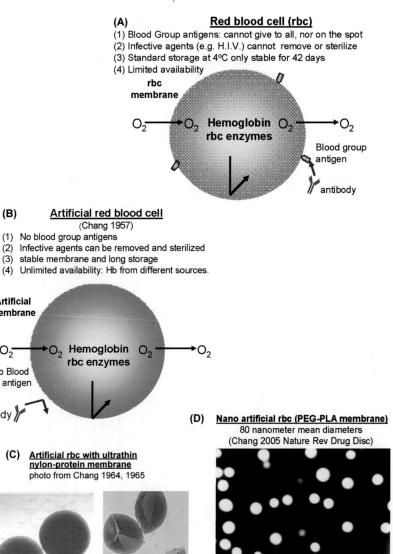

(A) <u>**Red blood cell (rbc)**</u>
(1) Blood Group antigens: cannot give to all, nor on the spot
(2) Infective agents (e.g. H.I.V.) cannot remove or sterilize
(3) Standard storage at 4°C only stable for 42 days
(4) Limited availability

rbc membrane

$O_2 \longrightarrow O_2$ **Hemoglobin** $O_2 \longrightarrow O_2$
rbc enzymes

Blood group antigen

antibody

(B) <u>**Artificial red blood cell**</u>
(Chang 1957)
(1) No blood group antigens
(2) Infective agents can be removed and sterilized
(3) stable membrane and long storage
(4) Unlimited availability: Hb from different sources.

Artificial membrane

$O_2 \longrightarrow O_2$ **Hemoglobin** $O_2 \longrightarrow O_2$
rbc enzymes

No Blood group antigen

antibody

(C) <u>**Artificial rbc with ultrathin nylon-protein membrane**</u>
photo from Chang 1964, 1965

hypertonic
hypotonic

(D) <u>**Nano artificial rbc (PEG-PLA membrane)**</u>
80 nanometer mean diameters
(Chang 2005 Nature Rev Drug Disc)

Fig. 1.1. (**A**) Characteristics of red blood cell. (**B**) Characteristics of artificial red blood cell. (**C**) Artificial rbc with ultrathin nylon-protein membrane. Spherical in hypotonic solution, becoming "crenated" in hypertonic solutions. Reversible when moved from one solution to another. (**D**) E/M of nanodimension (80 nanometer) artificial red blood cells containing Hb & rbc enzymes.

Artificial Cells

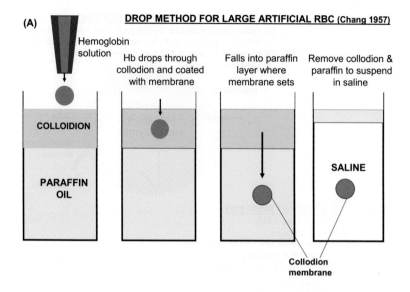

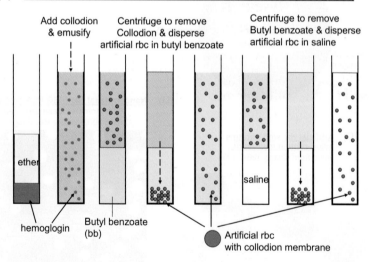

Fig. 1.2. Original (1957) method of preparing artificial cells (see Appendix I for details). (**A**) Drop method for preparing large artificial cells. Principle later extended for use in bioencapsulation of cells, stem cells, genetic engineered cells. (**B**) Emulsion phase separation method for preparing microscopic artificial cells (unlike the above, "collodion" prepared by removing most of alcohol and replaced with ether — see Appendix I). Principle extended to method for preparation of microscopic artificial cells and drug delivery systems and nanodimension artificial cells (Chang, *Nature Rev Drug Discovery*).

cells (Fig. 1.2). For the grading of this research project, the department asked for a sample of the artificial red blood cells. It was a relief when Professor Sekelj's analysis supported my oxygen dissociation curve and the research report was approved (Chang, 1957).

The extrusion drop method has become the basis for extension and modifications of methods for preparing artificial cells to encapsulate biological cells. The emulsion method has also become the basis for further extensions and modifications of the methods for the preparation of artificial cells of microscopic or nano dimension in nanotechnology. Because of this, the complete report including methods of preparation is reprinted in Appendix I in this monograph for interest and reference. The original project report has always been available in the McGill Medical Library. Part of this report was published in the 30th Anniversary Symposium volume (Chang 1988c).

1.3. Further Research and the First Routine Clinical Use of Artificial Cells

Over the next 5 years of my medical school and internship, the department continued to let me use the teaching laboratory space, providing me with summer salary support from the Faculty of Medicine. After this, since no one at McGill was carrying out related research, a Ph.D. committee was put together. In addition, Professor MacIntosh wrote a supporting letter for my submission to *Science* as the sole author of this work. This was under the condition that I use the title "semipermeable microcapsules" instead of my original pretentious title of "Artificial cells." *Science* accepted and published the paper (Chang, 1964). The Ph.D. thesis was also entitled "Semipermeable aqueous microcapsules" instead of "Artificial cells" (Chang, 1965). It is interesting to note that no one, including myself, knew precisely what "semipermeable microcapsules" meant. The outcome was that very few researchers followed up on this research for many years to come! The delay of others entering this line of research means that most of the earlier research in this area has been carried out in my laboratory.

In addition to the operating grant and career investigator award from the Medical Research Council of Canada (the present Canadian Institutes of Health Research), I was later awarded a large special program grant by the council. I took this as a challenge to show that all my findings about the potential implications of artificial cells can actually be put into clinical practice. With this, I was able to develop artificial cells which contain absorbents for use in hemoperfusion for the treatment of suicidal patients or patients involved in accidental use of sleeping pills and other medications. However, this being a very new approach, other physicians were initially concerned about medical liability. Thus, first worked out a way to scale up the preparation. After this, I carried out preclinical animal safety and efficacy studies as well as clinical treatment of patients. This required a lot of work and was highly time consuming; it was possible because of the unselfish support of my wife, Lancy, who knew the importance of this work for patients. Among many other things, she waited for more than 10 years before our first honeymoon. The result was hemoperfusion based on artificial cells, being carried out routinely around the world for the treatment of suicidal and accidental uses of medications in adult and pediatric patients.

In the meantime, I was completing an invited monograph for Charles C. Thomas Publisher and it contained a detailed description of methods. On becoming a tenured full professor in 1972, I immediately had this published (Chang 1972a). The advantage of having "tenure" is that it enabled me to safely use the title "Artificial cells" instead of "Semipermeable microcapsules" for this monograph! This self explanatory title and the detailed description of methods, combined with interests in the clinical use of artificial cells in hemoperfusion, led to a sudden surge of research activities in this area around the world. However, at that time the industries were mainly concerned with the development and use of artificial cells for hemoperfusion and drug delivery systems.

1.4. Importance of Progress in Parallel Areas of Biotechnology, Molecular Biology, and Regenerative Medicine

Most of my original ideas and basic research were related to enzyme and gene therapy, cell therapy, blood substitutes, regenerative medicine, nanomedicine and related areas. Developing these for actual clinical use required parallel developments in molecular biology and biotechnology that were not yet available. More recently, many groups around the world have made exciting progress in biotechnology, molecular biology, genetic engineering and related areas. The outcome is a recent new wave of research and development in artificial cells. Many groups around the world are now working on extensions and modifications of artificial cells for use in nanotechnology, nanobiotechnology, blood substitutes, regenerative medicine, gene therapy, cell/stem cell therapy and other areas (Orive *et al.*, 2003; Chang, 2005). Later chapters will describe some of these in more detail. This is now such a broad area that space will only allow for review of these areas with detailed examples based on research in my laboratory. Research in the many other centers will be presented by other authors in subsequent publications under the book series on "Regenerative Medicine, Artificial Cells and Nanomedicine."

1.5. Historical Milestones

Table I shows examples of the milestones of the first report of original ideas in artificial cells. This is not a complete listing and more details will be given in subsequent chapters.

Table 1.1 Artificial Cells (AC): Time Line of Ideas Since First Reported (1957)

1957 Chang	**First artificial cells prepared** with a synthetic membrane to replace RBC membrane and containing hemoglobin and red blood cell enzymes (emulsion phase separation, extrusion method or spray coating)
1964 Chang (*Science*)	Artificial cells (AC) containing enzymes, hemoglobin and cells formed by interfacial coacervation or interfacial polymerization to form membranes of polymer, crosslinked protein, polymer conjugated with protein, also crosslinked protein microspheres
1964, 1965 Chang	Nanobiotechnology: crosslinked protein (PolyHb) & conjugated Hb
1964,1965 Chang 1966: Chang *et al.*	Extrusion drop method for AC to encapsulate intact cells for immunoisolation in cell therapy
1965 Bangham *et al.*	Liquid crystal microspheres of multi-lamellar lipid (liposomes) as membrane model for basic research
1965, 1972a, 1973b Chang	AC for molecular sieve chromatography and separation
1965 Chang 1966 Chang *et al.*	AC with intracellular multi-compartments
1966 Chang	Silastic AC and microspheres containing protein
1966 Chang	AC containing magnetic materials and biological materials
1966, 1969a Chang	Ultrathin membrane AC containing adsorbents for hemoperfusion
1966 Clark & Gollan	Fluorocarbon as oxygen carrier
1967 Chang *et al.*	AC with polysaccharide complexed membrane for biocompatibility
1968 Chang & Poznansky (*Nature*)	Implanted enzyme AC for enzyme therapy in inborn error of metabolism (shown in congenital catalase-deficient acatalesemic mice)
1968 Bunn & Jandl	Intramolecularly crosslinked single Hb molecule
1968 Geyer *et al.*	Fluorocarbon effective in exchange transfusion in animal studies
1969d Chang 1972a Chang	AC with lipid-polymeric membrane or lipid-crosslinked protein membrane containing cyclic transport carrier (AC contains proteins)
1970–1975 Chang *et al.*	First clinical use of artificial cells in patients (in hemoperfusion)

(*Continued*)

Table 1.1 (*Continued*)

1971a Chang (*Nature*)	Implanted enzyme AC for lymphosarcoma suppression in mice
1971b Chang	Nanobiotechnology: glutaraldehyde crosslinked Hb into PolyHb. Later, others used this method for blood substitutes in patients
1972a Chang	First monograph on *Artificial Cells*
1972b Chang (*Lancet*)	AC hemoperfusion resulted in Grade IV hepatic coma patient recovering consciousness
1973 Gregoriadis	First use of liposomes to entrap enzymes and drugs. Led to extensive development of liposomes as delivery systems
1975h Chang	Paper discussing one shot vaccine using AC
1976a Chang	Biodegradable polylactide microcapsules and microparticles containing proteins and hormones
1976 Tam, Blumenstein & Wong	Soluble dextran conjugated hemoglobin
1976 Bonhard et al.	Develop glutaraldehyde crosslinked PolyHb as blood substitute
1977–1985 Chang with Campbell, Cousineau, Ilan, Grunwald, Wahl, Yu etc.	Artificial cells containing multienzyme systems with co-factor recyclying for multistep enzyme reactions
1978 Naito & Yokoyama	Developed perfluorodecalin as blood substitute towards clinical trials
1980 Lim & Sun (*Science*)	Alginate-polylysine-alginate AC encapsulated cells
1980 Rosenthal & Chang	AC membrane of lipid-protein-polymer containing Na^+K^+-ATPase
1980 Djordjevich & Miller	Lipid membrane AC encapsulated hemoglobin
1985 Mitsuno, Ohyanagi	Clinical trials of perfluorodecalin as red blood cell substitute
1986 Yuan & Chang	AC containing microsomes and cytosol
1986 Bourget & Chang	Oral enzyme AC for inborn error of metabolism (phenyketonuria rat)

(*Continued*)

Table 1.1 (*Continued*)

1986 Sipehia, Bannard & Chang	AC membrane that exclude small hydrophilic molecules but permeable to large lipophilic molecules
1986 Chang, Bourget & Lister	Novel finding of extensive enterorecirculation of amino acids leading to the use of oral enzyme AC therapy to selectively remove specific unwanted systemic amino acid
1988 Tsuchida's group 2002 Tsuchida *et al.*	Development and *in vivo* testing of synthetic heme complex either to liposome or to recombinant albumin as blood substitute
1989a Chang, 1989 Palmour *et al.*, Chang	Clinical use of oral enzyme artificial cells in a patient (patient with inborn error of metabolism: Lesch-Nyhan disease)
1989 Moss *et al.*	Clinical trials with glutaraldehyde crosslinked PolyHb
1990 Hoffmann *et al.*	Recombinant human hemoglobin
1994 Yu & Chang	Biodegradable polymeric membrane nanoartificial red blood cells
1994 Soon-Shiong *et al.*	AC encapsulated islet transplantation in a type 1 diabetic patient. Insulin independence reported
1996 Prakash & Chang (*Nature Med*)	Oral artificial cells containing genetically engineered cells lower systemic urea in a uremic rat model
1996 Aebischer, Lysagth *et al.* (*Nature Med*)	Polymeric fiber encapsulation of genetically modified xenogeneic cells for intrathecal delivery of CNTF in amyotrophic lateral sclerosis patients
1998 D'Agnillo & Chang (*Nature Biotech*)	Nanobiotechnology of crosslinking of Hb, catalase and superoxide dismutase to form soluble nanodimension PolyHb-CAT-SOD
1998 Tsuchida	Lipid AC vesicle Hb: developed and tested in animal towards clinical use
1999 Philips *et al.*	PEG-lipid membrane AC containing Hb increases circulation time
2000 Liu & Chang	AC coencapsulating hepatocytes and adult stem cells
2001 Lörh *et al.* (*Lancet*)	Clinical trial of AC microencapsulated cell-mediated treatment of inoperable pancreatic carcinoma in patients
2002 Gould *et al.*	The life-sustaining capacity of human polyhemoglobin in trauma surgery clinical trials
2002 Sprung *et al.*	The use of bovine polyhemoglobin in surgical patients: results of a multi-center, randomized, single-blinded trial

(Continued)

Table 1.1 (*Continued*)

2003 Chang, Powanda, Yu	PEG-PLA membrane nanodimension AC containing Hb and rbc enzymes
2004 Bloch *et al.*, Aebischer	Phase I clinical study for Huntington's Disease, using encapsulated cells engineered to secrete human ciliary neurotrophic factor
2004 Yu & Chang (*Melanoma Res J*)	Nanobiotechnological approach of PolyHb-tyrosinase: delays the growth of melanoma in a rat model
2006 Liu & Chang (*J Liver Trans*)	AC encapsulated bone marrow stem cells regenerate liver resulting in recovery and survival of rats with 90% of liver surgically removed

(Updated from Chang 2005, Nature Review Drug Discovery).

CHAPTER 2

Basic Principles

In this chapter, the basic principles of artificial cells are described and illustrated by a few of the numerous studies and references in the field. Specific areas with detailed examples and references will follow in the later chapters.

2.1. Basic Features of Artificial Cells

The initial research on artificial cells forms the basic principles of artificial cells, principles that have been extended for use in many areas by many groups. Indeed, as stated in the first monograph entitled *Artificial Cells* (Chang, 1972a): *"Artificial cell is not a specific physical entity. It is an idea involving the preparation of artificial structures of cellular dimensions for possible replacement or supplement of deficient cell functions. It is clear that different approaches can be used to demonstrate this idea."*

Basic features of early artificial cells

Earlier artificial cells have some of the simpler properties of biological cells (Fig. 2.1). Some examples of the basic features are:

(1) The membrane of an artificial cell separates its content from the outside. At the same time, the membrane can be prepared such that it can selectively allow different types of molecules to cross it. This ranges from membrane that does not allow any molecules to cross it to those that allow even very large molecules like proteins to cross it. In between these two extremes, there are artificial cell

Basic Principles of early Artificial Cells
(Chang 1957,1964)

intracellular

Like biological cells:
(1) content retained inside to
(2) act on permeant molecules
(3) release products
(4) intracellular environment
(5) extracellular environment

extracellular

ultrathin
artificial membrane

Types of early Artificial Cells
(Chang 1957 to 1966)

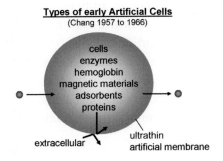

cells
enzymes
hemoglobin
magnetic materials
adsorbents
proteins

extracellular

ultrathin
artificial membrane

Present Status of Artificial Cells
(Chang 2005 Nature Rev Drug Disc)

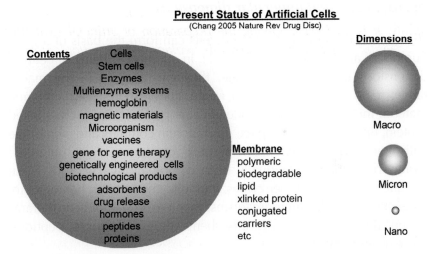

Dimensions

Contents
Cells
Stem cells
Enzymes
Multienzyme systems
hemoglobin
magnetic materials
Microorganism
vaccines
gene for gene therapy
genetically engineered cells
biotechnological products
adsorbents
drug release
hormones
peptides
proteins

Membrane
polymeric
biodegradable
lipid
xlinked protein
conjugated
carriers
etc

Macro

Micron

Nano

Fig. 2.1. *Upper*: Basic principles of early artificial cells. *Middle*: Different types of early artificial cells based on these basic principles. *Lower*: Present status of artificial cells with wide variations in contents, membrane material and dimensions.

membranes that restrict the movement of molecules according to molecular size, lipid solubility, affinity to carrier mechanisms, etc.

(2) Artificial cell membranes can be very thin, yet strong and have a large surface area. Thus, 10 ml of 20 μm diameter artificial cells has a total surface area of 2500 cm^2. This is the same as the total membrane surface area of an artificial kidney machine. In addition, an artificial cell membrane is 100 times thinner than that of an artificial kidney membrane. This means that smaller molecules can move across 10 ml of 20 μm diameter artificial cells 100 times faster than that across an artificial kidney machine (Chang, 1966). The microscopic size of artificial cells also allows material to diffuse rapidly inside the artificial cells.

(3) Artificial cells can contain the same biological material as biological cells. In addition, they are more versatile since adsorbents, magnetic materials, cells, drugs and other material can also be included separately or in combination (Fig. 2.1).

Present status of the basic features of artificial cells of macro, micron, nano and molecular dimensions

The general principles of artificial cells can form the basis of a large number of artificial systems (Fig. 2.1). In addition to being of cellular dimensions in the micron range, they can also be in the macro range, nano range or molecular range. Furthermore, the membrane material includes polymer, biodegradable polymer, lipid, crosslinked protein, lipid-polymer complex, lipid-protein complex and membrane with transport carriers. Artificial cells can contain an unlimited variety of material individually or in combinations (Fig. 2.1). These include cells, stem cells, enzymes, multienzyme systems, hemoglobin, magnetic materials, microorganisms, vaccines, genes for gene therapy, genetically engineered cells, adsorbents, drugs, hormones, peptides, proteins and others.

The following is a brief overview of some of examples that illustrate the basic principles. Later chapters contain detailed references and descriptions.

2.2. Nanotechnology and Nanobiotechnology

There is much recent interest in nanotechnology. Nanotechnology is a large and complex area that embraces many diverse approaches. One of these is to make the original artificial cells smaller using the same basic principles and method. This includes biodegradable nanoparticles, nanospheres and nanocapsules. Examples include nano artificial red blood cells with lipid membrane (Djordjevich and Miller, 1980) or biodegradable polymeric membranes nano artificial red blood cells (WP Yu and Chang, 1994) (Fig. 1.1). A later section will summarize other examples used in drug delivery systems.

Nanobiotechnology and artificial cells

Nanobiotechnology is the assembling of biological molecules into nanodimension structures, membranes with nanodimension thickness or nanotubules with nanodimension diameter.

The first nanobiotechnology approach reported is the crosslinking of hemoglobin into ultrathin polyhemoglobin (PolyHb) membrane with nanodimension thickness (Chang, 1964, 1965) (Fig. 2.2). This is used to form the membrane of artificial red blood cells (Chang, 1964, 1965). If the emulsion is made very small, then the whole submicron artificial cells can be crosslinked into PolyHb of nanodimension. Glutaraldehyde can crosslink hemoglobin to form soluble nanodimension PolyHb, each consisting of an assembly of 4–5 hemoglobin molecules (Chang, 1971b) (Fig. 2.2).

Two groups have independently developed this 1971 basic method of glutaraldehyde crosslinking for clinical use. One is glutaraldehyde human PolyHb (PolyHb) (Gould *et al.*, 1998; Gould *et al.*, 2002). Their phase III clinical trial shows that this PolyHb can replace blood lost in trauma surgery by keeping the blood hemoglobin at an acceptable level. The second PolyHb is glutaraldehyde-crosslinked bovine PolyHb which has been tested in Phase III clinical trials (Pearce and Gawryl, 1998; Sprung *et al.*, 2002). South Africa has approved this PolyHb for routine clinical use in patients. Unlike red blood cells, there is no blood group, and thus PolyHb can be given on the spot, without

NANOBIOTECHNOLOGY

Definition: assembling of biological molecules into nanodimension structures
(membrane thickness, nanotubule diameter or diameter of nanostructures)

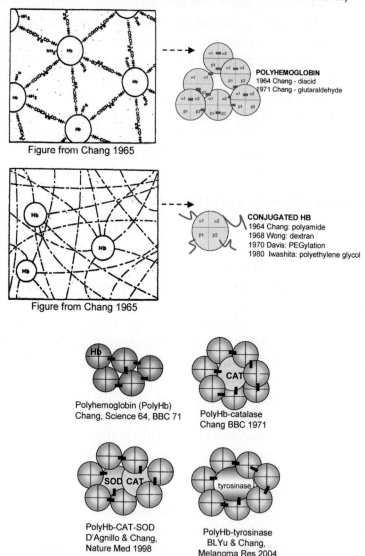

Figure from Chang 1965

POLYHEMOGLOBIN
1964 Chang - diacid
1971 Chang - glutaraldehyde

Figure from Chang 1965

CONJUGATED HB
1964 Chang: polyamide
1968 Wong: dextran
1970 Davis: PEGylation
1980 Iwashita: polyethylene glycol

Polyhemoglobin (PolyHb)
Chang, Science 64, BBC 71

PolyHb-catalase
Chang BBC 1971

PolyHb-CAT-SOD
D'Agnillo & Chang,
Nature Med 1998

PolyHb-tyrosinase
BLYu & Chang,
Melanoma Res 2004

Fig. 2.2. Nano artificial cells can be prepared in nanodimensions as membrane-enclosed nano artificial cells or by the use of nanobiotechnology to assemble biological molecules together into nanodimension structures. *Upper*: An example of assembling of biological molecules to form polyHb and conjugated Hb. *Lower*: Examples of different types of nanobiotechnology-based polyHb-enzymes.

waiting for typing and crossmatching in the hospital. They are also free from infective agents such as HIV, hepatitis C, bacteria, parasites and so on. Furthermore, whereas donor blood has to be stored at 4°C and is only good for 42 days, PolyHb can be stored at room temperature for more than a year. Thus, PolyHb can have important uses in a number of clinical conditions notably for surgery.

Nanobiotechnology and the assembling of hemoglobin with enzymes that remove oxygen radicals

As PolyHbs can be kept at room temperature and used immediately on the spot, they can have potential for use in treating severe bleeding (hemorrhagic shock). However, the process must be carried out fast because if delay occurs, the PolyHb alone might result in the production of oxygen radicals that cause tissue injury (ischemia-reperfusion injuries). Antioxidant enzymes normally present in red blood cells are not enough to prevent this problem. We use glutaraldehyde crosslinking to assemble a nanobiotechnology complex of PolyHb-SOD-CAT by crosslinking hemoglobin, superoxide dismutase and catalase (D'Agnillo and Chang, 1998) (Fig. 2.2). In this way, one can increase the antioxidant enzymes to a much higher level than those in red blood cells.

Obstruction of arteries due to clots or other causes can result in stroke or heart attack (myocardial infarction). Being a solution, PolyHb can more easily perfuse partially obstructed vessels. However, if there is a prolonged lack of oxygen, reperfusion with PolyHb alone may give rise to damaging oxygen radicals, resulting in ischemia-reperfusion injuries. Thus, in a rat stroke model, after 60 min of ischemia, reperfusion with PolyHb resulted in a significant increase in the breakdown of the blood-brain barrier and an increase in brain water (brain edema) (Powanda and Chang, 2002). On the other hand, polyHb-SOD-CAT did not result in these adverse changes (Powanda and Chang, 2002).

Nanobiotechnology for the assembling of hemoglobin with other enzymes

Abnormal microcirculation in tumor leads to a decrease in perfusion by oxygen carrying red blood cells (Pearce and Gawryl, 1998). PolyHb can more easily perfuse the abnormal microcirculation of tumors to supply oxygen needed for chemotherapy or radiation therapy. With a circulation half-time of 24 h, the effect can be adjusted to the duration of the chemotherapy or radiation therapy. When used together with chemotherapy, PolyHb decreases the growth of tumor and increases the lifespan in a rat model of gliosarcoma brain tumor (Pearce and Gawryl, 1998). We have recently crosslinked tyrosinase with hemoglobin to form a soluble PolyHb-tyrosinase complex (BL Yu and Chang, 2004) (Fig. 2.2). This has the dual function of supplying the needed oxygen and at the same time lowering the systemic levels of tyrosine needed for the growth of melanoma. Intravenous injections delayed the growth of the melanoma without causing adverse effects in the treated animals (BL Yu and Chang, 2004).

Conjugation of polymer with proteins

In the presence of diamine, sebacyl chloride crosslinks hemoglobin with polyamide to form conjugated hemoglobin (Chang, 1964,1965) (Fig. 2.2). This can be in the form of nanothickness artificial cell membranes or conjugated Hb nanospheres. An extension of this is the crosslinking of single enzyme or single hemoglobin molecule to soluble polymers (Tam *et al.*, 1976; Duncan, 2003; Li, Zhang & Liu, 2005; Winslow, 2006) (Fig. 2.2). Promising Phase II clinical trials are ongoing (Winslow, 2006). This extension is not nanobiotechnology since *conjugation of single biological molecule is not the same process as* assembling of biological molecules.

2.3.   Cell Homogenate, Organelle, Enzymes and Multienzyme Systems

Cell homogenate

As discussed in Chapter 1, artificial cells were first prepared by replacing the membrane of biological cells with artificial membranes (Chang, 1957). For example, an artificial membrane replaces the red blood cell membrane, but the content is the same as that in the original red blood cells. The same principle can also be applied to other types of biological cells. For example, we have prepared artificial cells containing hepatocyte microsomes and cytosol (Yuan and Chang, 1986).

Enzymes and enzyme therapy

A large array of enzyme systems is present in the cell homogenates. Simpler artificial cells can be made to contain only one enzyme (Chang, 1964, 1972) (Fig. 2.3). The enclosed enzyme would not leak

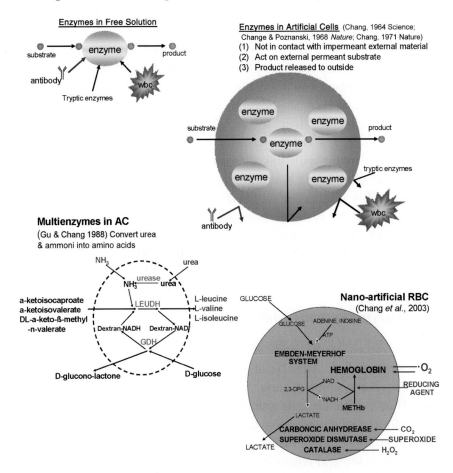

Fig. 2.3. *Upper left*: Problems related to injection of enzymes in free solution. *Upper right*: Enzymes inside artificial cells no longer have these problems. *Lower left*: Artificial cells can be prepared with multistep enzyme systems with recyling of cofactors. *Lower right*: Nano artificial red cells contain all the enzymes of red blood cells.

out, but can act on external permeant substrates. This would avoid protein sensitization, anaphylactic reaction, or antibody production with repeated injections (Fig. 2.3). Implanted urease artificial cells convert systemic urea into ammonia (Chang, 1964, 1965). Implanting artificial cells containing catalase can replace the defective enzyme in mice with a congenital catalase defect — acatalasemia (Chang and Poznanski, 1968). The artificial cells protect the enclosed enzyme from immunological reactions (Poznanski and Chang, 1974). Artificial cells containing asparaginase implanted into mice delay the onset and growth of lymphosarcoma (Chang, 1971).

Giving enzyme artificial cells by mouth obviates the need for repeated injections. For example, artificial cells containing urease and ammonia adsorbent can lower the systemic urea level (Chang, 1972a). In Lesch-Nyhan disease, enzyme defect resulted in the elavation of hypoxanthine to toxic levels. Given by mouth, artificial cells containing xanthine oxidase lowers the toxic systemic hypoxanthine levels in an infant with this disease (Chang, 1989; Palmour *et al.*, 1989). Phenylketonuria is a more common congenital enzyme defect. Artificial cells containing phenylalanine ammonia lyase, given by mouth, lower the systemic phenylalanine levels in phenylketonuria [PKU] in rats (Bourget and Chang, 1986). This leads to investigation into recombinant sources of this enzyme (Sarkissian *et al.*, 1999; Liu *et al.*, 2002).

Multienzyme systems with cofactor recycling

Most enzymes in the body function as multienzyme systems with cofactor recycling. After basic research on artificial cells containing multienzyme systems (Chang, 1985a), we looked into their possible use. Thus, artificial cells containing three different enzymes can convert metabolic wastes like urea and ammonia into essential amino acids (Gu and Chang, 1988) (Fig.2.3). The needed cofactor, NADH, can be recycled and retained inside the artificial cells by crosslinking to dextran or by using a lipid-polymer membrane. All the multienzyme systems in red blood cells can be included inside nanodimension artificial red blood cells (Chang *et al.*, 2003) (Fig. 2.3).

2.4. Artificial Cells Containing Intracelluar Compartments

Biological cells contain intracellular organelles. This allows separate compartments inside the cells to carry out specific functions more effectively. We have prepared artificial cells that also contain intracellular compartments (Chang 1965, 1972a; Chang *et al.*, 1966)

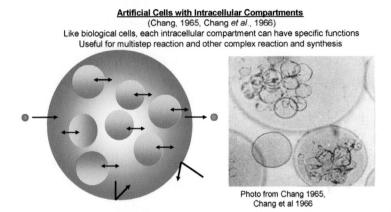

Artificial Cells with Intracellular Compartments
(Chang, 1965, Chang *et al.*, 1966)
Like biological cells, each intracellular compartment can have specific functions
Useful for multistep reaction and other complex reaction and synthesis

Photo from Chang 1965,
Chang et al 1966

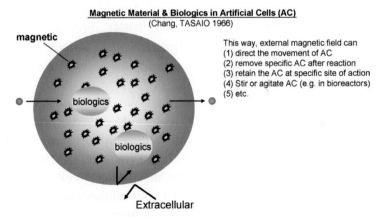

Magnetic Material & Biologics in Artificial Cells (AC)
(Chang, TASAIO 1966)

magnetic

This way, external magnetic field can
(1) direct the movement of AC
(2) remove specific AC after reaction
(3) retain the AC at specific site of action
(4) Stir or agitate AC (e.g. in bioreactors)
(5) etc.

biologics

biologics

Extracellular

Fig. 2.4. *Upper*: Artificial cells can be prepared with intracellular multi-compartments. *Lower*: Artificial cells containing biologics can also contain magnetic material allowing artificial cells to be site directed. Both principles are being extended by many groups and are being used in different areas of application and research.

(Fig. 2.4). Specific enzyme systems or other biologically active systems can be enclosed separately or in combination in each of these intracellular compartments to allow for more efficient stepwise functions.

2.5. Artificial Cells Containing Biologics and Magnetic Material

When magnetic material is included in artificial cells containing biological materials, one can use an external magnetic field to direct the artificial cells (Chang, 1966) (Fig. 2.4). This principle is now being used very extensively in bioreactors, in removing specific materials from a mixture as in diagnositcs kits, drug delivery systems and other areas of application.

2.6. Cells, Islets, Stem Cells, Genetically-engineered Cells and Microorganisms

Artificial cells containing cells

The first artificial cells containing intact biological cells were reported in 1964 based on a drop method (Chang, 1964), and it was proposed that "*protected from immunological process, encapsulated endocrine cells might survive and maintain an effective supply of hormone*" (1965, Chang *et al.*, 1966) (Fig. 2.5).

Chang asked Conaught Laboratory of insulin fame to develop this for use in islet transplantation for diabetes. Later, Sun and his collaborator from Conaught Laboratory developed Chang's original drop method (Chang, 1964,1965,1972a; Chang *et al.*, 1966), using alginate-polylysine-alginate (APA) for the artificial cell membranes (Lim and Sun, 1980). They showed that after implantation, the insulin secreting islets inside the artificial cells indeed remained viable and continued to secrete insulin to control the glucose levels of diabetic rats (Lim and Sun, 1980).

We have been studying the use of artificial cells containing liver cells (hepatocytes) for liver support. Implanting these increases the

Free Cell

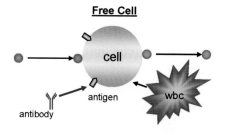

Cells in Artificial Cells
(Chang, 1964 Science; Chang 1965, Chang *et al.*, 1966)
(1)   Protected from immuno-rejection (antibody, wbc)
(2)   Oxygen & nutrients equilibrate rapidly into artificial cells
(3)   Secretion (e.g. insulin) controllable by permeant material (e.g. glucose)

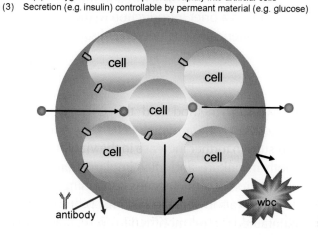

Cells in AC	**Stem Cells and Hepatocytes in AC**
(Chang 1965, Chang *et al.*,1966)	(Liu & Chang, 2000)

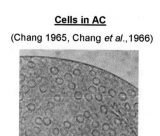 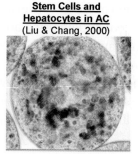

Fig. 2.5. *Upper*: Problems related to injection of free cells. *Middle*: Cells inside larger artificial cells no longer have these problems when injected. *Lower*: Artificial cells containing biological cells. Priniciple has been extended and used by many groups for bioencapsulation of islets, cells, genetically-engineered cells and stem cells (Chang, *Nature Rev. Drug Discovery*, 2005).

survival of acute liver failure rats (Wong and Chang 1986); lowers the high bilirubin level in congenital Gunn rats (Bruni and Chang, 1989); and prevents xenograft rejection (Wong and Chang, 1988). We developed a two-step cell encapsulation method to improve the APA method, resulting in improved survival of implanted cells (Wong and Chang, 1991a). Cell bioencapsulation for cell therapy has been extensively developed by many other groups especially using artificial cells containing endocrine tissues, hepatocytes, genetically-engineered cells and stem cells (Orive *et al.*, 2002; Chang, 2005). This is a very broad area that will be described in much more detail in the later chapters. Below is a brief introduction to the use of this principle for stem cells and genetically-engineered cells.

Stem cells

We used the two-step method and the coencapsulation of stem cells and hepatocytes into artificial cells (Liu and Chang, 2000) (Fig. 2.5). This results in further increase in the viability of encapsulated hepatocytes both in culture and after implantation (Liu and Chang, 2002). An implantation of artificial cells containing both hepatocytes-stem cells into Gunn rats lowers the systemic bilirubin levels, the low level being maintained for two months (Liu and Chang, 2003). Without stem cells, the implanted hepatocytes in artificial cells can only be maintained at the low level for one month. What is even more exciting is our recent finding using artificial cells containing only bone marrow stem cells and no hepatocytes. The control group of rats with 90% of their liver surgically removed did not survive. Unlike free stem cells, an intraperitoneal injection of artificial cells containing bone marrow stem cells alone results in long-term survival (Liu and Chang, 2006). Along with this, the livers regenerate and return to their normal weights.

Genetically-engineered cells

Many groups have carried out extensive research on artificial cells containing genetically-engineered cells. This more recent and very important area will be discussed in detail in a later chapter, including potential applications in amyotrophic lateral sclerosis, dwarfism, pain treatment, IgG_1 plasmacytosis, hemophilia B, Parkinsonism and axotomized septal cholinergic neurons, tumor suppression and other areas (Basic *et al.*, 1996; Tan *et al.*, 1996; Al-Hendy *et al.*, 1996; Okada *et al.*, 1997; Dalle *et al.*, 1999; Saitoh *et al.*, 1995; Hagihara *et al.*, 1997; Winn *et al.*, 1994; Aebischer *et al.*, 1996; Bloch *et al.*, 2004; Bachoud-Levi *et al.*, 2000; Xu *et al.*, 2002; Cirone *et al.*, 2002). To avoid the need for implantation, we studied the oral use of artificial cells containing genetically-engineered nonpathogenic *E. coli* DH5 cells to lower systemic urea in renal failure rats (Prakash and Chang, 1996; Chang, 1997).

2.7. Artificial Cells Containing Bioadsorbents

As mentioned earlier, the microscopic dimensions of artificial cells result in a large surface-to-volume relationship. This, together with the ultrathin membranes, allow artificial cells containing bioadsorbents to be much more effective when compared to standard hemodialysis in removing toxins and drugs from the blood of patients (Chang, 1966, 1969a, 1975g).

The most common routine application of this approach is the use of microscopic polymeric artificial cells encapsulating activated charcoal (Chang, 1969, 1973a,b, 1975g) (Fig. 2.6). Its use solves the major problems of release of embolizing particles and damage to blood cells when bioadsorbents are used without the artificial cell membranes (Fig. 8). The first successful application was in suicidal overdose patients (Chang *et al.*, 1973a,b). Since then, this has become a routine treatment worldwide for acute poisoning in adults and children, especially in cases of suicidal overdose (Chang, 1975b, 1975c; Winchester, 1988; Singh *et al.*, 2004; Lin *et al.*, 2004; Peng *et al.*, 2004; Lopez *et al.*, 2002; Kawasahi *et al.*, 2000; Lin *et al.*, 2002; Tominaga, 1997). The treatment is particularly useful in places where dialysis machines are not readily available.

The approach is also effective in removing toxic products in kidney failure patients (uremia), resulting in the relief of uremic symptoms (Chang *et al.*, 1971a; Chang, 1975g; Winchester, 1988). Components

Artificial Cells

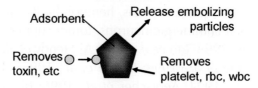

Free Adsorbent

Adsorbent

Release embolizing particles

Removes toxin, etc

Removes platelet, rbc, wbc

Adsorbent in Artificial Cells (AC)
(Chang, 1966 TASAIO)
(1) Separated from platelet, rbc, wbc
(2) Retain any particles and prevents embolizing
(3) Remove toxins that equilibrate rapidly into the AC

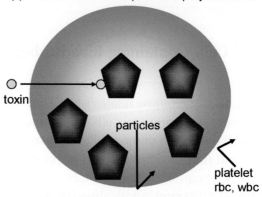

toxin

particles

platelet rbc, wbc

Artificial Cells containing Adsorbents
(Photo from Chang 1966 TASAIO)

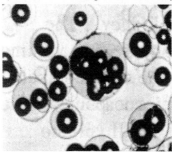

Fig. 2.6. *Upper*: Problems related to adsorbents when used to remove toxin from blood. *Middle*: Adsorbents inside artificial cells no longer have these problems and therefore can be used in hemoperfusion to remove toxins from blood. *Lower*: Adsorbent inside artificial cells. Since its first clinical use as artificial cells containing activated charcoal in hemoperfusion in patients, it is now a routine method for the treatment of acute poisoning.

for the removal of other uremic wastes need to be developed. The approach has also proved to be effective in removing toxic molecules in patients with severe liver failure, resulting in the recovery of consciousness of grade 4 hepatic coma patients (Chang, 1972b, 1975g; Gazzard *et al.*, 1974). Detoxification is only one of the functions of the liver, and this approach is being used as the detoxification component of hybrid liver support systems that are being developed (Liu *et al.*, 2001).

The success in the clinical uses of artificial cells containing bioadsorbents for detoxification has led to an increasing interest in research and development in many other areas. One of these is in artificial cells containing immuoadsorbents (Chang, 1980d).

2.8.   Research on Membrane Model Systems

Since the membrane compositions of artificial cells can be varied at will, they can serve as membrane model systems. Indeed, work in this laboratory shows that it is possible to vary the membrane properties as to porosity, thickness, charge, lipid content, protein content, mucopolysaccharide content, and polymer composition (Chang, 1964, 1965, 1972a). We have used artificial cells to study the biophysics of membrane transport (Chang and Poznansky, 1968c); the relationship of surface properties to survival in circulation (Chang, 1965, 1972a; Chang *et al.*, 1967b); and the relationship of physicochemical properties to the effect on coagulation and formed elements of blood (Chang *et al.*, 1967b; Chang, 1969a). Other workers have made use of these artificial cells in the study of the mechanical and electrical properties of membranes (Jay and Edwards, 1968; Jay and Burton, 1969; Jay and Sivertz, 1969).

Bangham and his co-workers prepared liposomes that are liquid crystal microspheres, each consisting of concentric shells of bimolecular lipid layers as a model system (Bangham *et al.*, 1965). Mueller and Rudin (1968) adopted the Chang procedure for the preparation of artificial cells (Chang, 1964) to prepare artificial cells having only a single bilayer lipid membrane. Since biological cell membranes consist of both protein and lipids, we have prepared artificial cells with membrane of lipid-polymer or lipid-protein for membrane transport studies (Chang, 1969d). This also includes the incorporation of macrocyclic molecules into the membrane as a carrier transport mechanism (Chang, 1969d). Another approach is to incorporate Na^+-K^+-ATPase into the lipid-polymer artificial cell membrane (Rosenthal & Chang, 1980). Since then, there have been very extensive studies on the inclusion of many different types of transport system or targeting agents or other materials into different types of artificial cell membranes (Torchilin, 2005; Chang, 2005).

2.9. Cell Physiology

In cell physiology, there is much research on the origin of cellular responses. One important area is the location of the triggering mechanisms and whether they are located at the cell surface, in intracellular organelles, or in the soluble constituents of the cytoplasm. It has been suggested (Chang, 1972a) that the enclosure of cell contents in artificial cells might be one way of studying this. It would be interesting to test the sites of action of certain hormones; changes in extracellular electrolytes; and substances causing the release of certain intracellular material. This would throw light on whether the substance acts on specific surface receptors, or only indirectly affects the intracellular material, or whether it directly affects the intracellular material.

2.10. Drug Delivery

Polymeric semipermeable microcapsules

Luzzi (1970a) used nylon membrane artificial cells as reported earlier (Chang, 1964) to microencapsulate drugs for slow release. Others have also extended this approach. However, the modern approaches in drug delivery systems are based on nanotechnology. This is to prepare much smaller artificial cells in nanodimensions.

Biodegradable polymeric artificial cells, nanoparticles, nanocapsules

Biodegradable membrane artificial cells have been prepared to contain enzymes, hormones, vaccines, and other biologicals (Chang, 1976a). The polylactide polymer can degrade in the body into lactic acid and finally into water and carbon dioxide. In the same study, variations in preparation can result in artificial cells that release insulin at different rates (Chang, 1976a). Biodegradable drug delivery systems are now used widely in different forms, ranging from microscopic to nanodimensions (LaVan *et al.*, 2002). They are also known as nanoparticles, nanocapsules, polymersomes, nanotubules, etc.

Liposomes — lipid membrane artificial cells

Bangham first reported the preparation of liposome consisting of microspheres of hundreds of concentric lipid bilayers — multi-lamellar (Bangham *et al.*, 1965). They used these as a membrane model for basic membrane research. When Gregoriadis completed his Ph.D. in biochemistry at McGill, he came to see me to discuss his future areas of research. I encouraged him to look into different ways of forming artificial cells for delivery of biologics. His subsequent research in England on the use of liposomes as drug delivery systems opened a whole new approach (Gregoriadis, 1976).

The large amount of lipid in the original multilamellar liposomes (Bangham *et al.*, 1965) limits the amount of drugs or biologics that can be enclosed. Thus, the basic principles and methods of artificial cells (Chang, 1957, 1964) were extended by researchers into an "ether evaporation method" to form single bilayer (unilamellar) lipid membrane liposomes (Deamer and Bangham, 1976). Much research and development since then have resulted in liposomes being used extensively as pharmaceutical drug carriers (Torchilin, 2005). The modern single bilayer lipid vesicles are actually single bilayer lipid membrane artificial cells. However, even now, there is still reluctance in using the term "artificial cells" and the term "liposome" continues to be used. However, some researchers and research groups are starting to call these "artificial cells."

2.11.   Other Systems

As suggested earlier (Chang 1974, 1965, 1972a), the general principles of artificial cells could be explored as follows. Artificial cells containing radioactive isotopes or antimetabolites might be used for intraarterial injection into tumors. In this case, some of the microcapsules might lodge at the tumor site, while others would be carried by lymphatic channels to act on tumor cells that have metastasized to the regional lymph nodes. Artificial cells containing radiopaque material would provide a contrast medium. Provided they can circulate readily in the bloodstream, they might be used as vehicles for contrast materials in angiography. Artificial cells containing highly magnetic alloys might provide a useful preparation for the measurement of blood flow in unopened vessels by electromagnetic techniques. If membranes of crosslinked protein can be made to retain the immunological characteristics of the protein, there might be a place for these in serological studies or target drug delivery.

CHAPTER 3

Oxygen Carriers Based on Nanobiotechnology

3.1. Introduction

As discussed in the last chapter, nanobiotechnology is the assembling of biological molecules into one of the following categories: nano-dimension structures, membranes with nanodimension thickness, and nanotubules with nanodimension diameter. The first nanobio-technology approach reported in the literature is the crosslinking of Hb into ultrathin polyhemoglobin (PolyHb) membrane with nanodimension thickness for artificial cell membrane (Chang, 1964, 1965) (Fig. 2.2). If the emulsions are made very small, then the whole submicron artificial cells can be crosslinked into PolyHb of nanodimension. Glutaraldehyde can crosslink Hb into soluble PolyHb of nanodimension (Chang, 1971b) (Fig. 2.2). This principle was later developed independently by two groups to produce oxygen carriers for the replacement of the oxygen carrying function of red blood cells in the final phases of clinical trials (Moss *et al.*, 1988; Gould *et al.*, 2002; Pearce & Gawryl, 1998; Pearce *et al.*, 2006; Sprung *et al.*, 2002). Oxygen carriers developed by one group are in routine clinical use in South Africa (Lok, 2001). There are also other types of Hb-based oxygen carriers that are not based on nanobiotechnology. This chapter discusses why nanobiotechnology has been used and how Hb-based oxygen carriers prepared using this approach compare with other types of Hb-based oxygen carriers.

3.2.   Hemoglobin as Oxygen Carrier

As shown in Fig. 1.1, red blood cell membranes contain blood group antigens, and typing and matching are needed before red blood cells can be transfused into patients. This results in delays in emergency situations. With the standard method, red blood cells can be stored for only about 42 days. Red blood cells cannot be sterilized to remove infective agents like hepatitis viruses, HIV and other potential emerging infective agents. Thus, red blood cell substitutes are being developed (Fig. 1.1). Red blood cell (rbc) contains Hb, antioxidant enzymes and a multienzyme system to prevent the conversion of Hb to nonfunctioning metHb. As such, preparing a clinically useful complete artificial red blood cell has been a very complicated process and will be discussed in the next two chapters. In the meantime, nanobiotechnology has allowed the development of simpler oxygen carriers containing only Hb. Although the simpler oxygen carriers are not as complete as the artificial red blood cell, some of them have more immediate possibility of clinical use.

Why do we need to modify Hb? Hb is a tetramer with two α subunits and two β subunits ($\alpha1\beta1\alpha2\beta2$) (Perutz, 1989). Although an excellent oxygen carrier, Hb extracted from the blood cells cannot be used for infusion because it is highly toxic to the kidney (Rabiner *et al.*, 1967). Even highly purified stroma-free Hb (SFHb), in which the rbc membrane stroma is removed, still showed toxicity to the kidney in humans (Savitsky *et al.*, 1978). It is now known that when free Hb is infused into the body, the tetramer ($\alpha1\beta1\alpha2\beta2$) breaks down into toxic dimers ($\alpha1\beta1$ and $\alpha2\beta2$) that cause renal toxicity and other adverse effects. The challenge is how to make use of the excellent oxygen carrying properties of Hb but at the same time eliminate its toxic effects. Figure 3.1 summarizes the four general methods of Hb modification.

3.3.   Modified Hb

Four types of modified Hb have been developed and tested in clinical trials. These are PolyHb based on nanobiotechnology, conjugated Hb, crosslinked tetrameric Hb and recombinant human Hb (Fig. 3.1).

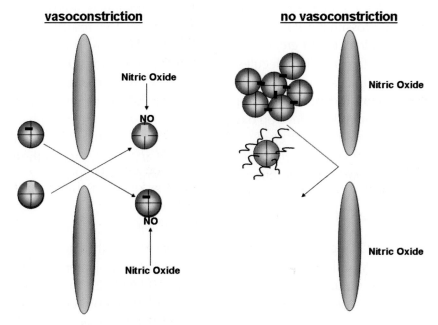

1ˢᵗ Generation modified hemoglobin (1ˢᵗ reports)

PolyHemoglobin (nanobiotechnology based)
1964 Chang (diacid)
1971 Chang (glutaraldehyde)

Conjugated Hb
1964 Chang (polyamide)
1968 Wong (dextran)
1970 Abuchowski & Davis (PEGylation)
1980 Iwashita (polyethylene glycol)

Crosslinked tetrameric Hb
1968 Bunn & Jandl
1979 Walder (diaspirin)

Recombinant Human Hb
1990 Hoffman

vasoconstriction   **no vasoconstriction**

Nitric Oxide

NO

NO

Nitric Oxide

Nitric Oxide

Nitric Oxide

Nitric Oxide

Fig. 3.1. First generation oxygen therapeutics PolyHb is based on nanobiotechnology.

PolyHb based on nanobiotechnology (Fig. 3.1)

In 1964, Chang first used bifunctional agents to crosslink the reactive amino groups of Hb to assemble Hb molecules together into PolyHb

(PolyHb). The first bifunctional agent used was sebacyl chloride (Chang, 1964, 1965).

$$Cl-CO-(CH_2)_8-CO-Cl + HB-NH_2 = HB-NH-CO-(CH_2)_8-CO-NH-HB$$

    sebacyl chloride   Hb   crosslinked PolyHb (Chang, 1964)

In 1971, another bifunctional agent, glutaraldehyde, was used to crosslink hemoglobin (10 gm/dl) and trace amounts of catalase (Chang, 1971). The reaction is as follows:

$$H-CO-(CH_2)_3-CO-H + HB-NH_2 = HB-NH-CO-(CH_2)_3-CO-NH-HB$$

    glutaraldehyde   hemoglobin   crosslinked PolyHb (Chang,1971)

In the 1971 glutaraldehyde procedure, crosslinking is adjusted so that the crosslinked PolyHb is in a soluble state. This results in less steric hindrance and greater ease of substrate diffusion. This method crosslinks the protein molecules to one another (intermolecular). It can also crosslink each protein molecule internally (intramolecular). Later, two groups independently applied the glutaraldehyde crosslinked PolyHb principle to develop oxygen carriers that are in ongoing phase III clinical trials (Moss *et al.*, 1988; Gould *et al.*, 2002; Pearce and Gawryl, 1998; Pearce *et al.*, 2006; Sprung *et al.*, 2002). One of these has been approved for routine use in South Africa (Lok, 2001).

Conjugated Hb (Fig. 3.1)

In 1964, Chang showed that in the presence of both diamine and Hb, sebacyl chloride can crosslink hemolgobin to polymers to form insoluble conjugated Hb (Chang, 1964, 1965) (Fig. 2.2). This method has been extended to produce (intramolecular) soluble conjugated Hb (Fig. 3.1) formed by the linking of one Hb molecule to soluble polymers, as prepared by Sunder in 1975; Wong in 1976; Iwashita in 1980; and the Enzon group in 1988. More recently, Winslow's group used an improved maleimide-polyethylene glycol conjugated Hb that with its water of hydration would not result in vasopressor effect (Winslow, 2006). This is now in an ongoing phase II clinical trial. Liu's group (Li *et al.*, 2005) has developed another type of PEG-conjugated Hb that is also in an ongoing clinical trial.

Intramolecularly crosslinked single tetrameric Hb (Fig. 3.1)

Bunn and Jandl in 1968 reported the use of a bifunctional crosslinker, bis(N-maleimidomethyl)ether, to crosslink Hb molecules internally (intramolecular). Walder *et al.* in 1979 reported the use of a 2,3-DPG pocket modifier, bis(3,5-dibromosalicyl) fumarate (DBBF), to intramolecularly crosslink the two α subunits of the Hb molecule. This approach prevents dimers formation and improves P_{50}, and has been extensively developed and tested in clinical trials. However, this did not progress further because of vasopressor effects and cardiac lesions in experimental animals (Nelson, 1998; Burhop and Estep; 2001).

Recombinant human Hb (Fig. 3.1)

Another very exciting area is the use of recombinant technology to produce recombinant human Hb from *E. coli* (Hoffman *et al.*, 1990). Recombinant human Hb is formed by the fusion of the two α subunits of each Hb molecule to prevent its breakdown into two half molecules (dimers). It also has a good P_{50}. This recombinant human Hb has been developed and tested in clinical trials (Freytag and Templeton, 1997). Vasopressor effects are observed in clinical trials. A new recombinant human Hb has been prepared that does not bind nitric oxide, thus eliminating the problem of vasopressor effect (Doherty *et al.*, 1998). The new recombinant human Hb is also a potential source of Hb for PolyHb and conjugated Hb and other future generation Hb-based blood substitutes.

Present status

As seen from Fig 3.1, most of the basic principles of modified Hb have been made available by the 1960s. Unfortunately, there was no public interest in these approaches then. The major efforts at that time were concentrated on the production of purified native Hb (SFHb) and perfluorochemicals. Savitsky *et al.*'s study in humans in 1978, showed that SFHb was unsuitable for use in humans. Shortly after this, it was found that perfluorocarbons available at that time had some adverse effects in humans and they did not carry enough oxygen in

Table 3.1 HIV Stimulated R & D for Modified Hb

POLYHb, CONJUGATED Hb & XLINKED TETRAMERIC Hb
Acharya, Agishi, Alayash, Bakker, Baldwin, Blumenstein, Benesch, Biro, Bonhard, Bucci, Burhop, Chang, D'Agnillo, DeAngello, DeVenuto, DeWoskin, Estep, Fagrell, Faivre, Feola, Fratantoni, Gawryl, Gould, Greenburg, Gulati, Hess, Hori, Hsia, Hughes, Intalietta, Iwashita, Jacobs, Jesch, Keipert, Klugger, Liu, Lowe, MacDonald, Magnin, Manning, Meinert, Messmer, Mckenzie, Moss, Nelson, Nose, Panter, Pearce, Powanda, Privalle, Przybelski, Pluira, Rausch, Seetharama, Sehgal, Sekiguchi, Sideman, Shorr, Su, Tsai, Valeri, Vandegriff, Winslow, Wong, Yang and many others

RECOMBINANT AND TRANSGENIC HEMOGLOBIN Casparl, Fronticelli, Hoffman, Kumar, Lemon, Looker, Olson, Shin, Shoemaker, Stetler and many others

ENCAPSULATED Hb (Artificial RBC)
Beissinger, Chang, Djordjevich, Farmer, Feuerstein, Gaber, Goin, Horinouchi, Hunt, Ikeda, Kondo, Kobayashi, Lee, Lutz, Miller, Nishiya, Ogata, Powanda, Phillips, Rabinovic, Rudolph, Sakai, Schmidt, Snohara, Su, Szebeni, Takahasi, Takaori, Takeoka, Tsuchida, Usuba, Yu and many others

many of the clinical applications. Thus, a few investigators started to turn to the earlier approach to approach modified Hb. This effort was suddenly intensified after 1986, when HIV in donor blood became a major public concern. As a result, extensive studies have been carried out by many groups on modified Hb (Table 3.1). The very extensive publications by these researchers can be found in the reference section of this book. This has led to clinical trials in all four types of modified Hb. Of these, nanobiotechnology-based PolyHb and conjugated Hb continue to show promise in clinical trials.

3.4. PolyHb in Clinical Trials

Gould *et al.* (2002) reported their phase III clinical trials on 171 patients using their glutaraldehyde crosslinked human polyHb with <1% unpolymerized Hb. They showed that this can successfully compensate for extensive blood loss in trauma surgery by maintaining the Hb level at 8 to 10 g/dl needed for safe surgery with no reported side effects. For example, transfusion of this PolyHb in patients with an Hb level as low as 2 g/dl, can raise the Hb level to within the 8 to

10 g/dl level, with the patients recovering from surgery. Their historical analysis showed that without transfusion, most patients with Hb levels of 3g/dl did not survive. This group has infused up to 10 liters of PolyHb into individual trauma surgery patients. They have carried out further clinical trials on its use in pre-hospital emergencies. These clinical trials have been completed and information on these trials and the protocol can be found on their website (www.northfieldlabs.com). In the USA, this product has been approved for compassionate use in patients and it is awaiting regulatory decision for routine clinical use.

Given that the supply of Hb from outdated donor blood is limited, a glutaraldehyde-crosslinked bovine polyHb with <4% unpolymerized molecular Hb, has been developed and tested in phase III clinical trials (Pearce and Gawryl, 1998; Pearce *et al.*, 2006; Sprung *et al.*, 2002). For example, they have carried out multicenter, multinational, randomized, single-blind, rbc-controlled phase III clinical trials in patients undergoing elective orthopedic surgery. A total of 688 patients are randomized 1:1 to receive either the polyHb or rbc at the time of the first perioperative rbc transfusion decision and 59.4% of the patients receiving polyHb required no rbc transfusion all the way to follow-up; 96.3% avoided transfusion with rbc on the first postoperative day and up to 70.3% avoided rbc transfusion up to day 7. This bovine polyHb has been approved for routine clinical use in patients in South Africa, a region with a higher incidence of human immunodeficiency virus (Lok, 2001). In North America, the PolyHb has been approved for compassionate use in patients. More details can be found on www.biopure.com.

Another way to solve the problem of the supply of Hb is the recent use of red blood cells from placentas that are discarded after birth as a source of human Hb for preparing glutaraldehyde human polyHb (Li *et al.*, 2006). Recombinant human Hb is another potential source of Hb to form polyHb (Doherty *et al.*, 1998; Olsen *et al.*, 2004; Fronticelli *et al.*, 2004).

Compared with donor red blood cells, nanobiotechnology-based polyHb is not a complete red blood cell substitute. It is only an oxygen carrier. Furthermore, its circulation half-time of about one day is much lower than that of donor rbc. On the other hand, polyHb has a number of advantages (Table 3.4).

Table 3.2 Comparing Four Types of Modified Hb

Type	Method of Preparation	Notes
PolyHb based on nanobiotechnology	Glutaraldehyde is the bifunctional agent being used to crosslink Hb intermolecularly to form soluble PolyHb polymers, each averaging 3–10 Hb molecules.	These PolyHbs with unpolymerized Hb removed do not cross the intercellular junction of the endothelial cell lining of blood vessels to remove nitric oxide; as a result, they do not have adverse vasopressor effects. Two types of PolyHb are in the final stages of phase III clinical trials and one of these is approved for human use in South Africa.
Conjugated Hb	PEG molecules are linked to each Hb molecule. The PEG and the water of hydration both contribute to a larger diameter PEG-Hb molecule.	If each of the PEG–Hb molecules with its added water of hydration approaches the required dimension, then there is no vasopressor effect as long as there are no free single Hb molecules. PEG–Hb is being tested in ongoing phase II clinical trials.
Crosslinked molecular Hb	Each Hb molecule is intramolecularly crosslinked to prevent the Hb from breaking down into half molecules (dimers).	These molecular-dimension Hb molecules cross the intercellular junction of the endothelial cell lining of blood vessels and remove nitric oxide needed for normal vasoactivity.
Recombinant molecular Hb	Recombinant human Hb with fusion of the two α subunits of each Hb molecule to prevent its breakdown into half molecules (dimers).	Vasopressor effects observed in clinical trials for the same reason as above. A new recombinant human Hb has been prepared that does not bind nitric oxide, thus eliminating the problem of vasopressor effects. This new type has still the disadvantage of being removed faster from the circulation but it is a potential source of Hb for PolyHb and conjugated Hb, and other future-generation Hb-based blood substitutes.

3.5.   Nanobiotechnology-based PolyHb Compared with other Modified Hb

This is summarized in Table 3.2.

Vasopressor effects

Not all modified Hb (Fig. 3.1 and Table 3.2) have been successful in clinical trials as vasopressor effects have been observed in some of the first generation modified Hb blood substitutes. For example, intramolecularly crosslinked Hb (Nelson, 1998; Burhop and Estep, 2001) and first generation recombinant Hb (Freytag and Templeton, 1997) blood substitutes contain 100% of molecular dimension modified Hb. Infusion of small volumes cause vasopressor effects and also increased smooth muscle contractions. With another type of polyHb that contains 36% molecular dimension modified Hb, significant vasoactivity and increased smooth muscle contractions could also be observed when larger doses are used. On the other hand, Gould *et al.* (2002) have not reported any vasopressor effects in their clinical trial using polyHb with <1% molecular dimension modified Hb even when large volumes of 10 liters are infused.

Theories why vasopressor effects are observed only in some types of modified Hb?

The hypothesis for the observed vasopressor effects is that the intercellular junctions of the endothelial lining of the vascular wall allow molecular dimension Hb to enter into the interstitial space (Chang, 1997; Gould *et al.*, 1998) (Fig. 3.1). There, Hb acts as a sink in binding and removing nitric oxide needed for maintaining the normal tone of smooth muscles. This results in constriction of the blood vessels and other smooth muscles, especially those of the esophagus and GI tract. Opponents to the hypothesis argue that one cannot compare the different types of modified Hb since there are major differences in the chemistry involved and in the oxygen affinity.

Effects of tetrameric Hb on vasoactivity and ECG (Fig. 3.2)

We therefore prepare nanodimension PolyHb each containing different percentages of molecular dimension Hb using the same glutaraldehyde crosslinking method and characterized the Hb to ensure that they all have the same oxygen affinity (BLYu,

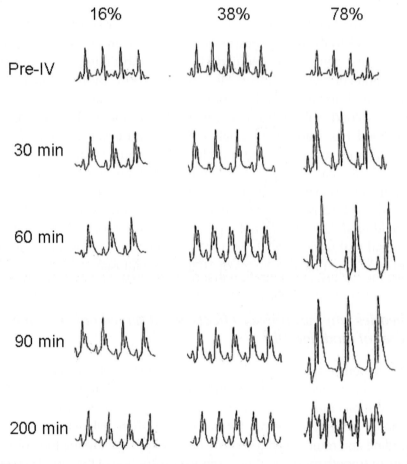

Fig. 3.2. Changes in electrocardiogram (ECG) after injection of 1/6 toploading with PolyHb containing 16%, 38% and 78% of tetramers, respectively at 30, 60, 90, and 120 min. Pre-IV: before intravenous injection; Post-IV: after intravenous injection. Preliminary study shows that similar toploading using PolyHb with 0.4% tetramers did not result in any changes. (BLYu, Liu & Chang, 2006.)

Liu and Chang, 2006). The result shows that the one with 0.4% molecular dimension modified Hb does not cause vasoconstriction nor changes in the electrocardiogram (e.g. 16%, 38% and 78%). With an increasing percentage of molecular dimensions modified Hb, there is an increasing degree of vasoconstriction and elevation of the ST segment of the electrocardiogram (Fig 3.2). ST elevation could be due to vasoconstriction resulting in decreased supply of oxygen to the heart and this may explain the observation of small subendocardial lesions in some primates and swine after infusion with one type of molecular dimension modified Hb (Burhop and Estap, 2001).

Based on our experimental results using PolyHb with 0.4%, 16%, 38% and 78% of tetramers, respectively, we carried out the following analysis. First, we calculate the total amount of tetrameric Hb (g/kg body weight) that is present in PolyHb with 0.4%, 16%, 38% and 78% of tetramers, respectively. Next, we calculate the total amount of tetramers needed to cause vasopressor effects or ECG changes. We then calculate the total amount of tetrameric Hb (g/Kg) present in 1%, 2%, 4%, 8%, 16%, 38% and 78% of tetrameric Hb for 1/6, 1/3, 1/2 and 3/3 blood volume replacement (Table 3.3). We plot this (Fig. 3.3). The top dashed line represents severe changes in ECG. The middle line is the level where the blood pressure increases markedly and this is also the highest concentration that does not cause marked changes in ECG, except for a slight elevation of the ST segment. The bottom solid line shows the level that did not cause an increase in arterial blood pressure (BLYu, Liu & Chang, 2006). The result shows that with 2% of

Table 3.3 Analysis of the Total Amounts of Tetrameric Hb (BLYu, Liu & Chang, 2006)

Blood Volume Replaced	PolyHb 1% Tetramer (g)	PolyHb 2% Tetramer (g)	PolyHb 4% Tetramer (g)	PolyHb 8% Tetramer (g)	PolyHb 16% Tetramer (g)	PolyHb 38% Tetramer (g)	PolyHb 78% Tetramer (g)
1/6	0.006	0.012	0.023	0.047	0.093	0.222	0.455
1/3	0.012	0.023	0.047	0.093	0.187	0.443	0.910
1/2	0.018	0.035	0.070	0.140	0.280	0.665	1.365
3/3	0.035	0.070	0.140	0.280	0.560	1.330	2.730

Artificial Cells

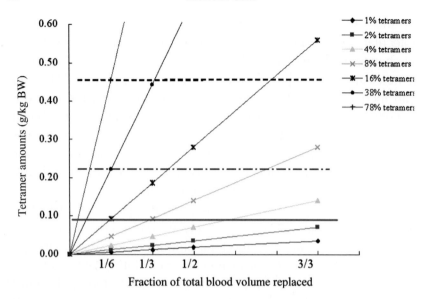

Fig. 3.3. The amount of tetrameric Hb (g/Kg body weight) in PolyHb with 1%, 2%, 4%, 8%, 16%, 38% and 78% of tetrameric Hb for 1/6, 1/3, 1/2 and 3/3 blood volume replacement. The top dashed line represents severe changes in ECG. The middle line is the level where blood pressure increases markedly and this is also the highest concentration that does not cause marked changes in ECG, except for a slight elevation of the ST segment. The bottom solid line shows the level that did not cause an increase in arterial blood pressure. (BLYu, Liu & Chang, 2006.)

tetrameric Hb, even replacing the total body blood volume, would not cause any vasopressor effects or changes in the ECG (Fig. 3.4).

Further support of the role of nitric oxide in vasopressor effects has come from Bucci's group who prepared large "zero-link" PolyHb (Matheson *et al.*, 2002). A second generation molecular dimension recombinant Hb that did not bind nitric oxide, also did not cause vasoconstriction (Doherty *et al.*, 1998). Since recombinant Hb crosses the intercellular junction and is removed quickly, their circulation time can be increased by crossliking to form nanodimension polyHb (Fronticelli *et al.*, 2004). Another approach in an ongoing clinical trial is to prepare maleimide-PEG conjugated Hb that with its water of hydration, would be too large to cross the intercellular junction and thus does not result in vasopressor effects (Winslow, 2006).

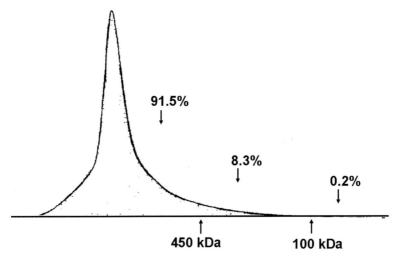

91.5%
↓

8.3%
↓

0.2%
↓

↑   ↑
450 kDa   **100 kDa**

Fig. 3.4. Analytical chromatograpy: Sepacryl S-300 (1.5 cm × 98 cm). Elution solution: 0.1 M Tris-HCl, pH 7.5, 0.154 NaCl; Elution velocity = 33.16 dl/h; Recorder velocity = 1 mm/min; Sens = ⟨5⟩. (Zolotareva & Chang, 2007.)

3.6. PolyHb with <2% Tetrameric Hb

The standard method of preparing polyHb usually results in about 7–10% of tetrameric Hb. Northfield has a proprietary method of preparing polyHb with <1% tetrameric Hb. Biopure can prepare their PolyHb with <4% tetrameric Hb. Our analysis described above (BLYu, Liu & Chang, 2006) shows that with 2% tetrameric Hb, there is neither vasoactivity nor change in ECG even with complete blood volume replacement. Therefore, we have designed methods for preparing PolyHb with <2% tetrameric Hb (Zolotareva & Chang, 2007). Increasing the degree of polymerization would decrease the % of tetrameric Hb. However, this also results in an increase in the molecular weight of the polyHb that tends to gel at a lower temperature or other conditions. Thus, instead of increasing the degree of polymerization, our final approach is to use preparative chromatography to remove the tetrameric Hb. Figure 3.4 shows the analytical chromatography of a preparation with 0.2% tetrameric Hb.

Table 3.4 Human Donor Blood vs. PolyHb

	Human Donor Blood	Nanobiotechnology-based PolyHb
Infective agents	Rare in some regions using costly screening tests, but more frequent in other regions. If a new unknown infective agent appears, it might take years to develop a screening test for blood to be safe.	Infective agents can be sterilized and removed.
Source	Limited availability.	Unlimited because, in addition to human Hb (Hb), bovine Hb and recombinant human Hb can be used.
Blood group antigens	Blood group antigens need typing and crossmatching.	No blood group antigens.
Delay for use	Delay in use due to need for typing and crossmatching.	Can be used immediately because no typing or crossmatching is needed; being tested in ambulances in ongoing phase III clinical trials.
Storage stability	Can be stored for 42 days with standard refrigeration at $4°C$.	Studies show that PolyHb can be stored for >1 year at room temperature.
Circulation time	About 60 days, depending on the length of storage.	Circulate effectively with a half-time of 24 h and are therefore good for short-term use, such as surgery and in emergencies. There are ways to increase its length of function, including: hemodilution, repeated infusion, and combination with erythropoietin.
Function	Complete RBC functions. In addition to carrying oxygen, RBC also have antioxidant enzymes and enzymes for preventing MetHb formation.	Function only as an oxygen carrier and are useful for several clinical uses. Other uses might require new generations of PolyHb (for example, PolyHb crosslinked to antioxidant enzymes). Future-generation nanodimension artificial RBC contain Hb and all of the enzyme systems of RBCs. Costs will increase with each new generation.
Vasopressor effect	No vasopressor effect.	To prevent vasopressor effect, single molecular-dimension Hb has to be eliminated.

For those entering this area of research for the first time, a major complaint would be the lack of any books containing laboratory procedures for the preparation and testing of blood substitutes. There is a lack of books written on procedures that readers can readily use in their laboratory for the preparation and testing of modified Hb. With this in view, a section on "Methods and Procedures" has been included at the end of most chapters.

3.7. Nanobiotechnology and Immunogenicity

Nanobiotechnology involves the assembling of biological molecules. In the case of polyHb, this is by polymerization of Hb molecules. Even though Hb is only weakly antigenic, it is generally thought that polymerization of protein might enhance its immunogenicity. In order to test this, we analyzed the *in vitro* and *in vivo* immunogenicity of crosslinking Hb into polyHb (Chang and Varma, 1987, 1988; Chang *et al.*, 1992; Hertzman, Keipert and Chang, 1986). Estep's group has reported on diaspirin-crosslinked Hb (Estep *et al.*, 1992). Bakker's group also studied the immunology of polymerized Hb (Bleeker *et al.*, 1995).

Antigenicity measured as antibody titers

The first basic study is to see whether crosslinking Hb into polyHb increases its antigenicity. Rats are immunized with either rat stroma-free Hb or rat polyHb in Freud's adjuvant. To account for cross species reactivity, two other groups of rats are immunized with either human stroma-free Hb or human polyHb. Antigens are labeled with ^{125}I. Figure 3.5 shows that rat Hb in Freud's adjuvant does not increase in the antibody titers when it is repeatedly injected subcutaneously into the rats (homologous). Even in the polymerized form (PolyHb), there is no increase in antibody titers using the very severe test of Freud's adjuvant. On the other hand, Hb from another species (heterologous) is slightly antigenic, resulting in a very small but significant increase in antibody titer. Crosslinking Hb from another species into PolyHb results in further increase in anitgenicity. However,

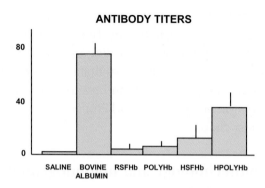

Fig. 3.5. Antibody titers in rats after repeated subcutaneous immunization of the solution in Freud's adjuvant. RSFHB — rat stroma-free Hb; RPOLYHB — rate polyHb; HSFHB — human stroma-free Hb; HPOLYHB — human polyHB.

even heterologous polyHb showed much lower antigenicity when compared with albumin (Fig. 3.5).

This low antigenicity agrees with the previous results that show Hb to be weakly antigenic. Furthermore, the effect of crosslinking of Hb into PolyHb in increasing immunogenicity is only apparent under heterologous conditions. Thus, the data suggest that within the same species, crosslinking of Hb into PolyHb does not increase its antigenicity. On the other hand, repeated subcutaneous injection of heterologous Hb or PolyHb in Freund's adjuvant resulted in antibody production.

Effects of infusion

Repeated subcutaneous injection with Freud's adjuvant is a very severe test that does not represent what happens in transfusion. We, therefore, carried out the following study to test the effect of the infusion (Chang and Varma, 1987, 1988; Chang *et al.*, 1992).

First, four groups of rats that had not received any Hb infusion before were used. Each group received one of the following infusions equivalent to 30% total blood volume: rat Hb; rat polyHb; Hb from another species (e.g. human); and polyHb from another species

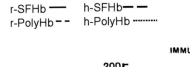

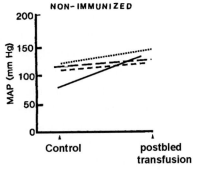
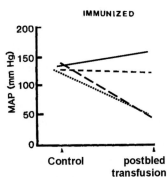

Fig. 3.6. (*Left*) Rat or human SFHb or PolyHb infused into rats that have not been immunized. No adverse effects. (*Right*) Rat or human SFHb or PolyHb infused into rats that have been immunized with the respective solutions. No adverse effects with rat SFHb or rat PolyHb. Adverse effects with heterologous sources, e.g. human SFHb or human PolyHb.

(humans). Infusion of any one of the four solutions did not result in any adverse effects (Fig. 3.6).

The next study was carried out in immunized rats as follows. The four groups of rats each received subcutaneously immunizing doses of one of the above four solutions in Freud's adjuvant. After this, the antibody titers were analyzed. The rats in each group then received infusion of the corresponding Hb solution. The volume was 30% of the total blood volume. Infusion of rat Hb or rat polyHb into rats that had received immunizing doses in Freud's adjuvant of the respective Hb and PolyHb, did not result in any adverse effects (Fig. 3.6). However, infusion of Hb or polyHb from another species into the immunized rats resulted in severe anaphylactic reactions (Fig. 3.6).

The above infusion studies showed that infusion of polyHb from the same species either for the first time or under the most severe immunization condition, did not result in any adverse response (Fig. 3.6). Infusion of polyHb from a different species for the first time also did not result in any adverse response. However, when polyHb from a different species was infused after the most severe immunization procedures, there were anaphylactic reactions. Thus, the result seems to show that polyHb from the same species would not have any immunological problems. It also shows that polyHb from a different species also does not produce any adverse effects when infused for the first time. This finding is supported by another group in actual clinical trials using large volumes of bovine PolyHb, which showed no adverse clinical immunology effects in humans (Pearce and Gawryl, 1998; Pearce *et al.*, 2006). What one cannot conclude from these results is how often can polyHb from another species be infused before there is immunological problem. Estep's group (Estep *et al.*, 1992), infused diaspirin crosslinked human Hb into Rheus monkeys. With 5 ml/kg monthly infusion, there was no production of IgG or IgM to the crosslinked Hb). Thus, this type of crosslinking did not result in any increase in antigenicity. They showed that five intravenous monthly infusions of human crosslinked Hb into a different species, Rheus monkey, did not result in any antibody production. This shows that the very severe test of subcutaneous injection of heterologous Hb with Freud's adjuvant is more likely to produce an immune response than intravenous infusion.

3.8. Bridging the Gap between Safety Studies in Animal and Humans

Response in animals is not always the same as for humans. This is especially in tests for hypersensitivity, complement activation and immunology. How do we bridge the gap between animal safety studies and use in humans?

In vitro screening test using human plasma before use in humans

We have devised a simple *in vitro* test tube screening test (Chang and Lister, 1990, 1992, 1993a, 1993b; Chang *et al.*, 1993b). This consists of adding 0.1 ml of modified Hb to a test tube containing 0.4 ml of human plasma or blood. The plasma is then analyzed for complement activation of C3 to C3a after incubation for one hour (Fig. 3.7). The use

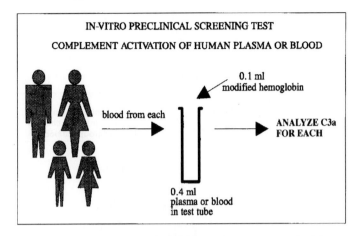

Fig. 3.7. *In vitro* screening test for PolyHb: This consists of adding 0.1 ml of PolyHb to a test tube containing 0.4 ml of human plasma or blood. The plasma was then analyzed for complement activation of C3 to C3a after incubation for one hour. Closest response to humans next to actual injection into the body. (Chang & Lister, 1993a.)

of human plasma or blood gives the closest response to humans next to an actual injection. If we want to be more specific, we can use the plasma of the same patient who is to receive the blood substitute.

Basic principles of screening test

Figure 3.8 summarizes the factors that initiate alternate and classical pathways leading to complement activation of C3 to C3a. Modified Hb may be contaminated with trace amounts of blood group antigen that can form antigen-antibody complex which can activate C3. Others include endotoxins, microorganisms, insoluble immune-complexes, chemicals, polymers, organic solvents and others (Chang, 1997). In the preparation or production of blood substitutes, different chemicals, reagents and organic solvents are used. These include crosslinkers, lipids, solvents, chemicals, polymers and other materials. Some of these can potentially result in complement activation and other reactions in humans. Other potential sources of problems include

Artificial Cells

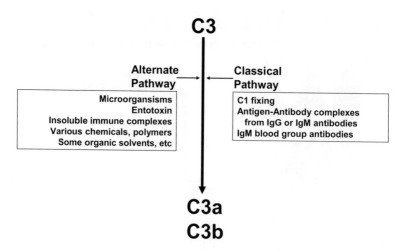

Fig. 3.8. Factors that activate complements in the classical and alternate pathways.

trace contaminants from ultrafilters, dialyzers and chromatography (Chang and Lister, 1994).

Use in research

Why wait for the completion of research, industrial production and preclinical animal studies? Why don't we do this test right at the beginning, during the research stage? If a new system is found to cause complement activation at this stage, one can avoid tremendous waste of time and money in further development, industrial production and preclinical animal study (Table 3.5) (Chang and Lister, 1994).

Use in industrial production

We have also used this preclinical test to help others in industrial production (Chang and Lister, 1994; Chang *et al.*, 1993). Thus in one of the earlier industrial scale-ups of polyHb, this *in vitro* screening test showed that certain batches caused complement activation. By conducting this test further, it was shown that complement activation was the result of using new ultrafiltrators. Re-used or washed

Table 3.5 Use of the *In Vitro* Screening Test

RESEARCH STAGE
POLYMERS, LIPIDS, CROSSLINKERS, CHEMICALS, SOLVENTS
AND OTHERS
Some of these can cause complement activation
Others may have different responses depending on the source of the material

ANIMAL SAFETY STUDIES
HUMAN RESPONSE NOT ALWAYS THE SAME AS ANIMAL
The *in vitro* screening test using human plasma or human blood is
useful in solving this problem without having to inject the material into
the human subject

INDUSTRIAL PRODUCTIONS
BATCH VARIATIONS, CONTAMINATIONS, CHEMICALS, ETC
A case in point: for unknown reasons, some batch causes complement activation.
Using the *in vitro* screening test, the cause was found to be the use
of new ultrafiltrators that release material that activate complements.
Repeated washing before use solves this problem

CLINICAL TRIALS AND USES
PRECLINICAL TRIAL FINAL SCREENING TEST
ANALYSIS OF VARIATIONS IN POPULATION RESPONSES

ultrafiltrators did not cause complement activation. Without this test, some batches could result in adverse effects of "unknown causes" in humans. Chromatography, ultrafiltrators, dialysis membranes and other separation systems are used extensively in the preparation of different types of blood substitutes. It is, therefore, important to screen for the possibility of trace contaminants that could cause complement activation. In the same way, different chemical agents and different reactants used in industrial production could be similarly tested (Table 3.5).

Correlation of in vitro complement activation to clinical symptoms

What are the clinical implications of C3a levels in the above *in vitro* complement activation screening test? (Table 3.5) There are clinical data in dialysis patients using different types of hemodialysis membranes. The result shows that clinical symptoms appear when

C3a is substantially elevated (Deane and Wineman, 1988). Increasing levels result in increasing severity of the symptoms. Symptoms include myalgia, chest tightness, fever, chill and others. C3a is a smaller molecule (M.W. 9000) than C3 (M.W. 180,000). Therefore, once it is formed, C3a equilibrates rapidly across the capillaries. Thus, in measuring C3a in patients one has to do this within very short intervals to catch the peak in C3a. The peak is reached in the first 15 min. After this, it declines rapidly to normal in 60 min. In following complement activation in patients receiving blood substitutes, one has to catch the peak. In the *in vitro* study, the C3a is retained inside the test tube and its concentration does not decline with time; therefore, the maximal level of C3a would be available.

Clinical trials and use in humans

The *in vitro* test may be useful in large scale screening for human response (Chang and Lister, 1994). For example, it could be used to study variation in production batches; to study individual variations; and to analyze the response of different human populations, especially with different disease conditions. It is important to note that all these could be done without ever introducing any blood substitutes into humans.

3.9. Design of Animal Study for Hemorrhagic Shock

Different types of modified Hb have been shown to be effective in the resuscitation of hemorrhagic shock. However, some of the results need to be further analyzed as discussed below.

Hemorrhagic shock model

When a small volume of blood is lost, the use of fluid to replace the volume is enough to maintain blood pressure and perfusion. It is only when a very large volume of blood is lost that modified Hb is needed (Table 3.6). Messmer (1988) has discussed volume replacement in hemorrhagic shock in some detail. Solutions for plasma replacement

Table 3.6 Validity of Hemorrhagic Shock Models

HEMORRHAGIC SHOCK MODELS
(1) **REMOVAL OF 30–40% BLOOD VOLUME**
* Needs volume replacement
* Does not need rbc replacement
An excellent model for testing volume replacement
<u>BUT</u> not a valid model for testing rbc replacement using blood substitutes
(2) **REMOVAL OF 67% BLOOD VOLUME (WIGGER'S MODEL)**
* Needs volume replacement
* Needs rbc replacement
Thus a valid model for testing rbc replacement using blood substitutes

include saline, Ringer's lactate, hypertonic saline, colloids and others. Most of the studies on volume replacement are based on bleeding 30–40% of the blood volume. These models are excellent for the study of volume replacement when red blood cell replacement is not needed. However, this model is not sensitive enough for testing the efficacy of red blood cell replacement in hemorrhagic shock.

Testing the effectiveness of blood substitutes for both volume and red blood cell replacement requires a more severe model (Table 3.6). Wigger's (1950) modified model involves two stages of bleeding, resulting in the removal of 2/3 of the total blood volume. We modified the model using a chronic tail-cannulation technique (Tabata and Chang, 1982b) to form an unanesthetized model (Chang and Varma, 1991). Another group used another chronic cannulation technique (Malcolm *et al.*, 1992) that works as well. Using these unanesthetized models, PolyHb blood substitutes are shown to be more effective than standard volume replacement using 3 volumes Ringer's lactate or 7.5% hypertonic-saline/6% dextran 70 plus 3 volumes Ringer's lactate (Keipert and Chang, 1985; Chang and Varma, 1992).

Experimental designs

Different experimental designs have been used. Some experimental designs are for a single blood replacement. They are not followed

Artificial Cells

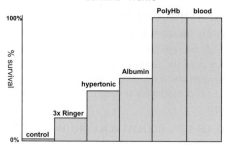

Fig. 3.9. Long-term survival and recovery of 67% blood loss replaced with different types of solution. Only blood and PolyHb resulted in long-term survival.

with other treatments or monitoring until recovery or death. This is more comparable to emergency replacement in major disasters or wars in the field, i.e. outside the hospital settings with no constant monitoring. Other experimental designs are for repeated infusions with continuous monitoring to maintain the blood pressure. This is more applicable to surgical or hospital settings. Some designs are more suitable for battlefield injuries. Furthermore, some experiments are acute experiments of a few hours. Here, only the vital signs are continuously monitored.

For example, we studied the effect of a single transfusion on the long-term survival of animals. The results (Fig. 3.9) showed that only whole blood and polyHb were effective for the long-term recovery of the animals. If we did not follow the long-term effects, the short-term response would have shown that stroma-free Hb is as effective as polyHb. Thus, proper experimental design is needed to show the effectiveness of using this for specific clinical conditions. Some applications only require short-term effectiveness with constant monitoring. Other applications may require longer-term effectiveness without the availability of constant monitoring or medical care. Thus, it is important not to make general conclusions regarding the effectiveness of blood substitutes in hemorrhagic shock or in other applications.

3.10.   General Discussions

Regional differences and the potential of unknown infective agents must be included in any discussions of the future prospect of blood substitutes. Also to be included is the degree of regulatory requirements. Blood substitutes are urgently needed in regions of the world where there are severe shortages of donor blood because of cultural or religious beliefs that cause people to be less willing to donate blood. They are also urgently needed in regions with higher incidences of infective agents like HIV and thus a higher potential for contaminated donor blood. It is less urgent in regions with a lower incidence of HIV and where costly screening tests are being used to screen out infective agents in donated blood. On the other hand, it is important to remember the past unexpected outbreaks of HIV and hepatitis C and the resulting contaminated donated blood that persisted for years until proper screening tests are developed. If this should happen again with some yet unknown agents, (e.g. avian flu and others), then it would again be disastrous if no blood substitutes, even first-generation blood substitutes, are immediately available. Past experience has shown that it takes many years to develop ideas on blood substitutes into products and that lack of basic information has resulted in much failure and delays. It is important to carry out basic research to gain important basic information needed for the simultaneous development of blood substitutes. In the meantime, two types of first-generation nanodimension polyHb are in the final stages of clinical trials in humans and one of these has been approved for routine clinical use in patients in South Africa. New nanodimension conjugated Hb is also being tested in clinical trials. Shortage of human Hb is being resolved by studies on recombinant human Hb, placenta Hb, bovine Hb and synthetic heme. Meanwhile, new generations of modified Hb are being developed that can modulate the effects of nitric oxide for those clinical applications that may have potential problems related to oxygen radicals. PolyHb can be crosslinked to an enzyme to suppress the growth of tumors. A further development is the use of PEG-lipids or PEG-biodegradable polymer membranes to prepare nanodimension artificial red blood cells containing Hb and complex enzyme systems.

3.11. Methods and Procedures

For those entering this area of research for the first time, a major complaint would be the lack of any books containing laboratory procedures for the preparation and testing of blood substitutes. This section describe the procedures that readers can readily use in their laboratory for preparing and testing modified Hb. The following is a simplified laboratory method for preparing polyHb. Also described are the experimental procedures for testing blood substitutes. Most of the basic laboratory principles are covered and these can form the basis for extension into new approaches. The method is suitable for preparing basic modified Hb for use in different laboratory research. However, it is not for use in preclinical or clinical applications. More detailed procedures are required in such settings, especially in the preparation of ultrapure stroma-free Hb. Other important aspects include stringent care in sterility; testing for endotoxin and complement activation; freedom from contaminating stroma fractions; and viral and bacterial inactivation. Those complicated and expensive procedures for practical use in larger industrial facilities are not included here.

Special method for preparation of PolyHb containing low levels of tetrameric Hb (0.2 to 2%)

Crosslinking reaction. This is carried out in a cold room at 4°C.

(1) Place 117 ml of a 8.55 g/dl Hb in a 500 ml flask. Add 40 ml 0.2 M sodium phosphate buffer, pH 7.4 and 4 ml 4 M NaCl.

(2) Add 1.33 ml of 1.3 M lysine monohydrochloride* in a 0.1 M phosphate buffer to the Hb solution (to a final molar ratio of 11.1:1). Fill the flask with nitrogen gas and seal opening. Place the flask in a shaker at 140 rpm at 4°C for 1 h.

(*1.3 M lysine monohydrochloride is prepared as follows:11.87 g (SIGMA or FISHER) + 50 mL 50 mM sodium phosphate buffer, pH 7.4;

this solution is kept at −20°C and refrozen immediately before being used.)

(3) Without stopping the shaker, slowly add 5 ml 0.5 M ice-cold degassed glutaraldehyde* (to a final molar ratio of M:M – 16.1:1). Fill the flask with nitrogen gas, seal the opening and continue shaking at 4°C for 24 h.

(*0.5 M glutaraldehyde is prepared as follows:10 mL 25% glutaraldehyde (SIGMA, G-5882, 10 × 10 mL) + 42.3 mL 50 mM sodium phosphate buffer, pH 7.4 and is kept at – 20°C. Defrost and degas this solution by Ultra Sone 30 min immediately before the reaction.)

(4) Without stopping the shaker, slowly add 31.6 g dry lysine to stop the reaction (until M:M ratio = 1118:1). Fill the flask with nitrogen, seal the opening and continue shaking at 150 rpm at 4°C for 1 h.

(5) Centrifuge at 16,000 g for 60 min [8000 rpm (JLA-8.1 rotor)] at 4°C.

Preparative chromatography. (For those who are preparing polyHb for *in vitro* studies and not interested in preparing polyHb with <2% tetrameric hemoglobin, the following complicated step is not needed.)

(1) Use 3 columns with Sephacryl S-300 (4.8 × 95 cm), equilibrated with saline. Load 3 g of polyHb to each column. The elution speed must be 100–120 ml/h. After the elution of the free volume (about 600 ml), continuously collect 10 ml fractions.

(2) Take 50 μL aliquots from each fraction to semimicro ultraviolet transparanted cuvetts; add 950 μL saline and measure the optical density at 280 nm.

(3) Take all the fractions corresponding to an optical density of more than 0.4000 until the eluate volume is 120–130 ml. Pool the 3 column fractions together. (Hb) is about 500–600 mg/dl (2000–2500 mg).

(4) Place the sample in a magnetic mixing ultrafiltrator (500 kDa cutoff 400 ml container) under nitrogen at 30 psi. Ultrafilter until the final volume reaches 50 ml or less.

(5) Dialyze twice in dialysis membrane tubing (M.W. cutoff 12–14 kDa, Spectrum Laboratories, Inc.) against 1.8 l Lactated Ringer's solution for 2 h.

(6) Centrifugate at 6,500 g [8,000 rpm (3745 rotor)] for 20 min at 4°C.

(7) Carry out sterile filtration with 0.22 μm filters with prefilter. Distribute the solution into 10 ml aliquots and keep in sterile condition under nitrogen at 4°C until use.

Prevention metHb formation. Additional steps can be taken to prevent increase in metHb during the preparative procedure. This includes those that have been described (Li *et al.*, 2005) or the use of carbon monoxide to form CO-Hb before crosslinking.

Analytical chromatography for molecular weight distribution. Sepacryl S-300 Column 1.5 × 98 cm; equilibrated with 0.1 M Tris-HCl, pH 7.5, 0.15 M NaCl. Load 1.5 mg polyHb in 300 μL elution solution with 5% glycerol. Sensitivity "5". Elution velocity = 32 ml/h. Recording velocity = 1 mm/min. Calculate percentage of area for: a) >445 kDa; b) 100–445 kDa; c) 100 kDa.

In vitro screening test using human plasma before use in humans

Laboratory procedure based on human plasma. Blood is obtained by clean venous puncture from human volunteers and put into 50 ml polypropylene (Sastedt) heparinized tubes (10 IU heparin/ml of blood). Immediately separate plasma by centrifugation at 5500 g at 2°C for 20 min and freeze the plasma in separate portions at −70°C. Do not use serum because coagulation initiates complement activation. EDTA should not be used as an anticoagulant, because it interferes with complement activation. Immediately before use, the plasma sample is thawed. 400 lambdas of the plasma are pipetted into 4 ml sterile polypropylene tubes (Fisher). 100 lambdas of pyrogen-free saline (or Ringer's lactate) for injection is added to the 400 lambdas of human plasma as control. One hundred lambdas of Hb or modified Hb are added to each of the other tubes containing 400 lambdas of human plasma. The reaction mixtures are incubated at 37°C at 60 rpm for 1 h in a Lab-Line Orbit Environ Shaker (Fisher Scientific, Montreal, Canada). After 1 h, the reaction is quenched by adding 0.4 ml of this solution into a 2 ml EDTA sterile tube containing 1.6 ml of sterile saline. The samples are immediately stored at −70°C until analyzed. The analytical kit for human complement C3a is purchased from Amersham, Canada. The method of analysis is the same as the instructions given in the kit, except for two minor modifications. Centrifugation is carried out at 10,000 g for 20 min. After the final step of inversion, the inside walls of the tubes are carefully blotted with Q-tips. The exact procedure and precise timing described above are important in obtaining reproducible results. The base line control level of C3a complement activation will vary with the source and procedure of obtaining the plasma. Therefore, a control baseline level must be used for each analysis. Furthermore, all control and test studies should be carried out in triplicates. Much practice is needed to establish this procedure before using the method for the first time. Reproducibility must be established before this test can be used.

Procedure based on blood from finger pricks. Instead of using plasma and to obviate the need to withdraw blood with a syringe, a simpler procedure involving obtaining blood from finger pricks can be carried out as follows.[90] Sterile methods are used to prick a finger. Blood is collected in heparinized microhematocrit tubes. The tubes are kept at 4°C, then used immediately as follows. Each blood sample used in the test contains 80 ul of whole blood and 20 ul of saline. Each test solution is added to a blood sample. Test solutions include saline (negative control) and Zymosan (positive control) or Hb. This is incubated at 37°C at 60 rpm. After 1 h of incubation, EDTA solution is added to the sample to stop the reaction. The analysis for C3a is then carried out as described. The test kit is based on ELISA C3a enzyme immunoassay (Quidel Co, San Diego, CA, USA).

Animal study on hemorrhagic shock and exchange transfusion

Animal model. Depending on the design of the experiment, the animal could be acutely cannulated under anesthesia. If a conscious animal experiment is needed, the animals are prepared for chronic cannulation as follows. Sprague-Dawley rats, 340 ± 40 g, from Charles River are randomly divided in groups of six ($n = 6$). They are allowed to acclimatize for four days. They then receive chronic cannulation as referred to above. During the 6–8 days of post-surgical period, the catheters are flushed daily with sterile heparinized saline. Body weights and hematocrits are followed and only those with the following values 6–8 days after surgery are used. These include average body weight of 361.14 ± 10.53 g and hematocrit of 37 ± 2.7%. In the case of conscious rats, they are partially restrained, allowing free access to the tails and catheters. In anesthetized rats, cannulation and experiment are carried out under general anesthetics. Each rat receives intravenous heparin, 150 IU/kg. One arterial cannula is connected to a statham pressure transducer for measuring of blood pressure and heart rates. These are recorded continuously on a Grass polygraph recorder. After steady baseline recording, lethal hemorrhagic shock is induced. This involves the removal of 67% total blood volume through the other arterial catheter at 0.5 ml/min, carried out in two stages. Removal of 36% of total blood volume is followed by a 10 min rest period. The second stage involves the removal of 31% of total blood volume. Throughout the bleeding, the blood pressure is maintained between 40 and 60 mm Hg. 67% of total blood volume is removed from each rat with an average time of 46.52 ± 2.71 min.

Resuscitation solutions. There are eight groups of rats with six rats randomly assigned to each group.
1. Controls — no resuscitation fluid.
2. Reinfusion of rats's own shed whole blood.
3. Ringer lactate solution. The volume is equivalent to the volume of shed blood.
4. Ringer lactate solution equivalent to three times the volume of shed blood.
5. Sterile human albumin 7 g% in Ringer's lactate solution. The volume is equivalent to the volume of shed blood.
6. Stroma-free Hb in Ringer's lactate solution. The volume is equivalent to the volume of shed blood.
7. PolyHb or other blood substitutes in Ringer's lactate solution. The volume is equivalent to the volume of shed blood.
8. 4 ml/kg (7.5g% NaCl/7g% Dextran 70 followed by three times the volume of shed blood of Ringer's lactate solution.

Immediately after shock induction, each rat receives one of the above intravenous resuscitation fluids at a rate of 0.5 ml/min. The mean arterial pressure and heart rates are recorded up to 30 min after completion of the infusion. Each rat is then returned to a separate cage. They are monitored for 14 days. Those rats that lived up to 14 days are considered as having "survived." The time of death is recorded for those that died before this time.

Exchange transfusion. This is one of the frequently used experimental procedures to test the efficacy of a given blood substitute in normovolumic hemodilution. Anesthetized male Sprague-Dawley rats with an average weight of 310 g are used. Cannulation of the femoral artery and femoral vein using 3 cm length PE10 tubing (0.28 mm inner, 0.61 mm outer diameter, Clay Adams) connected into a large diameter 20-cm length of PE50 tubing (0.58 mm inner, 0.97 mm outer diameter). These are filled with heparin (500 USP units/ml). Both catheters are connected to a pump (for example, Minipulse-2 Gelson Medical Electronics) fitted with tubing (0.89 mm inner). Rats are heparinized before exchange transfusion through the venous catheter at 60 units/100 g body weight. The polyHb solution is warmed to the body temperature of 37°C and infused intravenously into the femoral vein. Simultaneously, blood from the animal is removed from the femoral artery. The flow rate is maintained at 0.5 ml/min by adjusting the peristaltic pump. This is monitored by measuring the volume infused and the volume collected. Vital signs, hemodynamics and other measurements can be carried out as required.

CHAPTER 4

A Nanobiotechnologic Therapeutic that Transports Oxygen and Remove Oxygen Radicals: For Stroke, Hemorrhagic Shock and Related Conditions

4.1. Introduction

As discussed in Chapter 3, polyhemoglobin (PolyHb) is being used in surgical settings in the last stages of clinical trials in North America and in routine clinical use in South Africa. PolyHb promises to play an important role in surgery and the early stages of blood loss. However, in order to use blood substitutes for other clinical conditions like severe sustained hemorrhagic shock, stroke and so on, we have to consider the issue of oxygen radicals (Alayash, 2004; Chang, 1997a, 2006a). Red blood cells contain antioxidant enzymes like catalase (CAT), superoxide dismutase (SOD) and other enzymes, but in situations of prolonged and severe ischemia, even this is not enough. Furthermore, some PolyHbs are prepared from ultrapure Hb devoid of any red blood cell enzymes. Our aim is, therefore, to use nanobiotechnology to assemble these biological molecules into PolyHb-CAT-SOD, a therapeutic agent that not only contains enhanced antioxidant enzymes but also has the ability to carry oxygen.

4.2. Ischemia-Reperfusion

In prolonged severe hemorrhagic shock, stroke and other ischemic conditions, there is insufficient supply of oxygen resulting in ischemia.

Ischemia stimulates the production of hypoxanthine. Under normal conditions, xanthine dehydrogenase reduces purines to uric acid. During ischemia, this enzyme is converted into the xanthine oxidase form at a rate proportional to the duration of ischemia. When the tissue is again perfused with oxygen, reperfusion, xanthine oxidase converts hypoxanthine into superoxide. Via several mechanisms, superoxide results in the formation of oxygen radicals that can cause tissue injury. The enzymes in red blood cells help to prevent this to a limited extent. Thus, SOD converts superoxide into hydrogen peroxide that is in turn converted by CAT into water and oxygen. However, in severe and prolonged ischemia the normal amounts of these enzymes in the red blood cells are not enough to prevent ischemia-reperfusion injury. We have, therefore, prepared a PolyHb-CAT-SOD complex (PolyHb-CAT-SOD) with more CAT and SOD than are normally present in red blood cells. The idea is to have an oxygen carrier with enhanced antioxidant enzymes which can act as a therapeutic agent in ischemia-reperfusion.

4.3. PolyHb-CAT-SOD

This product is based on our earlier finding that polyHb can be prepared by crosslinking stroma-free Hb using a bifunctional agent, sebacyl chloride (Chang, 1964) (Fig. 2.2). As stroma-free Hb contained red blood cell enzymes, the crosslinking also resulted in a polyHb-enzyme complex. The first use of another bifunctional agent, glutaraldehyde, was to crosslink stroma-free Hb into polyHb, but with more CAT added to augment the CAT activity (Chang, 1971b) (Fig. 2.2). CAT crosslinked into the polyHb-CAT complex in this way is much more stable than the enzyme in free solution. Thus, the CAT activity of the polyHb-CAT remained at 98% after 43 days when stored at 4°C (Chang, 1971b). On the other hand, the activity of free CAT in solution fell to 10% in 35 days at 4°C. When stored at the body temperature of 37°C, polyHb-CAT retained 50% of its CAT activity after 7 days, whereas the activity of the free enzyme fell rapidly to 50% after 1 day.

We, therefore, used this principle to prepare PolyHb-CAT-SOD (D'Agnillo and Chang, 1998a) (Fig. 2.2). In this way, in addition to its antioxidant therapeutic component, PolyHb-CAT-SOD also acts as an oxygen carrier to supply the needed oxygen in ischemia. In the form of PolyHb-CAT-SOD, the enzymes stay in the circulation together with PolyHb that, in humans, has a half-life of 24 h. This is unlike the rapid removal of free SOD and CAT from the circulation (10 and 20 min, respectively). Furthermore, unless crosslinked to Hb, these enzymes are not located in close proximity to Hb, and are thus less likely to give adequate protection to Hb-initiated oxygen radicals. *In vitro* studies reveal that crosslinked Hb-SOD-CAT is effective in removing oxygen radicals and hydrogen hydroxide, and in minimizing damage to Hb and release of iron. Our animal studies show that PolyHb-CAT-SOD results in a significant decrease in oxygen radical related damage in an intestinal ischemia-reperfusion model as well as in a stroke model.

4.4. Enzyme Activities after Intravenous Injections

The following study shows that free SOD and CAT are removed rapidly within a few minutes from the circulation after intravenous infusion. On the other hand, after intravenous infusion of PolyHb-CAT-SOD, the enzyme activities remain in the circulation for hours with the same half-time as PolyHb (D'Agnillo and Chang, 1998a) (Fig. 4.1). This study includes the use of the following solutions injected into Sprague Dawley rats: 1) Free Hb, SOD, and CAT all in solution and not crosslinked; 2) PolyHb-SOD-CAT (prepared using a glutaraldehyde/Hb at a molar ratio of 8:1 glut:Hb); 3) PolyHb-SOD-CAT (8:1 glut:Hb) that has been ultrafiltered through 100 kDa cutoff membranes; and 4) PolyHb-SOD-CAT (16:1 glut:Hb) with a higher degree of polymerization.

Plasma Hb concentration (Fig. 4.1)

Free Hb shows a half-life of about 60 min (Fig. 4.1). A greater circulation time is achieved following polymerization (8:1 glut:Hb). Ultrafiltration does not increase the retention time since ultrafiltration through 100 kDa cut off membranes does not remove any significant fraction of tetrameric Hb. Greater polymerization (16:1 glut:Hb)

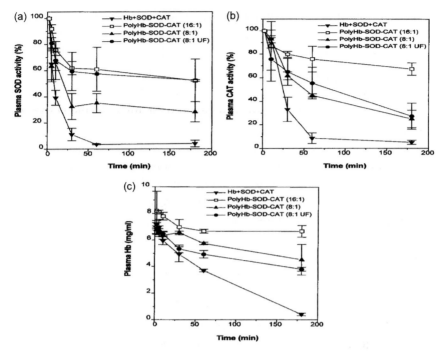

Fig. 4.1. Plasma circulation half-life in rats of (a) CAT (b) CAT (c) Hb in rats after I.V. injection of indicated solutions (300 mg Hb/kg). Data of each point on the chart obtained from samples collected from 3 separate animals. (D'Agnillo & Chang 1998a.)

decreases the fraction of free Hb molecules, resulting in a further increase in circulation time.

SOD activity in plasma (Fig. 4.1)

Free SOD is rapidly cleared from the circulation ($t_{1/2} \sim$ 4–6 min). PolyHb-SOD-CAT (8:1 glu:Hb) shows a greater retention of SOD activity with a biphasic clearance pattern. The rapid initial removal phase is likely related to some SOD that has not been crosslinked to PolyHb. Ultrafiltration removes the small fraction of non-crosslinked SOD, thus increasing the circulation half-time significantly. A higher degree of polymerization with 16:1 glutaraldehyde:Hb decreases the small fraction of non-crosslinked SOD, thus increasing the circulation half-time.

CAT activity in plasma (Fig. 4.1)

Free CAT is cleared rapidly ($t_{1/2} \sim 15$ min). PolyHb-SOD-CAT (8:1) shows a greater retention of CAT activity. Ultrafiltration of PolyHb-SOD-CAT (8:1) through 100 kDa membranes does not remove unbound CAT (mw of CAT 232,000), and there is no change in the circulation time compared with the unfiltered preparation. Greater polymerization (16:1) decreases the fraction of free CAT, resulting in an increase in the circulation time.

Discussions

Our present study shows that the circulation half-time of free SOD in solution is about 4–6 min. This may explain why in humans a single bolus injection of free SOD is ineffective against ischemia-reperfusion injuries. In theory, a continuous infusion of SOD and CAT can be given during and following the administration of PolyHb under conditions involving potential ischemia-reperfusion injuries. However, giving PolyHb-CAT-SOD would not require continuous infusion. Furthermore, when in the PolyHb-CAT-SOD configuration CAT and SOD are in the closest proximity to the Hb molecules, since PolyHb, like Hb, is a potentially reactive protein in ischemia-reperfusion injury. It is also a potential source of heme and iron release that can further increase oxygen radical formation in the Fenton reaction. Thus, in the form of PolyHb-CAT-SOD, the enzymes can act promptly to remove oxygen radicals released from reaction due to the effect of oxidants on Hb. As well, PolyHb has a circulation half-life in humans of 24 h; when crosslinked to polyHb, the enzymes circulate together with PolyHb for the same length of time. In this way, the PolyHb is at all times protected by the antioxidant enzymes. After one episode of ischemia-reperfusion in hemorrhagic shock, myocardial infarction or cerebral ischemia, further episodes can recur with varying degrees of severity and frequencies. When the enzymes are crosslinked together in the form of PolyHb-SOD-CAT, the enzymes would continue to circulate with PolyHb, thus preventing any potential problems at all times.

4.5. Tests in Animal Models for Stroke and Hemorrhagic Shock

The next step is to carry out a study using actual animal models of ischemia-reperfusion injuries comparing PolyHb-CAT-SOD with PolyHb. The three animal models chosen are ischemia-reperfusion of the intestine; ischemia-reperfusion in a combined hemorrhagic shock stroke model; and ischemia-reperfusion in an ischemia-reperfusion hind limb model. The first two are very sensitive to ischemia-reperfusion injuries, whereas the third one is not. As will be discussed in detail in a later section, after 60 min and 90 min of ischemia, respectively in the brain and intestine, reperfusion with PolyHb results in ischemia-reperfusion injuries or increase in oxygen radicals. On the other hand, PolyHb-CAT-SOD does not have such adverse effects on the intestine (Razak, D'Agnillo & Chang, 1997) or brain (Powanda & Chang, 2002). The hind limb is less metabolically active and reperfusion with PolyHb after 90 min of ischemia does not have any significant increase in oxygen radical production (D'Agnillo & Chang, 1997c).

4.6. Ischemia-Reperfusion of the Intestine

Severe hemorrhagic shock and the intestine

Since PolyHb has no blood group antigens and can be stored for a much longer time than donor blood even at room temperature, there is the potential of using this onsite for the treatment of hemorrhagic shock. However, if the ischemia is severe and prolonged before reperfusion, there could be ischemia-reperfusion injuries. In severe sustained hemorrhagic shock, there is severe vasoconstriction of some organs, especially the intestine and kidneys. Furthermore, the intestine is rich in xanthine oxidase that catalyzes the formation of superoxide. Thus, the intestine is very vulnerable to ischemia-reperfusion injury and

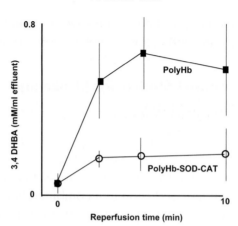

Fig. 4.2. Hydroxyl radical production assessed by the measurement of 3,4 dihydroxybenzoate in ischemic intestine perfused with 5 g/dl of PolyHb (■) or 5 g/dl of PolyHb-SOD-CAT (●). Intestinal effluent samples are collected during reperfusion at the indicated times. Data are presented as means ± SD. *denotes statistical significance; $P < 0.05$. (From Razack, D'Agnillo & Chang, 1997.)

therefore it would be a sensitive model for testing ischemia-reperfusion injury. In severe sustained hemorrhagic shock, reperfusion with red blood cells that contain the normal amounts of CAT and SOD may not be adequate to prevent ischemia-reperfusion. Since PolyHb-CAT-SOD can contain more enzymes than red blood cells, we tested this preparation for reperfusion in an ischemia-reperfusion intestine model (Razack, D'Agnillo & Chang 1997). Figure 4.2 shows that the use of PolyHb for reperfusion results in a significant increase in oxygen radicals. The use of PolyHb-CAT-SOD does not give rise to such an increase in oxygen radicals.

Test on effects of reperfusion on ischemic intestine

The details of the rat model and the measurement of oxygen radicals are described in Section 4.11 under "Methods and Procedures." After 90 min of ischemia, PolyHb or PolyHb-SOD-CAT (5 g/dl) containing 5 mM of 4-hydroxybenzoate (4HB) is reperfused at 0.8 ml/min for 10 min without recirculation. Outflows from the portal veins are collected, centrifuged, and the spun perfusates are stored at $-80°C$ for later analysis of hydroxyl radicals. In this analysis, hydroxyl radical reacts with 4HB to form 3,4-dihydroxybenzoate (3,4-DHBA). High performance liquid chromatography measures the 3,4-DHBA produced and thus the amounts of hydroxyl radicals generated.

Reperfusion using PolyHb resulted in significant increases in the level of 3,4-DHBA corresponding to increases in hydroxyl radicals (Fig. 4.2). On the other hand, infusion of PolyHb-CAT-SOD only results in a minimal increase in hydroxyl radicals as measured by the production of 3,4-DHBA (Fig. 4.2).

4.7. Test in Combined Hemorrhagic Shock and Stroke Rat Model

Prolonged hemorrhagic shock and stroke model

Stroke is common in North America, claiming some 500,000 victims each year. The most common cause is arteriosclerosis, which results in narrowing and obstruction of blood vessels. Red blood cells, being 8 micron in diameter, have difficulty flowing through these partially obstructed vessels to supply the needed oxygen to the brain (Fig. 4.3). PolyHb, being a solution, can perfuse partially obstructed vessels better than red blood cells to supply the needed oxygen. However, when an ischemic brain is perfused with an oxygen-rich solution, reactive oxygen species will be generated, resulting in ischemia-reperfusion injury. These oxygen radicals can lead to increased leakiness of the blood-brain barrier, brain edema and cell death (Fig. 4.3).

The therapeutic agent, PolyHb-SOD-CAT, has the dual function of an oxygen carrier and the ability to remove oxygen radicals. Thus, we used a transient global ischemia-reperfusion rat brain model to compare the use of this PolyHb-SOD-CAT solution with four other oxygen carrying solutions: 1) PolyHb; 2) PolyHb with CAT and CAT added as solution without crosslinking to polyHb; 3) purified Hb; and 4) oxygenated saline solution. Reperfusions are initiated 1 h after ischemia and the effects on blood-brain barrier and brain edema are followed for 6 h. Reperfusion with PolyHb-SOD-CAT solution did not result in brain

Artificial Cells

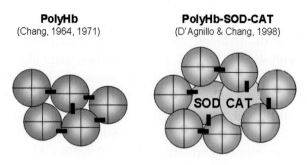

PolyHb
(Chang, 1964, 1971)

PolyHb-SOD-CAT
(D'Agnillo & Chang, 1998)

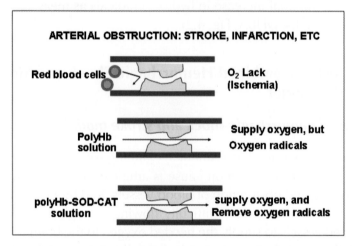

ARTERIAL OBSTRUCTION: STROKE, INFARCTION, ETC

Red blood cells

O₂ Lack
(Ischemia)

PolyHb
solution

Supply oxygen, but
Oxygen radicals

polyHb-SOD-CAT
solution

supply oxygen, and
Remove oxygen radicals

Fig. 4.3. Stroke is common and in North America and about 500,000 persons suffered from this each year. The most common cause is arteriosclerosis which results in narrowing and obstruction of blood vessels. Red blood cells being 8 micron in diameter have difficulty flowing through these partially obstructed vessels to supply the needed oxygen to the brain. PolyHb being a solution can perfuse partially obstructed vessels better than red blood cells to supply the needed oxygen. However, when an ischemic brain is perfused with an oxygen-rich solution, reactive oxygen species would be generated resulting in ischemia-reperfusion injury. Unlike PolyHb, PolyHb-SOD-CAT has the dual function of an oxygen carrier and the ability to remove oxygen radicals. (Figure adapted from Chang, *Nature Rev Drug Discovery*, 2005i.)

edema and the effect on blood-brain barrier is minimal. On the other hand, when compared with the sham control, the use of the other four solutions resulted in brain edema and disruption of the blood-brain barriers (BBB). Theses studies show that in this transient global ischemia-reperfusion rat brain model, the PolyHb-SOD-CAT solution can carry oxygen to ischemic brain tissue without causing significant ischemia-reperfusion injury.

Rat model of combined hemorrhagic shock and stroke

This is a combined model of hemorrhagic shock and bilateral carotid arteries occlusion transient global ischemia-reperfusion rat model (Smith *et al.*, 1984). Since the brain is particularly sensitive to ischemia-reperfusion injuries, we used this rat model for the studies described in the rest of this Section (Powanda and Chang, 2002). The detailed procedure for preparing this model is described in Section 4.11 under "Methods and Procedures." Male Sprague-Dawley rats (235–255 g) are anesthetized with intraperitoneal injection of pentobartbital (Somnotol, 65 mg/kg). Blood is withdrawn from the left femoral-artery cannula to maintain the mean blood pressure at 30–35 mm Hg. Then both the common carotid arteries are temporarily occluded with microvascular clips to cut the circulation to the brain.

Duration of stroke and reperfusion injuries

First, we carried out studies to analyze the relationship between the duration of ischemia before reperfusion and the severity of ischemia-reperfusion injuries using an ischemia time of 0 (control), 20, 30, 40 and 60 min, respectively (Powanda and Chang, 2002). After one hour of reperfusion, one of the oxygen carriers containing no enzymes is infused into the rats in each group: oxygenated saline, stroma-free Hb, PolyHb and sham control. The effect of the duration of ischemia before reperfusion on the blood-brain barrier is measured using Evans blue extravasations. Under normal conditions, the tight intercellular endothelial junctions of the brain capillaries do not allow the albumin-bound dye, Evans blue, to enter from the circulating blood. Ischemia-reperfusion injuries disrupt the blood-brain barrier, resulting in leakiness of the normally tight intercellular endothelial junctions to the large albumin-bound dye molecule, Evans blue.

Figure 4.4 summarizes the results obtained. The results show that for an ischemic duration up to 30 min, reperfusion does not significantly

Artificial Cells

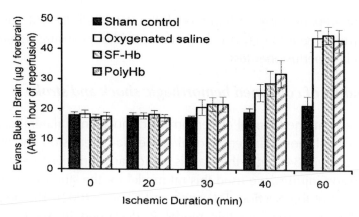

Fig. 4.4. Reperfusion after 0, 20, 30, 40 or 60 min of ischemia. 60 min of ischemia resulted in the most significant increase in Evans blue extravasations compared with sham control ($P < 0.01$). (Powanda & Chang, 2002.)

increase the disruption of blood-brain barrier, as shown by Evans blue measurement. An ischemia of 40 min significantly increases the disruption of the blood-brain barrier, with increased extravasations of Evans blue. One hour of ischemia produced the most significant effect and this is chosen for the following ischemia-reperfusion studies.

Effects of different solutions on reperfusion injuries

Ischemia is terminated in 1 h by removing the bilateral carotid arteries occlusion and restoring the blood pressure by replacing the lost blood, using one of the solutions of each of the following groups: oxygenated saline, ultrapure hemoglogin, ultrapure Hb containing solutions of CAT and CAT, PolyHb, PolyHb-SOD-CAT and sham control. Animals in the sham control group are subjected to the same surgical procedure as described above, with the exception that ischemia is not induced and no infusion is carried out.

Disruptions of blood-brain barrier

This is based on the increased leakiness of the blood-brain barrier with resulting influx of Evans blue into cerebral tissues. Compared with the other solutions, there is significantly less increase in the cerebral

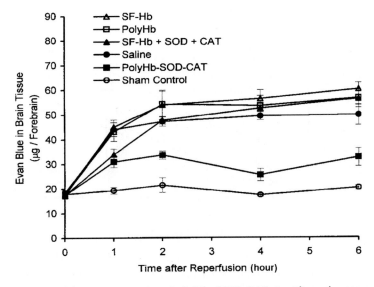

Fig. 4.5. Evans blue extravasation. PolyHb-SOD-CAT significantly attenuated the severity of BBB disruption as compared to saline, SF-Hb, SF-Hb + SOD + CAT, and PolyHb. Statistical significance is $P < 0.01$. (From Powanda & Chang, 2002.)

tissue Evans blue levels in the rats receiving Poly-SOD-CAT after one hour of ischemia, when followed for 2, 4 and 6 h (Powanda & Chang, 2002) (Fig 4.5). In the other groups, by 2 h after reperfusion, there are already significant increases in cerebral tissue Evans blue in the groups receiving: oxygenated saline ($47 \pm 2\,\mu g$), PolyHb ($54 \pm 4\,\mu g$), stroma-free Hb ($54 \pm 6\,\mu g$); and stroma free-Hb containing solutions of SOD and CAT ($48 \pm 2\,\mu g$ and $56 \pm 4\,\mu g$ by the 6th h) (Fig 4.5). It is interesting to note that the solutions of SOD and CAT do not have the same protective effect as when they are crosslinked into PolyHb-CAT-SOD. This is likely because in free solution, SOD and CAT are rapidly removed from the circulation within minutes. On the other hand, the enzymes in PolyHb-CAT-SOD circulate with PolyHb for hours.

Brain edema

Cerebral edema is measured as changes in cerebral water content. There is no significant difference between the group that received

Artificial Cells

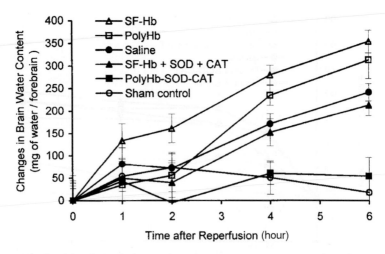

Fig. 4.6. Brain edema: changes in brain water content. The changes in brain water content of PolyHb-SOD-CAT treated animals are not significantly different from that of the sham control. The increase in water contents of saline, SF-Hb, SF-Hb + SOD + CAT, and PolyHb are significantly different from that of the sham control and PolyHb-SOD-CAT group by the 4th h continued to increase and thereon with time. Statistical significance is $P < 0.01$.

PolyHb-SOD-CAT one hour after ischemia and sham control group when followed for 2, 4 and 6 h, respectively (Powanda and Chang, 2002) (Fig 4.6). On the other hand, significant increases in cerebral water content are observed for those groups that received the following solutions after one hour of ischemia: PolyHb (234.8 ± 21.9 mg); free-Hb plus SOD and CAT in free solution (152.3 ± 30.2 mg); oxygenated saline (171.3 ± 23.0 mg); and SF-Hb (279.8 ± 21.8 mg). Except for PolyHb-CAT-CAT, all the other solutions like oxygenated saline, SF Hb and polyHb resulted in cerebral edema. Again, the results show that CAT and SOD have to be crosslinked into PolyHb-CAT-SOD to be effective in protecting against ischemia-reperfusion.

4.8. Discussion

The above studies show that in ischemia-reperfusion, the use of PolyHb-SOD-CAT resulted in significantly less production of oxygen radicals as compared with PolyHb. This difference, as is expected, is much more obvious in the case of ischemia-reperfusion of the intestine or brain. The intestine and brain are organs that are particularly sensitive to ischemia-reperfusion injury. On the other hand, the hind limb is much less sensitive to ischemia-reperfusion injury even when PolyHb is used.

The attenuation in hydroxyl radical generation with PolyHb-SOD-CAT shows promise for its potential role as a protective therapeutic agent in clinical situations of ischemia and oxidative stress. Another area of potential application would be in organ transplantation. A Korean group (Chang, Lee *et al.*, 2004a and 2004b) has done this in experimental kidney and liver transplantations. Yet another area of potential application would be the prevention of reperfusion injury associated with the states where there is a period of no perfusion. One example is cardiopulmonary bypass (CPB) with circulatory arrest. Ischemia-reperfusion injury has been noted to occur after CPB in children undergoing surgery for congenital heart disease. However, further research is needed to study these potential applications.

4.9. Preparation and Characterization

Method of preparation

Typically, each gram of Hb is crosslinked with 3000 U of SOD and 300,000 U of CAT (D'Agnillo & Chang, 1998a). The ratio of Hb to SOD and CAT (in mg/ml) is 1 : 0.009 : 0.0045. The very small amount of enzymes results in minimal change in the relative amount of Hb. Thus, there is minimal change in the oxygen carrying capacity of PolyHb-CAT-SOD when compared with PolyHb. The ratio for the enzyme can be varied over a very wide range; thus, a much higher amount of enzymes can be included when needed. The detailed method is discussed under "Methods and Procedures" in Section 4.11.

Molecular characterization of PolyHb-CAT-SOD

A molecular weight distribution analysis is performed using two methods. One is based on gel filtration chromatography on a Sephadex G-200 column equilibrated with 0.1 M Tris-HCl, pH 7.5. Another is

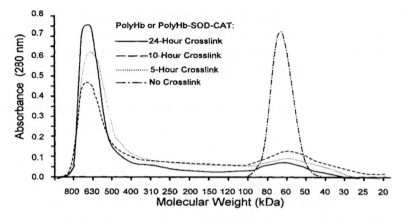

Fig. 4.7. Molecular distribution of crosslinked and non-crosslinked ultrapure Hb: Elution profiles of 5-, 10-, 24-h crosslinked PolyHb, PolyHb-SOD-CAT and non-crosslinked bovine SF-Hb are obtained by running 1 ml of the 10× diluted samples on Sephadex G-200 1.6 cm × 70 cm column (VT = 105.5 ml, Vo = 38 ml, equilibrated with 0.1 M Tris-HCl, pH 7.5, and eluted at 11.5 ml/h. (Powanda & Chang, 2002.)

based on HPLC molecular weight gel filtration analysis on a BioSep Sec S-3000 column (5000–700,000 kD exclusion limits, Phenomenex, Torrance, CA). Since only trace amounts of antioxidant enzymes are used in relation to Hb, the molecular weight distribution of PolyHb-SOD-CAT is the same as that of PolyHb (Fig. 4.7). The ratio of Hb to SOD and CAT (as mg/ml) is typically 1:0.009:0.0045. Only 5% of the enzyme is not crosslinked into the PolyHb-CAT-SOD (Powanda & Chang, 2002) (Table 4.1). There is no CAT or SOD enzyme activity in

Table 4.1 Enzyme Activity after Crosslinked into PolyHB-CAT-SOD

Sample	%SOD Activity Retained	%CAT Activity Retained
Hemoglobin + SOD + Catalase	100	100
PolyHB-SOD-CAT 8:1 Glut:Hb	90	99
PolyHB-SOD-CAT 16:1 Glut:Hb	85	95
PolyHB 8 or 16:1 Glut:Hb	<1	<1

the PolyHb prepared from ultrapure Hb. Crosslinking CAT, SOD and Hb with a glutaraldehyde: Hb molar ratio of 8:1 retains 90% of SOD activity and 99% of CAT activity. A higher molar ratio of 16:1 leads to a slightly lower enzyme activity as shown in Table 4.1.

Molecular Weight	MW Component (%)	SOD Activity (%)	Catalase Activity (%)
PolyHb-SOD-CAT sample			
>450 kDa	77	83 ± 3	82 ± 2
100 to 450 kDa	16	7 ± 2	11 ± 1
<100 kDa	7	4 ± 1	5 ± 1
PolyHb sample			
>450 kDa	77	0	0
100 to 450 kDa	16	0	0
<100 kDa	7	0	0

Oxygen dissociation curve

Figure 4.8 summarizes the results of the oxygen dissociation analysis for the 24-h crosslinked PolyHb ($n = 6$), the 24-h crosslinked PolyHb-SOD-CAT ($n = 6$) and the ultrapure Hb ($n = 6$). The P_{50} for polyHb is 26.1 ± 0.2 mm Hg, while that of PolyHb-SOD-CAT is 24.7 ± 0.2 mm Hg, and for ultrapure Hb, 27.8 ± 0.0 mm Hg.

Effects of crosslinking procedure on metHb formation

We have carried out an analysis of metHb formation (Quebec & Chang, 1995). During the crosslinking process, there is minimal formation of metHb when Hb is crosslinked with SOD and CAT (Fig. 4.9) (Quebec & Chang, 1995).

When crosslinking is carried out with Hb alone, there is a marked formation of metHb during the preparative procedure. Thus, PolyHb contains higher starting levels of metHb compared with PolyHb-SOD-CAT. This shows that crosslinking Hb with SOD and CAT may also provide oxidative protection during the preparation of PolyHb (Quebec & Chang, 1995).

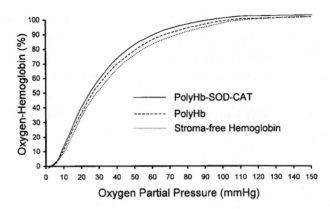

Fig. 4.8. Oxygen-Hb dissociation: 24 h of crosslinking with glutaraldehyde did not significantly affect Fig. 4.8. Oxygen dissociation curve of SF-Hb. PolyHb and PolyHb-SOD-CAT P_{50} values are 26.1 ± 0.2 mm Hg and 24.7 ± 0.2 mm Hg, respectively. SF-Hb P_{50} value is 27.8 ± 0.0 mm Hg. All oxygen-Hb dissociation curves are presented as averages of 6 trials ($n = 6$). (From Powanda & Chang, 2002.)

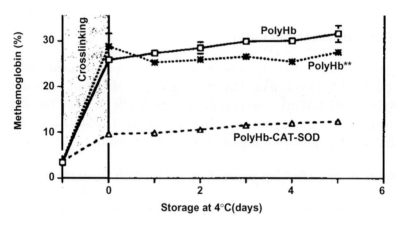

Fig. 4.9. Comparison of effects of crosslinking reaction on metHb formation: PolyHb and PolyHb-CAT-SOD. (From Quebec & Chang,1995.)

4.10.   *In Vitro* Studies of the Antioxidant Properties of Crosslinked HB-SOD-CAT

We found that PolyHb-SOD-CAT is effective in scavenging superoxide, H_2O_2 and oxygen-derived free radicals. This effect is dependent on the concentration of SOD-CAT crosslinked with Hb. Oxidative challenge with H_2O_2 resulted in minimal changes in the absorbance spectra of PolyHb-SOD-CAT. With PolyHb, there are spectral changes reflecting the formation of metHb and heme degradation. Furthermore, the amount of iron released, after incubation with H_2O_2, is significantly reduced when PolyHb-SOD-CAT is used as compared to PolyHb. These results show that crosslinked PolyHb-SOD-CAT decreases oxidative reactions. More detailed description follows.

Scavenging of superoxide ($O_2^{\bullet-}$)

Unlike PolyHb, PolyHb-CAT-SOD effectively removes superoxide that is enzymatically-generated. In this study, the level of superoxide is measured as the rate of reduction of cytochrome c by superoxide (Fig. 4.10). The higher the level of superoxide, the faster will the superoxide reduce cytochrome c. The detailed method of analysis

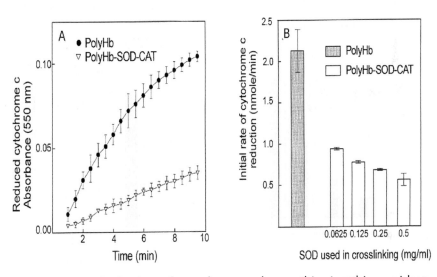

Fig. 4.10. (**A**) Reduction of cytochrome c by xanthine/xanthine oxidase-generated superoxide, in the presence of PolyHb or PolyHb-SOD-CAT (5 μM). The reaction was started by adding xanthine oxidase. (**B**) The initial rate of cytochrome c reduction of either PolyHb (5 μM) or PolyHb-SOD-CAT (5 μM) prepared with different concentrations of superoxide dismutase. Error bars represent the standard error for triplicate measurements.

is given in Section 4.11 under "Methods and Procedures." PolyHb is not effective in removing superoxide as shown by a cytochrome c reduction rate of 2.13 nmoles cyt. c/min. On the other hand, PolyHb-CAT-SOD effectively removes superoxide, and the lowered superoxide is shown by the much lower cytochrome c reduction rate of 0.56 nmoles cyt. c/min. (Fig. 4.10). Hb does not interfere with the reaction, since the rate of reduction in PolyHb mixtures is the same as that in the buffer alone. Increasing the SOD concentrations in the PolyHb-CAT-SOD increases the rate of superoxide removal (Fig. 4.10).

Scavenging of hydrogen peroxide (H_2O_2)

In the reaction mixtures containing 25, 50, 100 or 200 μM of H_2O_2, addition of PolyHb did not result in any significant decrease in the H_2O_2 level (Fig. 4.11). However, the addition of PolyHb-CAT-SOD markedly reduces the H_2O_2 level (Fig. 4.11) The H_2O_2 scavenging activity increases with increasing concentrations of CAT in the PolyHb-CAT-SOD preparation (Fig. 4.11).

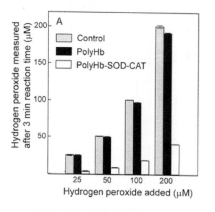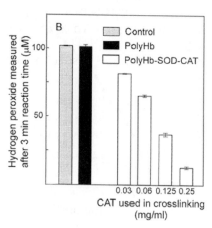

Fig. 4.11. (**A**) Peroxidase-catalyzed measurement of H_2O_2 in 3 ml reaction volumes containing added H_2O_2 and either PolyHb (5 μM) or PolyHb-SOD-CAT (5 μM). (**B**) Hydrogen peroxide measured, 3 min after the addition of 100 μM H_2O_2, in either PolyHb (5 μM) or PolyHb-SOD-CAT (5 μM) prepared with different CAT concentrations. Error bars represent the standard error for triplicate measurements. (From D'Agnillo & Chang, 1993.)

Effects of oxidative challenge on the degradation of Hb and iron release in PolyHb and PolyHb-CAT-SOD

Oxygen radicals can oxidize the Fe^{2+}-heme component of Hb to Fe^{3+}-heme that can stimulate the production of superoxide $(O_2^{\bullet-})$.[9] Furthermore, excessive oxidative damage to Hb leads to the release of iron from heme. In the presence of free iron, $O_2^{\bullet-}$, H_2O_2 rapidly react to produce hydroxyl radical ($^{\bullet}OH$) (i.e. Fenton reaction).

$$O_2^{\bullet-} + Fe^{3+} \rightarrow Fe^{2+} + O_2$$
$$H_2O_2 + Fe^{2+} \rightarrow Fe^{3+} + OH^- + {}^{\bullet}OH$$
$$O_2^{\bullet-} + H_2O_2 \rightarrow OH^- + {}^{\bullet}OH + O_2$$

Hydrogen peroxide can also react with Hb to produce ferrylHb. These can promote cellular injury by reacting with carbohydrates, nucleic acids and proteins. These reactive species can also readily abstract methylene hydrogen from polyunsaturated fatty acids resulting in lipid peroxidation. We carry out the following studies to determine whether PolyHb-CAT-SOD can prevent these effects on PolyHb.

Absorbance spectra and Hb degradation

We use absorbance spectra to follow the effects of superoxide and hydrogen peroxide on the Hb components of PolyHb and PolyHb-CAT-SOD (D'Agnillo & Chang, 1998a) (Fig. 4.6). Xanthine-xanthine oxidase is used to generate superoxide (Fig. 4.12 ABC). For PolyHb, this results in a stepwise decrease in absorbance at 540 and 577 nm, showing that Hb (Fe^{2+}) is oxidized to metHb (Fe^{3+}) (Fig. 4.12A). Increase in metHb (Fe^{3+}) is shown as a stepwise increase at the absorbance peak of 630 nm (Fig. 4.6A). PolyHb-CAT-SOD with a SOD/CAT ratio of 0.01 prevents this oxidation (B), while a SOD/CAT ratio of 100 produces no protective effect (Fig. 4.12C). A glucose-glucose oxidase system is used to generate hydrogen peroxide (Fig. 4.12 DEF). The ferryl (Fe^{4+}) intermediate is shown by the appearance of new peaks at 545 and 580 nm (Fig. 4.12D). PolyHb-CAT-SOD with a SOD/CAT ratio of 0.01 reduces this oxidation (Fig. 4.5E), while a SOD/CAT ratio of 100 produces no protective effect (Fig. 4.12F).

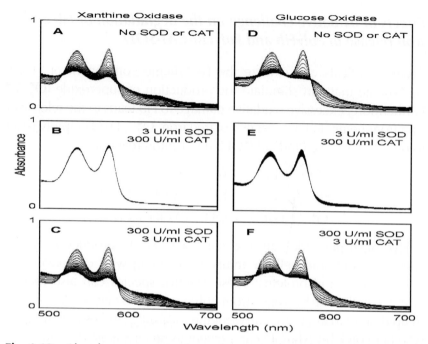

Fig. 4.12. Absorbance spectra of reaction mixtures containing 50 μM of Hb and 100 μM of xanthine, and 10 mU/ml of xanthine oxidase (*left*) or 10 mM of glucose and 10 μg/ml glucose oxidase (*right*). SOD/CAT in each mixture are shown. Each panel represents a series of spectra collected over a period of 20 min at 1 scan/min. In the absence of CAT or even the addition of 3 U/ml of CAT, there is still inadequate protection from the effects of hydrogen peroxide. 300 U/ml of CAT to have significant protective effects. (D'Agnillo & Chang, 1998a.)

In a separate study (Fig. 4.13), hydrogen peroxide is added to PolyHb (10 mM) or PolyHb-SOD-CAT (10 mM), and the absorbance spectra (450–700 nm) is followed (Fig. 4.13). For PolyHb, the spectral changes reflect the oxidation of ferrous (Fe^{2+})-heme to ferric (Fe^{3+})-heme (Fig. 4.13AB). The rate of degradation increases with increase in hydrogen peroxide added in amounts from 10 to 100 to 500 μM. The absorbance spectra of PolyHb-CAT-SOD are minimally affected, indicating that these reactions are minimized due to the elimination of H_2O_2 (Fig. 4.13CD). Similar results are observed following

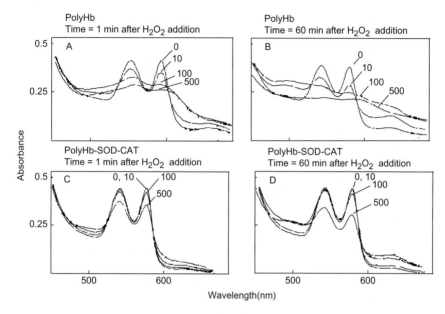

Fig. 4.13. Absorbance spectra of PolyHb and PolyHb-SOD-CAT (10 μM) following H_2O_2 addition of 0, 10, 100, and 500 μM. (From D'Agnillo & Chang, 1993.)

oxidative challenge with exogenous superoxide (O_2-) produced from xanthine/xanthine oxidase.

Release of iron

The next step is to study the impact of oxidative challenges on iron released from PolyHb and PolyHb-CAT-SOD. The methods are discussed in Section 4.11 under "Methods and Procedures." PolyHb (15 μM) or PolyHb-SOD-CAT (15 μM) are incubated with different concentrations of H_2O_2 for 1 h at 37°C. We use the Ferrozine assay method to measure the release of iron. For PolyHb, the addition of H_2O_2 leads to the release of up to 37% of the total iron in PolyHb (Fig. 4.14A). On the other hand, for PolyHb-CAT-SOD, less than 1% of the total iron is released for the same amount of H_2O_2. It is important

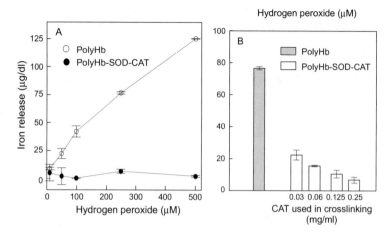

Fig. 4.14. (**A**) The release of iron, measured by ferrozine assay, from PolyHb (15 μM) or PolyHb-SOD-CAT (15 μM) incubated with H_2O_2 for 1h at 37°C. (**B**) Iron release from either PolyHb (15 μM) or PolyHb-SOD-CAT (15 μM) prepared with different CAT concentrations, following incubation with 250 μM H_2O_2. Error bars represent the standard error for triplicate measurements. (From D'Agnillo & Chang, 1993.)

to note that decreasing the amount of CAT crosslinked to the PolyHb-CAT-SOD will decrease the protective effects on the release of iron during H_2O_2 incubation (Fig. 4.14B).

Lipid peroxidation and salicylate hydroxylation

Other important mechanisms are related to lipid peroxidation and salicylate hydroxylation. As shown in Fig. 4.15 (left), adding 50 μM or 500 μM of hydrogen peroxide to PolyHb in the presence of lipid resulted in very significant increases in lipid peroxidation. On the other hand, the addition of 500 μM of hydrogen peroxide to PolyHb-SOD-CAT causes significantly less lipid peroxidation (D'Agnillo & Chang, 1998b). Similarly, the addition of 500 μM of hydrogen peroxide to PolyHb causes salicylate hydroxylation (Fig. 4.15 right). However, this does not happen for the case of PolyHb-SOD-CAT.

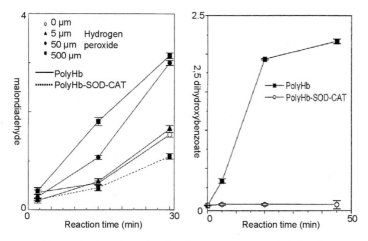

Fig. 4.15. Effects of PolyHb and PolyHb-SOD-CAT on (*left*) lipid peroxidation as shown by the production of malondiadehyde and (*Right*) on salicylate hydroxylation as shown by the product 2,5 dihydroxybenzoate. (From D'Agnillo & Chang, 1998b.)

4.11. Methods and Procedures

Preparation of PolyHb-SOD-CAT

Materials: Purified bovine Hb is purchased from Biopure Corporation, Boston, MA. SOD from bovine erythrocytes (EC 1.15.1.1, 3000 units/mg stated activity) and CAT from beef liver (EC 1.11.1.6, 72,575 units/mg stated activity) are purchased from Boehringer Mannheim. Glutaraldehyde (25%) is obtained from Polyscienes, Warrington, PA in one of the studies (D'Agnillo and Chang, 1998). Bovine milk xanthine oxidase, xanthine, and glucose oxidase are obtained from ICN Biomedicals, Aurora, OH. Over the years, there are differences in specific activity in commercially available enzymes. Other chemicals of reagent and HPLC grade are obtained from BDH, Toronto, ON. Cytochrome c from horse heart (type III), 4-aminoantipyrine, horseradish peroxidase type IV (EC 1.11.1.7), ferrozine (0.85%), and the iron standard (500 g/dl) are obtained from Sigma. The Hb assay kit is purchased from StanBio Labs. All the other reagents are of analytical grade.

Preparative procedures for PolyHb and PolyHb-SOD-CAT:

Reaction mixtures are prepared containing Hb (7g/dl), SOD (1050 U/ml), and CAT (21,000 U/ml) in 50 mM sodium phosphate, pH 7.4 (D'Agnillo and Chang, 1998). The final ratio is 1g Hb:15,000 U SOD:300,000 U CAT. For ischemia-reperfusion studies (Powanta and Chang, 2002), SOD (2,100,000 U/dl) and CAT (2,100,000 U/dl) in 0.1 M sodium phosphate buffer (PBS) (pH 7.4). For the PolyHb mixtures, an equivalent volume of buffer replaces enzyme addition. Prior to the start of crosslinking, lysine is added at a molar ratio of 10:1 lysine:Hb. Crosslinking reaction is started with the addition of glutaraldehyde at a molar ratio of 8:1 or 16:1 glutaraldehyde:Hb. Glutaraldehyde is added in four equal aliquots over a period of 10–15 min. Non-crosslinked preparations consist of the identical mixtures except that a buffer is added instead of glutaraldehyde. After 3.5 h at 4°C under aerobic conditions with constant stirring, the reaction is stopped with excess lysine at a molar ratio of 100:1 lysine:Hb. Solutions are dialyzed in physiological saline solution and passed through sterile 0.2 μM filter. Aliquots (300 μl) of the 8:1 crosslinked preparation are concentrated using 100 kDa microconcentrators (Amicon, Beverly, MA). Samples are centrifuged at 3000 g for 25 min at 23°C. After the first spin, the retentate is collected, diluted, and spun again. The final retentates are diluted and pooled. The Hb concentration is determined by cyanometHb measurement at 540 nm.

Measurement of SOD activity

SOD activity is determined by cyctochrome c reduction assay. One unit of SOD is defined as the amount of enzyme that inhibits the rate of cytochrome c reduction by 50% under specified conditions. The reagent solution consisted of xanthine (50 μM), cytochrome c (10 μM), and CAT (500 units/ml) in 50 mM potassium phosphate buffer, 0.1 mM EDTA, pH 7.8. CAT added to the reagent solution prevented interference due to the Hb-hydrogen peroxide interactions. The reaction volumes are prepared containing the test sample (25 μl of 0.6 g/dl), reagent solution (1.45 ml), and xanthine oxidase (25 μl of 0.2 units/ml) to start the reaction. The initial rate of cytochrome c reduction is monitored at 550 nm using a Perkin-Elmer Lambda 4B Spectrophotometer.

Measurement of CAT activity

CAT activity is assessed by measuring the decomposition of hydrogen peroxide at 240 nm. The spectrophotometer is zeroed with two cuvettes (1 ml) containing the test sample (10 μl of 0.6 g/dl) in 50 mM potassium phosphate buffer, pH 7.0. In the experimental samples, hydrogen peroxide (250 μl of 30 mM) is added instead of a buffer, and the absorbance at 240 nm is monitored for 36 sec. The disappearance rate of hydrogen peroxide is used as a measure of CAT activity.

Hydrogen peroxide scavenging

Reaction volumes (3 ml) containing the horseradish peroxidase/4-aminoantipyrine/phenol reagent solution (1.2 ml), PolyHb or PolyHb-SOD-CAT (5 mM), water, and hydrogen peroxide are prepared. Identical mixtures containing no H_2O_2 served as blanks. After allowing the mixture to stand for 3 min at 22°C, the absorbance at 505 nm is recorded. Hydrogen peroxide and the reagent solution participate in a peroxidase-catalyzed reaction to form a dye which can be measured at this wavelength.

Ratio of SOD/CAT and Hb oxidation

Reaction volumes are prepared containing 50 μM (as heme) in 50 mM potassium phosphate, pH 7.4 and varying ratios of SOD/CAT (0, 1, 100), with the final enzyme activities of 3 or 300 units. Samples also contained either xanthine (100 μM)/xanthine oxidase (10 mU/ml) or glucose (10 mM)/ glucose oxidase (10 μg/ml). Reactions (total volume 1 ml) are started with the addition of the oxidase enzymes at 23°C. The absorbance spectra (500–700 nm) are recorded at 1 scan/min for 25 min using a Perkin-Elmer Lambda B spectrophotometer.

Measurement of FerrylHb

Hydrogen peroxide (5, 50 and 500 μM) is added to reaction volumes containing 50 μM PolyHb or PolyHb-SOD-CAT in 50 mM potassium phosphate, pH 7.4. Xanthine (100 μM)/xanthine oxidase (10 mU/ml) is used as the oxidation system in separate experiments. At given time intervals, excess CAT is added to remove residual hydrogen peroxide in both the oxidation systems. Sodium sulfide (2 mM) is added to the mixtures, and the absorbance is measured at 620 nm. Blank samples consisted of measuring absorbance in similar mixtures except that water is added instead of hydrogen peroxide.

Iron release study

Hydrogen peroxide is added to 50 μM of PolyHb or PolyHb-SOD-CAT in 50 mM potassium phosphate, pH 7.0. At specified time intervals, the reaction is terminated by adding excess CAT. After 2–3 min, ascorbic acid (0.5 ml of 0.02%) is added and mixed for 5 min. Trichloroacetic acid (0.5 ml of 20%) is then added to precipitate proteins. The 1.5 ml mixture is centrifuged, and the supernatant (1 ml) is added to an ammonium acetate buffer (0.45 ml) and a ferrozine reagent (50 μl). The absorbance of iron color complex is measured at 560 nm. The amount of iron released is calculated by measuring the absorbance of an iron standard (500 μg/dl), (treated as described above) against H_2O blank.

Measurement of iron: The first step is to incubate PolyHb (15 mM) or PolyHb-SOD-CAT (15 mM) in hydrogen peroxide (total volume, 0.5 ml) for 60 min at 37°C. CAT is added to remove the residual H_2O_2, then ascorbic acid (0.5 ml of 0.02%) is added and mixed for 5 min. Trichloroacetic acid (0.5 ml of 20%) is then added to precipitate proteins. The 1.5 ml mixture is centrifuged, and the supernatant (1 ml) is added to an ammonium acetate buffer (0.45 ml) and a ferrozine reagent (50 ml). The iron color complex is measured at 560 nm. The amount of iron released is calculated by measuring the absorbance of an iron standard (500 mg/dl) (0.5 ml), (treated as described above) against blank (0.5 ml H_2O).

Circulation time studies

Fasted male Sprague-Dawley rats (250–275 g, Charles River, St-Constant, QC) are anesthetized with 65 mg/kg sodium pentobarbital i.p. (Somnotol, Becton Dickinson, NJ). Following a 15 min stabilization period, a bolus infusion of the test solution (300 mg Hb/kg rat) is given via the femoral vein. Femoral venous samples (0.5 ml) are centrifuged at 15,600 g for 5 min. Recovered plasma is stored at −80°C.

Rat model of intestinal ischemia-reperfusion

Fasted male Sprague-Dawley rats (280–310 g) are anesthetized with intraperitoneal injection of pentobarbital (Somnotol, 65 mg/kg). This is followed by midline laparotomy. The left and right renal pedicles are mobilized, isolated and ligated. The abdominal aorta is isolated and cannulated just proximal to the bifurcation using catheter-over-the-needle technique (24 gauge angiocath, Critikon, Tampa, Fl) and fixed in place by ligature of the aorta and vena cava distal to the catheter insertion. The abdominal inferior vena cava (IVC) is then isolated just proximal to this ligation and cannulated and affixed in a similar manner with a 22 gauge angiocath. At this point, midline sternotomy is performed and the thoracic IVC and descending aorta are isolated and ligated. Intestinal ischemia is taken to begin at this point. A closed circuit for intestinal and liver perfusion is created with the aortic cannula as the inlet and the IVC cannula as the outlet (Fig. 2). The system is flushed with 20 ml of heparinized (10 units/ml) lactated Ringer's solution delivered by a peristaltic pump at 0.8 ml/min. After this, the portal vein is isolated and cannulated with a 20 gauge angiocath and fixed in place by ligature of the vessel. After a 90 min ischemic time, PolyHb or PolyHb-SOD-CAT (5 g/dl) containing 5 mM of 4HB is perfused at 0.8 ml/min for 10 min without recirculation. Portal vein effluent samples are collected, centrifuged, and the centrifuged perfusate is stored at −80°C for later analysis by high performance liquid chromatography.

Measurement of oxygen radical

Hydroxyl radical generation is assessed by an aromatic hydroxylation technique with 4-hydroxybenzoate (4-HB). Reaction of hydroxyl radical with 4-HB produces 3,4-dihydroxybenzoate (3,4-DHBA).

Combined hemorrhagic shock and transient global cerebral ischemia-reperfusion

Procedure for rat model: Global cerebral ischemia is induced by bilateral common carotid arteries (BCCA) occlusion with hemorrhagic hypotension (to lower the mean arterial blood pressure) (Flothe and Otting, 1984). Male Sprague-Dawley rats (235–255 g) are anesthetized with intraperitoneal injection of pentobartbital (Somnotol, 65 mg/kg). The rectal temperature is maintained at 38°C during the experiments. The right femoral artery is cannulated to allow continuous recording of the arterial blood pressure. The left femoral vein and artery are catheterized for intravenous injection and induction of hypotension, respectively. The left and right common carotid arteries are isolated and ready for later initiation of global cerebral ischemia. The animals are allowed to rest for 30 min. After this, 5 ml of blood is withdrawn from the left femoral-artery cannula to reduce the mean blood pressure to 35 mm Hg; and both the common carotid arteries are temporarily occluded with microvascular clips to prevent circulation flow to the brain. During the occlusion period, the mean blood pressure is maintained by withdrawing or reinfusion of blood.

Baseline studies on blood-brain barrier: Baseline studies are first carried out to analyze the duration of ischemia needed to result in significant ischemia-reperfusion effects when followed by reperfusion. The duration of ischemia studied is: 0, 20, 30, 40 and 60 min. This is followed by reperfusion using one of the following "baseline" solutions: 1) oxygenated saline solution; 2) polyHb solution; and 3) Hb solution. One hour later, the ischemia-reperfusion effect is measured by Evans blue extravasation. This is compared with a Sham-control where only surgery is carried out but without hypotension or occlusion of the common carotid arteries (no ischemia) and no infusion (no reperfusion).

Studies on blood-brain barrier and brain edema: Based on the result obtained, an ischemia duration of one hour is chosen for the following detailed studies. Ischemia is terminated 1 h after the occlusion of the common carotid arteries by removing the microvascular clips. This is followed by the isovolumic replacement of the lost blood using one of the following solutions at a rate of 0.8 ml/min: 1) oxygenated saline solution; 2) Hb solution; 3) PolyHb solution; 4) solution containing a mixture of Hb, SOD and CAT; and 5) PolyHb-SOD-CAT. After preset periods of 1, 2, 4 and 6 h following reperfusion of each of the above solution, the animals are sacrificed and the effects on blood-brain barrier and brain edema measured. The results are compared with Sham-control groups, where only surgery is carried out but without occlusion of the common carotid arteries (no ischemia) and no infusion (no reperfusion).

Blood-brain barrier and brain water content measurement: We use the method of Chan *et al.* (1991) with the following modifications. Evans Blue, 0.1 ml (4%) in 0.9% saline solution is administered through the left femoral veins. For animals in the 1-, 2-, 4- and 6-h experimental groups, Evans Blue is injected at 0,1, 2, 4 and 6 h after reperfusion and allowed to circulate for 30 min. After this, a thoracotomy is carried out on each animal, followed by clamping of the descending limb of the aorta and inferior vena cava with microvascular clips. The left and right external jugular veins are then cut and each animal is perfused with 5 ml of 0.9% saline at room temperature through the aorta via the left ventricle using a peristaltic pump at a rate of 0.8 ml/min. The brains of the animals are quickly removed to obtain the forebrain. Each forebrain is weighed to obtain the wet weight and then dried at 105°C for 1 week, then reweighed to obtain the dry weight. To measure the extravasation of Evans Blue, the forebrain is then placed in 2 ml of 0.1 M PBS, pH 7.4, for another week. The samples are then homogenized in PBS and centrifuged at 1000 g for 5 min. Samples of 0.7 ml of the supernatants are mixed with an equivalent volume, 0.7 ml, of 12% (w/v) trichloroacetic acid. The mixtures are incubated at 4°C for 18 h, then centrifuged at 1000 g for 30 min. The Evans Blue concentration of the supernatants is quantitatively measured at 610 nm for absorbance. The water content of each forebrain is calculated as the difference between the forebrain wet weight and its dry weight (water content = wet weight – dry weight). The changes in water content of the forebrain, as presented in Fig. 4.6, is calculated as the difference between the forebrain water contents obtained from 1-, 2-, 4- and 6-h groups and that of the 0-h group. Change in water content of forebrain = Water content$_i$ – Water content$_0$ (where i = 1, 2, 4, and 6 h; 0 = zero hour).

CHAPTER 5

Nanotechnology-based Artificial Red Blood Cells (RBCs)

5.1. Introduction

As discussed in the previous two chapters, investigators have successfully used the basic principles of nanobiotechnology-based blood substitutes to develop clinically useful polyHb and conjugated Hb (Chang, 1964, 1965, 1972a). Are we now ready to ask the next question? Is it possible to develop the more complete artificial red blood cells (Chang 1967, 1964) for clinical use in patients?

In nature, hemoglobin and complex enzyme systems are retained inside red blood cells (Fig. 5.1). In this way, they can carry out their functions: to transport oxygen and carbon dioxide, remove oxygen radicals and prevent hemoglobin from being oxidized into methemoglobin. It has been showed as far back as 1957 (Chang, 1957), that the contents of red blood cells can be enveloped in artificial red blood cells using ultrathin polymer membranes (Fig. 5.1). In this form, the hemoglobin and the enzyme systems normally present in red blood cells are retained inside the membranes. At the same time, oxygen and carbon dioxide can equilibrate rapidly across the membranes to interact with the enclosed materials (Chang, 1957, 1964). The membranes of the artificial red blood cells would not become fragile after prolonged *in vitro* storage. The artificial red blood cells could be transfused without cross-matching, since they do not have any blood group antigens. In order to develop this basic idea of a complete artificial red blood cell for practical applications, researchers from

Artificial Cells

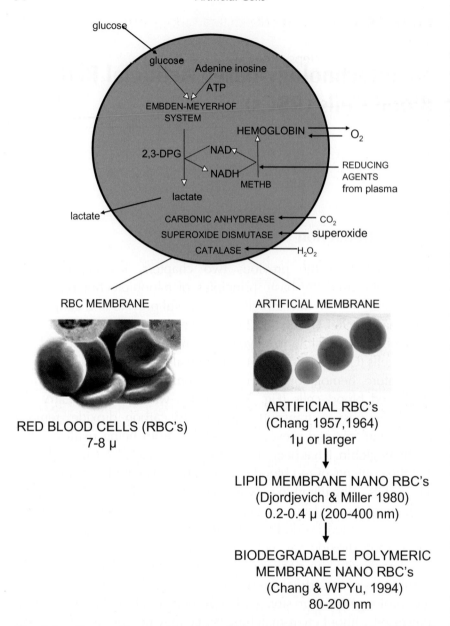

Fig. 5.1. *Upper*: Composition of red blood cells and artificial red blood cell. *Lower left*: Red blood cells. *Lower right*: First artificial rbc's of 1 micron or larger diameters; first lipid membrane nanodimension artificial rbc's, first nano dieamension biodegradable polymeric membrane artificial rbc's.

around the world have devoted a tremendous amount of effort to the research.

There are three steps in the development of a complete artificial rbc (Fig. 5.1), i.e. developing: 1) a micron dimension artificial red blood cell; 2) a submicron lipid membrane artificial red blood cell; 3) a nanodimension biodegradable membrane artificial rbc.

5.2. Micron Dimension Artificial RBCs with Ultrathin Polymeric Membrane

The first attempt at preparing an artificial red blood cell involved replacing the red blood cell membrane with an ultrathin polymeric membrane (Chang, 1957). This results in an artificial rbc with the same contents as those in a rbc, but with an artificial membrane instead of the red blood cell membrane.

Hemoglobin

The content of rbc, hemolysate, enveloped in a butyl benzoate coated collodion membrane artificial cell, contains a combination of oxyhemoglobin and methemoglobin. The oxyhemoglobin retains its ability to combine reversibly with oxygen (Fig 5.2) (Chang, 1957).

Chang and Poznansky (1968a) showed that if additional catalase is enclosed along with the red blood cell hemolysate, there is a marked reduction in methemoglobin formation. In the case of nylon membrane artificial cells, the preparative procedure results in the formation of a much higher proportion of methemoglobin (Chang, 1964, 1965). Red cell hemolysate enclosed in silastic rubber artificial cells (Chang, 1966) remains unaltered and can combine reversibly with oxygen. A Japanese group has also supported these findings. They showed that hemoglobin microencapsulated within polymer membranes like polystyrene, ethyl cellulose and dextran stearate (Sekiguchi and Kondo, 1966) would combine reversibly with oxygen (Toyoda, 1966).

Carbonic anhydrase

Carbonic anhydrase, present at high concentration in rbc's, plays an important role in the transport of carbon dioxide. Carbonic anhydrase

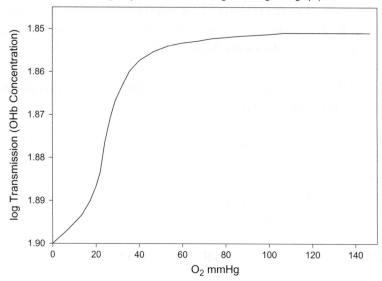

Fig. 5.2. Oxygen dissociation curve of collodion membrane artificial rbc's (scanned from Chang, 1957).

of the red blood cell hemolysate enclosed in artificial cells with collodion, nylon, or cross-linked protein membranes, retains its activity (Fig. 5.3) (Chang, 1964, 1965).

The carbonic anhydrase activity was measured as the rate of fall of pH when carbon dioxide was bubbled through a buffered suspension. Artificial rbc's and rbc's both accelerate the rate of fall of pH (Fig. 5.3). Acetazolamide completely inhibited the catalytic activity of both rbc's and artificial rbc's with membranes of nylon, cross-linked protein, or collodion. Samples of buffered saline that had been in contact with their own volume of artificial rbc's for 24 h showed no detectable enzyme activity, indicating that the enclosed carbonic anhydrase had not leaked out during this period. Carbonic anhydrase in artificial cells have about 75% of the activity of the enzyme in rbc's. Thus,

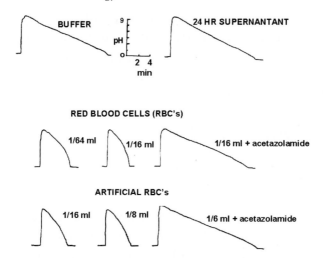

Fig. 5.3. Carbonic anhydrase activity measured as rate of fall of pH as carbon dioxide is bubbled into the respective solution (modified from Chang, 1964).

carbonic anhydrase in artificial rbc's acts efficiently in catalyzing carbon dioxide.

Catalase

Catalase is another important component of mammalian rbc's. We have analyzed the catalase activity of collodion membrane artificial cells containing hemolysate (Chang, 1967; Chang and Poznansky, 1968a). Both its natural substrate, hydrogen peroxide, and its chemical substrate, sodium perborate, have been used. Catalase enclosed in collodion membrane artificial cells acts efficiently *in vitro*. It also acts efficiently *in vivo* for enzyme replacement therapy in mice with a congenital defect in their red blood cell catalase (Chang and Poznansky, 1968a). This will be described in detail in the next chapter.

Immunological studies

Human ABO blood group antisera cause marked agglutination of both incompatible human rbc's and heterogeneous rbc's. On the other hand, these antisera do not cause any agglutination of artificial rbc's containing rbc hemolysate, even though the hemolysate has been obtained from incompatible human rbc's or heterogeneous rbc's (Chang, 1972a). Thus, an artificial red blood cell membrane does not appear to possess the blood group antigens normally present in biological cell membranes.

5.3. Intravenous Injection

Polymeric membrane artificial RBCs 1 micron and larger

The fate of intravenously injected nylon membrane artificial cells containing hemolysate has been studied (Chang, 1964, 1965, 1972a). A suspension of $5\mu \pm 2.5\mu$ diameter (mean $\pm$ standard deviation) nylon membrane artificial rbc's was infused intravenously over a period of 1.5 min, and samples were obtained from the anesthetized cat during and following the infusion.

The arterial level rose steeply during infusion, but upon discontinuation of the infusion, the level fell rapidly (Fig. 5.4). Thus, most of the artificial cells successfully completed one or more circulations, after which they were removed rapidly from the bloodstream. Experiments using ^{51}Cr-labeled nylon membrane artificial cells showed that most of them were removed by the lung, liver, or spleen. Collodion membrane artificial cells are also removed rapidly from the circulation. Thus, it appears that nylon and collodion membrane artificial cells smaller than rbc's are rapidly removed from the circulation. During the infusion, there is a fall in arterial blood pressure and an increase in venous pressure and other signs compatible with pulmonary embolism. These artificial rbc's do not have the deformability of rbc's that allows the latter to pass through capillaries of smaller diameters.

Can artificial rbc's survive longer if they are made much smaller in order that they can pass through the capillaries? Unfortunately, it has long been known that even submicron foreign particles are removed very rapidly from the circulation (Halpern *et al.*, 1958). After all, this is part of the body's defence against circulating microorganisms and other foreign particles. Thus, we looked into another factor, the surface

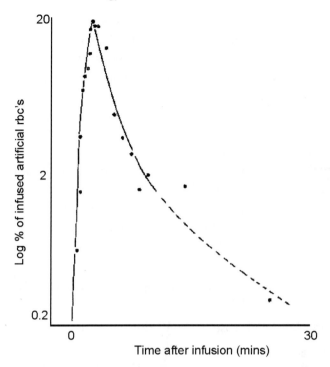

Fig. 5.4. Circulation time of artificial rbc's after infusion (from Chang 1965, 1972a).

property of artificial cells (Chang, 1964, 1965, 1972a). Red blood cells and other formed elements of blood have a negative surface charge due to the presence of n-acetylneuraminic acid containing mucopolysaccharide on the red blood cell membrane (Cook *et al.*, 1961). Danon and Marikovsky (1961) found that the younger rbc's were about 30% more strongly charged than the older ones obtained from the same blood sample. All this information led us to look into the effects of surface properties on circulation time for rbc's, foreign particles, and artificial rbc's (Chang, 1965,1972a).

Neuraminic acid and circulation time of red blood cells

We first studied the effects of removing neuraminic acid from rbc's on their survival in the circulation (Chang 1965,1972a; Chang and

MacIntosh, 1964). Experiments on the survival of autologous ^{51}Cr-labeled rbc's were carried out using four dogs (Fig. 5.5). First, ^{51}Cr-labeled rbc's obtained from each of the dogs were reinjected. They stayed in the circulation with the usual lifetime of about three to four months for dog rbc's. The second injection consisted of rbc's obtained from each of the dogs and then treated with neuraminidase to remove neuraminic acid. The procedure for the use of neurominidase to remove neuraminic acid follows that of Cook *et al.* (1961). Removal of neuraminic acid resulted in the rapid removal of these rbc's from the circulation, with the level falling exponentially with a half-time of only about 2 h (Fig. 5.5). The third injection consisted of normal ^{51}Cr-labeled rbc's obtained from each of the dog. The circulation time

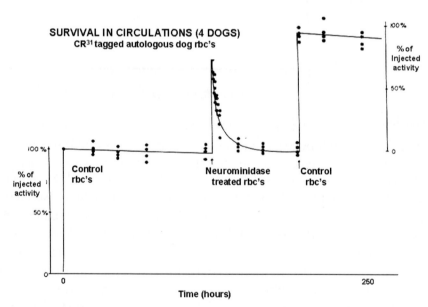

Fig. 5.5. Circulation time of reinfused autologous dog rbc's, tagged with [51]Cr. Data from four dogs: each point represents the radioactivity of one blood sample from a limb vein. In each case, the first and third tests were with normal rbc's from the same dog. The second test was with neuraminidase-treated rbc obtained previously from the same dog. Ordinates give percentage of injected label in circulation at time of sampling. Note rapid disappearance of neuraminidase-treated rbc's. (From Chang, 1965, 1972a.)

was not significantly different from that for the first injection. These results led to our suggestion that the survival of rbc's may be related to the neuraminic acid present in the membranes (Chang, 1965, 1972a; Chang and MacIntosh, 1964).

Effects of surface charge of polymeric particles on circulation time

These results led to the investigation of a simpler model system (Chang, 1965,1972a). Polystyrene latex particles (Dow Chemical Co.) with diameters of $2.05\mu \pm 0.018\mu$ (mean $\pm$ S.D.) were diluted 1: 100 in saline. A similar suspension was prepared using polystyrene latex sulfonated for 30 sec by Ag_2SO_4 and H_2SO_4, and then washed with saline (Chang, 1965, 1972a). With the sulfonated polystyrene latex, the initial counts were close to the expected number. Then the level declined, first rapidly and then more slowly. In the case of the control polystyrene latex particles, there constituted only 10 to 15% of the injected particles in the circulation in the initial count $1/2$ min after injection. The level then fell much more steeply, with a half-time of about 15 to 30 sec. Further experiments showed that the sulfonated polystyrene latex particles were not trapped to any extent by the lung, but were removed by the reticuloendothelial system. The control polystyrene latex particles, on the other hand, were trapped fairly efficiently by the lung, and very efficiently by the reticuloendothelial system.

Artificial rbc's with sulfonated nylon membrane

The above results led us to prepare artificial cells with sulfonated nylon membranes (Chang, 1965, 1972a); We used sulfonated diamine such as 4,4'-diamino-2,2'-diphenyldisulfonic acid (Chang, 1964, 1965, 1972a; Chang *et al.*, 1966). Sulfonated nylon membrane artificial cells of varying surface negative charges can be prepared by adding different amounts of the sulfonated diamine (e.g. 3 mM/liter) to the alkaline 1,6-hexanendiamine solution, and then the standard procedure for preparation of nylon membrane artificial cells as described in Appendix II can be used. The reaction is shown in Fig. 5.6.

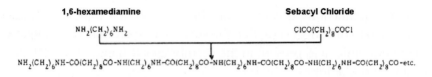

NYLON MEMBRANE

4,4'-diamino-2,2-diphenyldisulfonic acid   **Sebacyl Chloride**

NYLON MEMBRANE WITH STRONG ACIDIC GROUP

Fig. 5.6. *Upper:* schematic representation of nylon artificial cell membrane. *Lower:* schematic representation of sulfonic nylon artificial cell membrane (Chang, 1965, 1972a).

Sulfonated nylon membrane artificial cells prepared as described in the foregoing have a negative surface charge compared to that of rbc's. Although sulfonated nylon artificial cells are also removed rapidly, they survived in the circulation significantly longer than nylon membrane artificial cells (Fig. 5.7). In addition, the sulfonated nylon membrane artificial cells are less liable to be trapped in the pulmonary circulation and the reticuloendothelial system compared with nylon membrane artificial rbc's.

Polysaccharide incorporation

Since natural rbc's and the endothelium of vessels are covered by mucopolysacchride, further experiments were carried out to incorporate a complex containing a mucopolysaccharide (heparin) into the collodion membrane of artificial cells (Chang *et al.*, 1967; Chang, 1970). Preliminary results showed that while both types were removed rapidly from the circulation, those with a heparin-complexed membrane survived significantly longer.

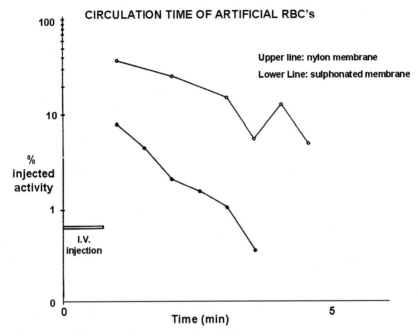

Fig. 5.7. Circulation time of ^{51}Cr labeled artificial rbc (mean diameter, 5μ) after intravenous infusion into cats; arterial sampling. *Upper line*: sulfonated nylon membrane artificial rbc's. *Lower line*: nylon membrane artificial rbc's. Survival estimated as percentage of radioactivity. (From Chang, 1965, 1972a.)

Summary

The *in vitro* results showed that artificial rbc's with the required *in vitro* properties to substitute rbc's can be prepared. However, the single major problem is the extremely short circulation time of the artifical rbc's. We found that they have to be less than 1 micron in diameter to avoid being trapped in the lung capillaries (Chang, 1965, 1972a). Even then, they are removed by the reticuloendothelial system. We attempted to look at why the rbc's circulated for so long, by using neuraminidase to remove sialic acid from the red blood cell membrane. We showed that the removal of sialic acid resulted in the rapid removal of the rbc's from the circulation (Chang 1965, 1972a, Chang & MacIntosh, 1964a) (Fig. 5.5). This observation led us to prepare artificial rbc's with modified surface properties.

This included the use of other synthetic polymers and cross-linked protein membranes (Chang 1965, 1972a); membrane with negative surface charge and polysaccharide surface as sialic acid analogues (Chang *et al.*, 1976, Chang 1970); and lipid-protein and lipid-polymer membranes (Chang 1969d, 1972a). Some of these improved the circulation time. However, the circulation time was still not sufficiently long for practical applications.

5.4.   Submicron Lipid Membrane Artificial RBCs

We prepared larger lipid membrane artificial cells where the lipid is supported in the form of lipid-protein membrane and lipid-polymer membrane (Chang 1969d, 1972a). In 1980 Djordjevich and Miller reported that they could prepare submicron 0.2 micron-diameter artificial rbc's using lipid membrane vesicles to encapsulate hemoglobin (Fig. 5.8). This increased the circulation time significantly, although it was still rather short. Many investigators have since carried out research to improve the preparation and the circulation time. Examples of the many investigators are Beissinger; Domokos; Farmer; Gaber; Hunt; Kobayashi; Miller; Phillips; Rudolph; Schmidt; Szeboni; Takahashi; Tsuchida; Usuuba, and many others. Improvements include modifications of surface properties such as surface charge, and the use of sialic acid analogues. The most successful approach to improve the circulation time is to incorporate polyethylene-glycol (PEG) into the lipid membrane artificial rbc's (Fig. 5.8), resulting in a circulation half-time of more than 30 h (Philips *et al.*, 1999).

Rudolph's group has carried out a very extensive study that has contributed much to this area (Rudolph, 1994, 1995, Rudolph *et al.*, 1997). Tsuchida's group has for many years carried out research and development leading to commercial development and preclinical animal studies in preparation for clinical trials (Tsuchida, 1994, 1998; Tsuchida *et al.*, 1988, 1992). In North America, lipid membrane artificial red blood cell's are called lipid encapsulated hemoglobin (LEH). In Japan, they called hemoglobin vesicles (HbV).

Lipid encapsulated hemoglobin (LEH) is a very large area that will be best left for later book volumes of our book series "Regenerative Medicine, Artifical Cells and Nanomedicine" This way, each specialist

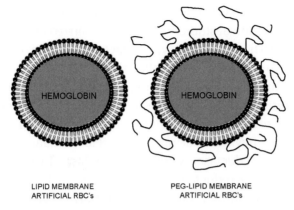

LIPID MEMBRANE
ARTIFICIAL RBC's

PEG-LIPID MEMBRANE
ARTIFICIAL RBC's

Fig. 5.8. *Left:* Bilayer lipid membrane that forms the membrane of lipid encapsulated hemoglobin (lipid membrane artificial rbc's). *Right:* Incorporation of polyethylene-glycol (PEG) to the lipid membrane resulted in marked improvement in the circulation time of hemoglobin PEG-lipid vesicles (PEG-lipid membrane artificial rbc's).

in a special area can present in detail his or her own specialized area. Furthermore, there will not be any copyright infringement by the author of this monograph. The following is a brief overview of LEH.

Composition and preparation

LEH lipid membrane consists of cholesterol and phospholipids. One of the aims is to increase the encapsulation efficiency of hemoglobin within a minimal amount of lipid. This can be done by using anionic lipids. Furthermore, inclusion of anionic lipid significantly increases the circulcation time (Awasthi *et al.*, 2004). However, some anionic liposomes may interact adversely *in vivo* with the complement (Szebeni, 1998). On the other hand, it is important to remember that the same type of lipid produced by different manufacturers may have different abilities in terms of activating the complement (Chang and Lister, 1990). This further supports the view that there is a need to use the C3a screening test for each lipid before its use in preparing LEH (Chang and Lister, 1990). The use of a new synthetic anionic lipid, 1,5-diplamitoyl-l-glutamate-N-succinic acid may solve the problem of complement activation (Sou *et al.*, 2003).

Tsuchida's group has been scaling up the production of PEG-LEH for preclinical and clinical studies. Briefly, they carried out pasteurization at 60°C for 10 h to achieve viral inactivation. The conversion of hemoglobin to carbonylhemoglobin (HbCO) increases the stability of hemoglobin during pasteurization. During pasteurization, all the other proteins including red blood cell enzymes are denatured and precipitated, resulting in a pure Hb solution. This also removes the red blood cell enzymes that prevent metHb formation, for instance methemoglobin reductase system and catalase. Thus, they tried to co-encapsulate the reductant-like glutathione of homocysteine and catalase with the purified Hb. They could also regulate the oxygen affinity by co-encapsulating allosteric effectors. After pasteurization, the Hb solution is passed through a series of ultrafiltrators to remove any virus, and then concentrated by ultrafiltration to 40g/liter. After adjustment for electrolytes, this is then encapsulated into the PEG-LEH. HbCO is then covereted back to oxyHb by illuminating the LEH with a visible light under O_2 The group was able to use the extrusion method to prepare PEG-LEH at a pH of 7.0–7.4. Their mixed lipids contained dipalmitoylphosphatidylcholine (DPPC) as the main component. Using a number of progressing steps of extrusion, they finally obtained PEG-LEH with a diameter of 250 ± 20 nm with a Hb/lipid ratio of 1.7–1.8.

The Hb inside PEG-LEH, like the red blood cell Hb, does not exert a colloid osmotic pressure. Thus, human serum albumin, hydroxyethly starch or dextran has to be added to the suspending solution. They chose to use recombinant human albumin since there is no immunological problem and infective agents are absent. For purpose of storage, they use LEH with PEG on the surface to prevent aggregation and deoxygenate the hemoglobin to prevent metHb formation. The group reported that their PEG-LEH can be stored at 4°C and 23°C for 2 years. PEH-LEH stored at 40°C became unstable after one year.

Methemoglobin

Reducing agents and glucose cannot cross the LEH lipid membrane. Thus, reducing agents from the plasma cannot cross the PEG-Lipid membrane to reduce the formation of methemoglobin. Furthermore, all the red blood cell enzymes are removed during the pasteurization procedure. Even if the methemoglobin reductase system can be added, it will require glucose as a substrate to recycle the cofactors needed in the enzymatic conversion of methemoglobin back to hemoglobin. As a result, hemoglobin in LEH is slowly converted to methemoglobin after infusion at body temperature. Increase in circulation time would not serve any useful purpose if the enclosed hemoglobin is converted to methemoglobin in long circulating PEG-LEH. As a result, many studies has have been carried out especially by Tsuchida's group to resolve this problem. As mentioned above, Tsuchida's group added antioxidants like glutathione or homocysteine and catalase to hemoglobin before the latter is encapsulated into the PEG-LEH. To slow down metHb formation during storage, the LEH can be deoxygenated by perfusing it with nitrogen gas and storing it at 4°C.

5.5.   LEH in Animal Studies

Circulation time after infusion

As discussed under polymeric membrane artificial rbc's above, size is very important as far as circulation time is concerned. LEH larger that 200 nm are removed rapidly. Those under 200 nm are less likely to be removed by the reticuloendothelial system. However, those of about 60 nm are removed quickly by the liver, most likely through the fenestrated hepatic endothelium (Liu *et al.*, 1992).

Thus, up to a point, a smaller size is important in maintaining LEH in the circulation. However, decreasing the size of LEH would also decrease the encapsulation efficiency of hemoglobin. In this regard, Philips *et al's* (1999) study showing that PEG-lipid membrane artificial rbc's can have a markedly increased circulation time is very important. The use of PEG-lipid membrane will allow for the use of larger LEH, with a better efficiency for encapsulating hemoglobin. Thus, a size range of 210–275 nm is possible with PEG-lipid membranes since these still maintain a good circulation time (Awasthi *et al.*, 2003). Larger than this and the LEH would be removed rapidly. Inserting

PEG-linked phosphatidylethanolamines (PEG-PE) into the outer layer of LEH, will, improve the encapsulation efficiency of hemoglobin and the circulation time (Awasthi *et al.*, 2003). Thus, PEG-LEH infused into rats shows a circulation half-time of 30 h. Most of the LEH are removed by the reticuloendothelial system (RES), liver and spleen. Compared to humans, rats have a highly active RES; thus, it has been suggested that a $T_{1/2}$ of 12–20 h in rats is equivalent to 40–60 h in humans (Woodle *et al.*, 1995).

Similar results were obtained with PEG-lipid membrane artificial rbc's prepared by Tsuchida's group (Saiki *et al.*, 2001; Sou *et al.*, 2005). Their study using radioisotope-labeling showed that the circulation half-life was 32 h in rats with a dose of 14 ml/kg (Sou *et al.*, 2005). Gamma camera imaging showed that they were removed mainly by the liver, spleen and bone marrow. They were found in the phagosomes of the macrophages one day after infusion (Sakai *et al.*, 2001), but cannot be observed after seven days.

Safety of lipid membrane artificial rbc's

As mentioned earlier, Tsuchida's group at Waseda University is developing their PEG-lipid membrane artificial cells for commercial scale up in preclinical animal studies towards clinical trial. They have collaborated with Kobayashi at Keio University to carry out detailed safety and efficacy studies in animal (Kobayashi *et al.*, 1997, Izumi *et al.*, 1997; Saikai *et al.*, 1997, Saikai *et al.*, 2004a, Saikai *et al.*, 2004b, Saikai *et al.*, 2004c, Saikai *et al.*, 2004d; Yushizi *et al.*, 2004). The following is a summary of their results. For copyright reasons, many of the excellent figures could not be included in this monograph, but are available in the referred papers.

In their study, HbV (LEH) were suspended in 5 g/dl human serum albumin. The resulting colloid osmotic pressure was 20 mmHg and the viscosity, 3 cP, both similar to whole blood. They studied the effects of daily repeated infusion of HbV by following serum biochemistry (Sakai *et al.*, 2004a, Sakai *et al.*, 2004b, Saikai *et al.*, 2004d). Indicators of liver function like albumin, ALT, AST and LDH stayed within the normal ranges. Bilirubin and ferric ion remained at a low level. The lipid concentration increased, but returned to its original level in 2 wk, indicating that the lipid component of the HbV were metabolized in the body. They found that infusion of these 250 nm diameter PEG-lipid membrane artificial cells did not result in any vasopressor effects. This lends further support to the hypothesis that vasopressor effects is due to single tetrameric Hb crossing the intercellular junction of the endothelial cell lining of blood vessels.

Efficacy in exchange transfusion, hemorrhagic shock and hemodilution

The group carried out exchange transfusion of 90% of the blood volume in rats and studied oxygen transport by inserting needle oxygen electrodes into their renal cortex and skeletal muscle. When albumin alone was used for the 90% exchange, there was a marked decrease in both the arterial blood pressure and renal cortical oxygen tension followed by the death of the rats. On the other hand, all the rats survived when the 90% exchange transfusion was carried out using their artificial lipid membrane artificial rbc's suspended in human serum albumin. Both the arterial blood pressure and renal cortical oxygen tension remained unchanged. They also studied the use of recombinant human serum albumin (rHSA) instead of human serum albumin to suspend the HbV. HbV suspended in this way was found to be safe and effective as priming volumes for 50% hemodilution in cardiopulmonary bypass. All the rats survived with continuing growth as shown by body weight gains. The group reported no irreversible changes in plasma biochemistry, hematology and histopathology.

They also studied the use of rat HbV for resuscitation in hemorrhagic shock in rats (Sakai *et al.*, 2004c; Yoshizu *et al.*, 2004). In anesthetized rats, removing 50% of the total blood volume resulted in hypotension and metabolic acidosis. Infusion of either HbV in rHSA or autologous rbc's in rHSAA restored the blood pressure. However, rHSA alone with HbV did not restore the arterial blood pressure and some of the rats in this group died.

Summary

The single major obstacle to the practical realization of the original artificial rbc's (Chang, 1957, 1964, 1972a) was the low circulation time. With extensive research by many groups, the use of PEG-lipid membrane to form 250 nm-diameter PEG-lipid membrane artificial rbc's has increased the circulation time to a clinically useful level. This is an important and major step in the development towards a more complete artificial rbc for clinical use.

5.6. Nanodimension Biodegradable Polymeric Membrane RBCs

Introduction

Submicron PEG-lipid membrane artificial rbc's as described above, have solved the major barrier of short circulation time. The lipid encapsulated hemoglobin (LEH) has potential uses in a number of clinical conditions. However, in some other clinical conditions one may have to look at the effect of the lipid membrane of the LEH especially regarding the following two aspects.

(1) The smaller the particle, the larger will be the total surface area in the same volume of suspension. Thus, a 1 ml suspension of 200 nm-diameter LEH would have more than 10 times the total surface area of a 1 ml suspension of rbc's. Since both have a bilayer lipid membrane, the total amount of lipid would, at least be 10 times higher for the 1 ml suspension of 200 nm lipid membrane artificial rbc's compared to rbc's. A large amount of lipid can decrease the phagocytic function of the reticuloendothelial systems (RES). In traumatic hemorrhagic shock, the RES needs to be efficient in removing contaminating microorganisms. Furthermore, in ischemic reperfusion, lipid may induce lipid peroxidation. Therefore, we are carrying out a study to see if we can cut down as much as possible the amount of lipid needed for the membrane of nano artificial rbc's.

(2) Another problem is methemoglobin formation. For example, reducing agents present in the plasma of the circulating blood can reduce the rate of methb formation in polyHb or conjugated Hb. However, these reducing agents cannot cross the lipid membrane of LEH. Glucose that is needed for the methemoglobin reductase system inside red blood cell's also cannot cross the lipid membrane of LEG. Thus, research to find a way to prevent metHb formation in long circulating LEH is an ongoing process. For biodegradable membrane nano artificial rbc's, we can use membranes that are permeable to reducing agents and small molecules like glucose.

Nanodimension biodegradable polymeric membrane artificial rbc's

Our aim is, therefore, to study how to prepare nanodimension artificial rbc's with the following properties:

1. Contains little or no lipid in the membrane.
2. Persist in the circulation after infusion for a sufficiently long time.
3. Be stable in storage.
4. Remains stable after infusion for the duration of its function as a blood substitute — but it has to be biodegraded soon after the completion of its action in the body.
5. The membrane material and its degradation products have to be nontoxic.
6. In addition to hemoglobin, the nano artificial cells should contain important rbc enzymes like superoxide dismutase, catalase, carbonic anhydrase and methemoglobin reductase, as shown in Fig. 5.1.
7. The membrane should be permeable to reducing agents and/or glucose.

We have been using biodegradable polymer, e.g. polylactic acid, to prepare micron dimension artificial cells for the encapsulation of hemoglobin, enzymes and other biologically active material since 1976 (Chang, 1976a). With the availability of methods to prepare artificial cells of nanometer dimensions, we started to prepare hemoglobin nanocapsules of less than 0.2 micron mean diameter using polylactic acid (PLA) membrane, PEG-PLA membrane and other biodegradable polymers (Chang & WPYu, 1992, 1997a, 1998, 1999, 2001; Chang *et al.*, 2003a, 2003b; WPYu & Chang, 1994, 1996).

Safety and reasons for selection of polylesters as the biodegradable polymer membrane

Polymers for forming nano artificial rbc membrane have to be non-toxic and can be degraded in the body into nontoxic degradation products. Polyesters like polylactic acid, polyglycolic acid and their copolymers polylactoglycolides finally degrade into lactic acid and glycolic acid which are normal human metabolites. The rate of degradation can be adjusted by variations in the molecular weight, particle size and ratios of the copolymer. Other potentially useful biodegradable polyesters include Poly(hydroxybutyric acid) and its copolymer with hydroxyvalerianic acid, polyorthoesters, polyanhydrides and others. Potential toxicity can come from residue monomers and chemicals used in polymerization to form nanocapsules. The solvent evaporation method (Chang, 1976a) for preparing polylactide membrane artificial cells and microspheres is an extension of the original method for preparing artificial rbc's (Chang, 1957). This method obviates the need for monomers or chemical reagents and nontoxic solvents can be used.

Since 1982, there has been extensive clinical experience gained in the use of polylactic acid, polyglycolic acid and polylactoglycolides in humans as surgical sutures or implants with no adverse or toxic effects. The amounts used are many folds greater that needed for the ultrathin membrane of the nano artificial rbc's. Polyglycolic acid (Dexon-S) suture has been used routinely in surgical practice since 1982 with no reported toxicities. Meadows *et al.* (1993) show that grafting polylactic acid (PLA) granules is as successful as using decalcified freeze-dried bone allograft. There is no reported toxicity in many other surgical implantations of polylactide or polyglycolide in amounts many folds greater than what we plan to use for nano artificial red blood cell membrane. The basic approach of preparing polylactic acid artificial cells or microspheres (Chang, 1976a) for drug devliery has been used clinically. For example, Jaquel *et al.* (1995) gave repeated injections of Poly(DL-lactate-coglycolide) microspheres containing Octreotide in 100 acromegalic patients.

As mentioned above, long term experiences in humans show that the polyesters and their degradation products are safe. We also have long term experience in research on biodegradable polymer microencapsulation using polylactides, having initiated this area in 1976 (Chang, 1976a). Thus, we decided to look into the use of polylactic acid as the biodegradable membrane for nano artificial rbc's. Details of the method of preparation will be described later.

5.7. Characterization of Nano Artificial RBCs

Electromicroscopic appearance

A Philips EM300 electron microscope, at 60 KV, was used. The sample was treated by standard negative staining. A typical electron micrograph for the biodegradable polymer hemoglobin nanocapsules prepared with d,l-PLA is shown in Fig. 5.9. They are spherical and homogeneous. Their diameter in the micrograph ranges from 40–120 nm, with a mean diameter of 80 nm. The membrane thickness is 5–15 nm.

Size distribution of polylactide membrane artificial rbc's

The diameter and size distribution of the biodegradable hemoglobin are determined by using the Nicomp Size Analyzer (Model 370). The instrument operates by light scattering. The average particle size of the biodegradable nanocapsules containing hemoglobin is dependent on the formula used for preparation. A typical size determined using the Nicomp Sizer Analyzer is presented in Fig. 5.10. A unimodal distribution is obtained for all the samples. With different preparation processes and different polymers, the mean diameters of the biodegradable hemoglobin nanocapsules can be shown to be as low as 74 nanometers.

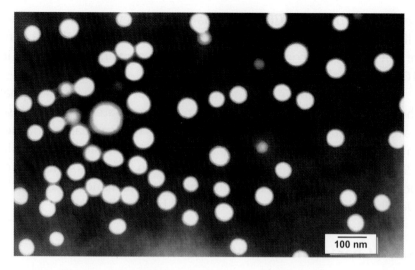

Fig. 5.9. Electromicroscopy of polylactide membrane nano artificial rbc's. Mean diameter is 80 nanometer (from WPYu & Chang, 1996).

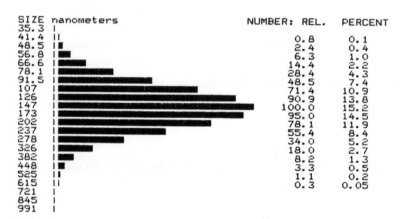

Fig. 5.10. Size distribution of PLA nano artificial rbc's determined using the Nicomp Size Analyzer (Model 370). An unimodal distribution is obtained. The mean diameter for this sample is about 100 nm. (From WPYu & Chang, 1996).

Other properties

The characteristics of the preparation containing 10.97 gm/dl of hemoglobin prepared with d,l-PLA is listed in Table 5.1. Nanocapsules can be prepared with up to 15 gm/dl hemoglobin concentration.

Table 5.1 Characteristics

Hemoglobin concentration: 10.97 g/dl (up to 15 g/dl)
Polymer concentration: 1.2 g/dl
Phospholipid concentration: 0.6 g/dl
Specific gravity (22°C): 1.0047
Viscosity (37°C): 3.7 to 3.8 cp

Table 5.2 Steady Shear Viscosity (22°C, cp)

Shear Rat (1/sec)	Viscosity of Preparation
45	6.5
90	6.5
225	5.0

Steady shear viscosity determination

Steady shear viscosity of the suspension of the nano artificial rbc's was measured with a Wells-Brookfield Syncro-Lectric Microviscometer (Model LVT) equipped with a 0.80° cone (Model CP-40). Shear rates were from 45 to 450 s^{-1} at 22°C. The steady shear viscosity of the biodegradable polymer hemoglobin nanocapsules is shown in Table 5.2.

5.8.   Safety and Efficacy of Nano Artificial RBCs

Amount and fate of PLA membrane

By assuming that all the polymer used in the solvent evaporation method is incorporated into the nano artificial red blood cell membrane, the maximal amount of polymer per 500 ml of suspension of nano artificial rbc's is shown in Fig. 5.11. The total amount of membrane material is much less that the total membrane material in LEH. Furthermore, the membrane material of PLA nano artificial rbc's as shown in Fig. 5.11 is made up mostly of biodegradable polymer with a minimal amount of lipid. Since polymer is stronger than lipid and is also porous, much less membrane material is required.

Artificial Cells

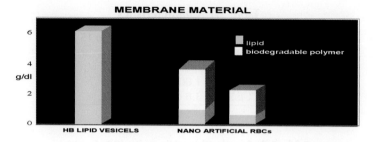

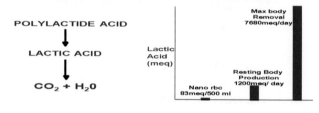

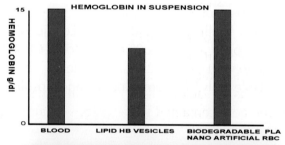

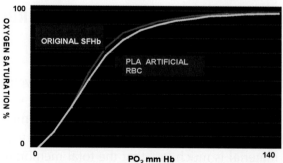

Fig. 5.11. *Top*: Amount of membrane material in LEH and PLA nano rbc's, *2nd from top*: Fate of polylactide membrane in PLA nano rbc's compard to PLA metabolism. *3rd from top*: Hb concentration reported. *Bottom*: Oxygen dissociation curve of PLA nano rbc's compared to SF Hb used in preparation.

Polylactic acid is degraded in the body into lactic acid and then to carbon dioxide (Fig. 5.11). These are all normal body metabolites. However, in hemorrhagic shock much lactic acid can be produced. Therefore, it is important to analyze how much lactic acid is produced in the degradation of PLA nano artificial rbc's. Polylactide is degraded into lactic acid and then water and carbon dioxide (Fig. 5.11). For a 500 ml suspension of PLA nano artificial rbc's, the total lactic acid produced is 83 mEq. This is far less than the normal resting body lactic acid production of 1000–1400 mEq/day) (Fig. 5.11). The maximal body capacity to break down lactic acid is 7080 mEq/day. Thus, 83 mEq is equal to about 1 to 2% of this. Furthermore, the polylactic acid in the PLA nano artificial rbc's is biodegraded over at least two days, and therefore, there is an even smaller amount released per day for each unit of PLA nano artificial rbc.

Efficiency in nanoencapsulating hemoglobin

At present, the hemoglobin nanoencapsulation efficiencies range from 13 to 29% of the starting quantities of hemoglobin, depending on the polymer used. A higher amount of hemoglobin is encapsulated with poly(D.L.)lactic acid. Nano artificial rbc's prepared with poly(D.L.)lactic acid contained fewer defects in the membrane than that prepared using poly(L)lactic acid. The ratio of hemoglobin: phospholipid : polymer for the preparation with hemoglobin of 10.97 gm/dl suspension is 15 : 1 : 2. Higher Hb concentration of up to 15 gm/dl suspension has also been prepared more recently (Fig. 5.11). This approaches the hemoglobin concentration in whole blood.

Oxygen affinity, Hill coefficient and Bohr effect

We use the TCS Hemoxanalyser (TCS Medical Products Co., U.S.A.) to measure oxygen affinity. The result for PLA artificial rbc's with 10.97 gm/dl hemoglobin, is shown in Fig. 5.11. There is no significant difference between hemoglobin nano artificial rbc's and the original bovine hemoglobin used for the preparation. Hill coefficient is 2.4 to

Artificial Cells

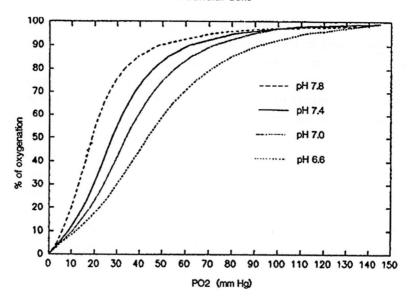

Fig. 5.12. Effects of pH on oxygen affinity of PLA nano artificial rbc's. (From WPYu & Chang, 1996.)

2.9. These results show that the procedure of preparation does not have adverse effects on the hemoglobin molecules. In the physiological pH range, oxygen affinity of PLA nano artificial rbc's changes with pH (Fig. 5.12) and the Bohr effect is -0.22 to -0.24.

Enzymes and multienzymes

We have earlier carried out basic research on artificial cells containing multienzyme systems with cofactor recycling (Chang, 1987b). A number of enzymes normally present in rbc's have been encapsulated within nano artificial rbc's and they retain their activities (Chang *et al.*, 2003a, 2003b). For example, we have encapsulated the red blood cell methemoglobin reductase system into the nano artificial rbc's (Chang *et al.*, 2003a, 2003b) and showed that this can convert methemoglobin to hemoglobin. For PLA nano artificial rbc's, the membrane can be made permeable to glucose and other small molecules. This allows us to prepare nano artificial rbc's containing the methemoglobin reductase system to function as shown in Fig. 5.1. External glucose can diffuse into the nano artificial rbc's. Products of the reaction can diffuse out and therefore do not accumulate in the nano artificial rbc's to inhibit the reaction. Furthermore, reducing agents from the outside can enter the nano artificial cells to help in preventing metHb formation. These studies are discussed in more detail below.

Nano artificial rbc's containing metHb reductase system

By increasing the circulation time, we can increase the *in vivo* functioning of the hemoglobin in the nano artificial rbc's. However, with the increase in circulation in the body at 37°C, there would be a steady increase in methemoglobin. Oxidation of hemoglobin to methemoglobin inside rbc's is prevented by the enzyme systems of the red blood cell (rbc).

PLA nano artificial rbc's can contain all the enzymes of the red blood cell (Fig. 5.1). This is done by extracting from red blood cell all the enzymes and hemoglobin. In preparing nano artificial rbc's, there is usually an increase in the metH level (Fig. 5.13). After this, when suspended in Ringer lactate at body temperature, metHb rose rapidly (Fig. 5.13). With the addition of 100 mg/dl glucose and 0.02 mM NADH, instead of increasing, metHb level decreased with time. This preliminary result is exciting because it shows that we only need to encapsulate fresh red blood cell contents with the normal amount of methemoglobin reductase system. In this way, 100 mg glucose (can be available as blood glucose) and 0.02 mM NADH in a suspending medium not only prevent metHb formation, but can convert preformed metHb back to Hb. Through further optimization of the NADH concentration, this can be increased further. Unlike NADH, the larger cofactor NADPH is not permeable across the nanocapsules. Since NADPH is not permeable, it can be enclosed inside the PLA nano artificial rbc's for recycling. This would obviate the need to supply external cofactor and only a continuous supply of glucose from the circulating blood is needed.

Artificial Cells

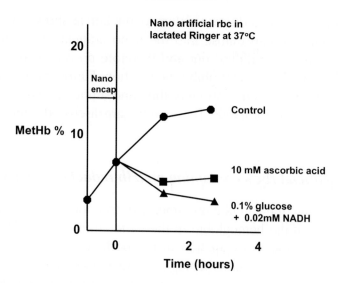

Fig. 5.13. Nanoencapsulation increases MetHb level. After this when incubated at 37°C metHb increases quickly. Addition of a reducing agent, ascorbic acid prevents the increase in MetHb. Addition of glucose and NADH decreases MetHb further.

Nano artificial rbc's permeable to reducing factors

There are reducing agents in the plasma that prevent methemoglobin formation. Examples include ascorbic acid and glutathione. These can cross the membrane of the PLA nano artificial rbc's. In our study, we find that ascorbic acid, glutathione or methylene blue can cross the membrane of PLA nano artificial rbc's. When nano artificial red blood cell is suspended in Ringer lactate with a 10 mM concentration of ascorbic acid, there is no increase in metHb as long as there is enough supply of ascorbic acid. (Fig. 5.13). Thus, it would appear that Hb in the nano artificial rbc's could be exposed to reducing factors in the circulating plasma that prevent the formation of metHb. This is an additional and simpler means of preventing the oxidation of hemoglobin into methemoglobin.

Enzymes and multienzymes in PLA nano artificial rbc's

Other red blood cell enzymes present in the red blood cell extract, like carbonic anhydrase, catalase and superoxide dismutase are also included in the nano artificial rbc's (Fig. 5.1). In addition, higher concentrations of red blood cell enzymes like catalase and superoxide dismutase can be included. Other non-red blood cell enzymes can also be included.

5.9.   Circulation Time of Nano Artificial RBCs

Circulation half-life of PLA artificial rbc's

The *in vitro* results obtained so far appear promising and encourage us to continue to carried out detailed *in vivo* studies. We started with studying the circulation time. Circulation half-life of PLA nano artificial rbc's is evaluated in anesthetized male rats. Each rat received 1/3 of its blood volume via intravenous top loading. PLA artificial red blood cell is removed very rapidly from the circulation.

As a result, we investigated different ways of changing the surface properties of the PLA nano artificial rbc's as described below. This includes different ways of incorporating PEG into the PLA membrane.

Different types of PEG-PLA copolymers for nano artificial rbc's

Figure 5.15 summarizes the results of the circulation time in rats for polyHb and different types of PLA and PEG-PLA nano artificial rbc's (Chang *et al.*, 2003b). PLA nano artificial rbc's are removed very rapidly from the circulation. PolyHb circulates with the same half-time as that normally obtained in rats. Depending on the method used, PEG-PLA nano artificial rbc's (Fig. 5.14) have significantly higher circulation time than PolyHb. The details are as follows.

Anesthetized rats received a 30% blood volume toploading by intravenous infusion. Figure 5.15 shows that glutaraldehyde crosslinked polyhemoglobin (10 gm/dl) resulted in maximal non-red blood cell Hb concentration of 3.35 gm/dl, falling to half its maximal

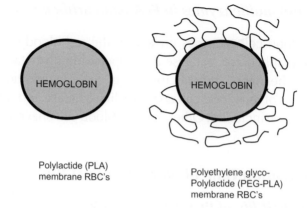

Fig. 5.14. *Left:* PLA membrane that forms the membrane of nano artificial rbc's. *Right:* Incorporation of polyethylene-glycol (PEG) in the PLA membrane results in marked improvement of circulation time of PEG-PLA membrane nano artificial rbc's.

concentration of 1.67 gm/dl in 14 h. From here on, PolyHb (17:1) is used as the basis for comparison with all nano artificial rbc preparations.

Four factors, individually or in combination, should be considered when preparing the different types of PEG-PLA nano artificial rbc's. These are studied in regard to the maximal systemic Hb attained and the time of maintaining a systemic non-red blood cell Hb level equal to that of polyHb (17:1), 1.67 gm/dl. The four factors are as follows.

(i) Use of different degrees of polymerized Hb.
(ii) Effects of higher M.W. PLA.
(iii) Effects of concentrations of the PEG-PLA copolymer.
(iv) Crosslinking of the newly formed PEG-PLA Hb nanocapsules.

(1) *Effects of molecular weight distribution of polyHb*
    used in the PEG-PLA nano artificial rbc's

The use of PolyHb (10:1) for PEG-PLA nano artificial rbc's results in a maximal non-red blood cell Hb level of 3.05 ± 0.03 S.D. gm/dl.

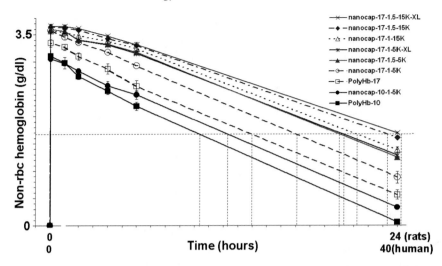

Fig. 5.15. Comparison of the maximal systemic non-rbc hemoglobin reached after infusion of different preparations and the time to reached a given non-rbc hemoglobin level. The time for PolyHb-17 to reach a non-rbc hemoglobin level of 1.67 gm/dl is 14 hours in rats equivalent to 24 hours in human. The time for different types of nano artificial red blood cells to reach this non-RBC hemoglobin concentration is used to calculate the equivalent time for human (Chang *et al.*, 2003b).

The non-red blood cell Hb falls to 1.67 gm/dl in 12.3 h (Fig. 5.15). Calculations based on body weight, blood volume, plasma volume and dilution factors show that the maximal non-red blood cell hemoglobin concentration for nano artificial rbc's should have been 3.6 gm/dl. This seems to show that a significant part (about 16%) has been removed nearly immediately on infusion.

We then used a more refined method to improve the degree of polymerization to markedly reduce the amount of tetrameric hemoglobin (PolyHb 17:1). This was then used for preparing PEG-PLA nano artificial rbc's. Two min after infusion, the maximal non-rbc Hb was 3.58 ± 0.04 S.D. gm/dl. This was significantly higher than the 3.05 gm/dl (S.D.= 0.04) above. This also approached the maximal possible initial non-red blood cell Hb concentration. Furthermore, it took 17.1 h in rats for the non-red blood cell Hb level to fall to 1.67 gm/dl. The next step was to further improve this in the following step-wise incremental design until we reached a maximal concentration of 3.66 ± 0.03 S.D. gm/dl and 24.2 h to fall to the level of 1.67 gm/dl in rats.

(2) *Effects of higher concentration of PEG-PLA copolymer combined with polyHb (17:1)*

The same method as above with PolyHb (17:1) used, but with a 1.5 times higher PEG-PLA concentration. This resulted in a thicker membrane with better membrane stability, which in turn resulted in a further increase in circulation time (Fig. 5.15). Thus, in two min after infusion, the maximal non-red blood cell Hb was 3.60 ± 0.01 S.D. gm/dl. Furthermore, it took 20.0 h in rats for the non-red blood cell Hb level to fall to the level of 1.67 gm/dl. We also used twice the concentration of PEG-PLA polymer to further improve the stability of the membrane. However, the Hb nanocapsules formed this way tended to aggregate and therefore was not tested in animal studies.

(3) *Effects of higher molecular weight PLA for the PEG-PLA copolymer*

We looked at the use of a higher molecular weight PLA to increase the stability. For this we replaced the 5K PLA in (1) above with a 15K PLA to form the PEG-PLA copolymer. This significantly increases the circulation time of the preparation as compared to method (1). Thus, in two min after infusion, the maximal non-red blood cell Hb was: 3.57 ± 0.05 S.D. gm/dl. The slope of the disappearance was also much more gradual, but what is more important is that it took 21.2 h in rats for the non-red blood cell Hb level to fall to 1.67 gm/dl.

(4) *Higher PEG-PLA concentration combined with higher molecular weight PLA*

We next looked at combining the use of a higher molecular weight PLA (15K) as in (3) above with a 1.5 times higher concentration of the polymer as in (2) above. This is a combination of the following 3 factors:

(a) Using polyhemoglobin with low percentage of crosslinked tetrameric hemoglobin.
(b) Using a 1.5 x concentration of the PLA-co-PEG copolymer.
(c) Using a higher molecular weight PLA (15K).

This resulted in a further significant improvement. In 2 min after infusion, the maximal non-red blood cell Hb was: 3.6458 ± 0.02 gm/dl. The slope of the disappearance was also much more gradual, and it took 23.3 h for the non-red blood cell Hb level to fall to the level of 1.67 gm/dl (Fig. 3).

(5) *Effect of crosslinking the newly formed Hb nanocapsules*

The method described in (1) above using polyHb (17:1) was modified by adding glutaraldehyde to the newly form PEG-PLA nano artificial cells. This approach has been used earlier to stabilize both the larger Hb microcapsules and the protein inside (Chang, 1971b). The polymerization was stopped by adding 2 M of lysine (at molar ratio of lysine/hemoglobin = 100:1) after 24 h. This approach also increased the circulation time to the same degree as when using 1.5 times concentration of the polymer in (2) above. Thus, in 2 min after infusion, the maximal non-red blood cell Hb was higher: 3.60 ± 0.01 S.D. gm/dl. The slope of the disappearance was also much more gradual, but what is more important is that it took 20.3 h in rats for the non-red blood cell Hb level to fall to 1.67 gm/dl (Fig. 5.15).

(6) *Effects of the combination of all four factors to prepare Hb nanocapsules*

Finally, we combined all the above four factors as follows:

(a) Using the polyhemoglobin (17:1) with a low percentage of single crosslinked tetrameric hemoglobin.
(b) Using a 1.5 x concentration of the PLA-co-PEG copolymer.
(c) Using a higher molecular weight PLA (15K).

(d) Crosslinking of the newly formed Hb nanocapsules with glutaraldehyde.

In 2 min after infusion, the maximal non-red blood cell Hb was higher at 3.6583 ± 0.03 S.D. gm/dl (Fig. 5.15). The slope of the disappearance was also gentler and it took 24.2 h in rats for the non-red blood cell Hb level to fall to 1.67 gm/dl.

Analysis of results

All the results obtained are summarized in Fig. 5.15. and Table 5.3.

After a 30% blood volume toploading using polyHb (10 gm/dl), the best polyHb-17 can only attain a maximal Hb concentration of 3.35 gm/dl. The best PEG-PLA artificial red blood cell's, on the other hand, can reach a maximal Hb concentration of 3.60 gm/dl. This cannot be explained by polyHb having more colloid osmotic pressure than Hb inside the nanocapsules and resulting in some hemodilution. After all, in those PEG-PLA Hb nanocapsules that are removed more rapidly than PolyHb (17:1), e.g. nanocap-10-1-5K, the maximal Hb reached for nanocap-10-1-5K is lower than that for PolyHb-17. The more likely explanation is that for the best PEG-PLA Hb nanocapsules, they all remain in the circulation when the first samples are taken at 2 min after infusion. For nanocap-10-1-5K and PolyHb, a small fraction must have been removed within 2 min after infusion.

Table 5.3 Analysis of Results

Preparation	Max Hb	Time to 1.67g/dl (h)	
PolyHb-10	3.10 gm/dl	10.4 (rats)	17 (humans)
PolyHb-17	3.35 gm/dl	14.0 (rats)	24 (humans)
Nanocap-10-1-5K	3.05 gm/dl	12.3 (rats)	21 (humans)
Nanocap-17-1-5K	3.58 gm/dl	17.1 (rats)	29 (humans)
Nanocap-17-1.5-5K	3.60 gm/dl	20.0 (rats)	34 (humans)
Nanocap-17-1-5K-XL	3.60 gm/dl	20.3 (rats)	35 (humans)
Nanocap-17-1-15K	3.57 gm/dl	21.2 (rats)	36 (humans)
Nanocap-17-1.5-15K	3.65 gm/dl	23.3 (rats)	39.9 (humans)
Nanocap-17-1.5-15K-XL	3.66 gm/dl	24.2 (rats)	41.5 (humans)

Relevance for clinical application in humans

After a 30% blood volume toploading, the best polyHb can maintain a systemic hemoglobin level of 1.67 gm/dl for 14 h in rats (equivalent to about 24 h in humans). In the case of the best PEG-PLA nano artificial rbc's a similar toploading can maintain a much longer systemic Hb level, reaching 1.67 gm/dl after 24.2 h in rats (Fig. 5.15). If one uses the rat results in this study for polyHb of 14 h and its clinical equivalent of about 24 h, we might calculate and extrapolate this as follows. For the best PEG-PLA nano artificial rbc's with the ability to maintain a systemic Hb level reaching 1.67 gm/dl after 24.2 h, it is likely equivalent to about 41.5 h in humans (Fig. 5.15). Compared to humans, rats have a highly active RES; thus, it has been suggested that a $T_{1/2}$ of 12–20 h in rats is equivalent to 40–60 h in human (Woodle *et al.*, 1995). If this is the case, then the best type of PEG-PLA nano artificial rbc's may have an equivalent circulation time in humans of more than 60 h. This is a significant and important increase that would allow longer function after infusion and thus decrease the need for donor blood and further increase the avoidance of donor blood. A further advantage of PEG-PLA nano artificial rbc's containing polyHb instead of Hb is that even if PolyHb leaks slowly after infusion it would continue to act and, unlike tetrameric Hb, would not cause adverse effects. Indeed, PEG-PLA nano artificial cells could be useful carrier for other modified Hb, including recombinant Hb, PEG conjugated Hb to prolong further their circulation time, inclusion of enzyme systems and avoid direct external exposure.

5.10. General

As discussed in Chapter 4, successes in nanobiotechnology-based polyHb and conjugated Hb stimulate research into the next generation systems. This is also the case in nanobiotechnology-based artificial rbc's. Thus, research in micron dimension artificial rbc's has led to the successful development of lipid encapsulated hemoglobin. This has encouraged us to look into the next step towards a further generation of nano artificial rbc's in the form of PEG-PLA nano artificial rbc's. We are only at the early stage of this research. In the meantime, it is important to complete the development of the lipid encapsulated hemoglobin for clinical trials.

5.11. Method of Preparation of Biodegradable Polymeric Nano Artificial RBCs

Organic phase

Dissolve 100 mg (d.l)-polylactic acid (MW 25000) (Polysciences, Warrington, PA) in 8 ml acetone. Dissolve 50 mg hydrogenated soybean phosphatidylcholine (Avanti Polar Lipids, Alabaster, AL) in 4 ml ethanol with help of Ultrasonic water bath (very low power). The two solutions are mixed (added either one to another), and used as organic phase.

Aqueous phase

Take 0.04 ml of Tween 20, mix with 25 ml 15 g/dl hemoglobin.

Preparation of PLA nano artificial red blood cells

Slowly inject (8 ml/min). (The injection head is made with a 0.2 ml pipette tip). The organic phase integrates into the aqueous phase under magnetic stirring [with Therm-O-Swirl Stirrer (Precision Scientific Co., Chicago) setting to 6)], under 4°C. Nanoparticles are formed immediately, and the suspension is stirred continually for 15 min. The suspension prepared comes up to 37 ml.

The organic solvent is partly removed from the above suspension prepared by rotary evaporator under vacuum at 20°C for about 10 min. 33 ml of the suspension is obtained (i.e. 4 ml of organic solvent removed). The remaining suspension is mixed with 15 ml of 0.9% NaCl. Then the organic solvent and free hemoglobin are removed by ultrafiltration (by Amicon ZM 500,000 membrane, MW cut off 500,000). The suspension is repeatedly washed by 0.9% NaCl by ultrafiltration.

The detailed washing process is as follows:

a. The original 33 ml of Hb nanoparticles suspension plus 15 ml 0.9% NaCl, are concentrated to 10 ml.

b. 10 ml of suspension plus 40 ml 0.9% NaCl are concentrated to 6 ml.

c. 6 ml suspension plus 18 ml 0.9% NaCl are concentrated to 8 ml.

Preparation of PEG-PLA nano artificial red blood cells

The procedure for the preparation of the new PEG-PLA copolymer is as follows. 1/½g DL-PLA (M.W. 5000 or 15,000) and 0.75 g methoxypolyethylene glycol (M.W. 2000 or 5000) are dried under vacuum overnight. 5 ml acetone is added. The mixture is heated to 180°C for 2 h. under nitrogen. After adding 10 µl of stannous-2-ethylhexanoate, the mixture is heated at 180°C for another 3 h under nitrogen. This is soluble in acetone but not in water.

This PEG-PLA is used for the preparation of nano artificial cells. The procedure is similar to the procedure for the preparation of PLA nano artificial rbc's described above, with the following change. 150 mg PEG-PLA copolymer is dissolved in 8 ml acetone. 50 mg hydrogenated soybean phosphatidylcholine is dissolved in 4 ml ethanol. The two solutions are mixed to form the organic phase.

CHAPTER 6

Use of Enzyme Artificial Cells for Genetic Enzyme Defects that Increase Systemic Substrates to Toxic Levels

6.1. Introduction

General

The basic principle on the use of artificial cells containing enzymes in enzyme therapy has been discussed in Chapter 2 (Fig. 2.3). This principle has been studied for use in enzyme therapy for genetic enzyme defects, removal of unwanted metabolites and treatment of amino acid-dependent tumors (Fig. 6.1). This chapter describes examples of the use of enzyme artificial cells in three genetic enzyme defects: Acatalasemia, Lesch-Nyhan disease and phenylketonuria (Fig. 6.1). Other terminologies for genetic enzyme defects include inborn errors of metabolism, hereditary enzyme defects and congenital enzyme deficiency. The cause of each of these conditions is the congenital defects of an enzyme or an enzyme system. This results in the accumulation of the substrate for the enzyme to an unacceptable or toxic level.

Substrate → // **DEFECTIVE ENZYME** // → Product

The accumulation can be inside the cells as in the case of storage diseases, or throughout the body fluid. There is already an extensive literature on inborn errors of metabolism (Scriver, 1989); thus, this

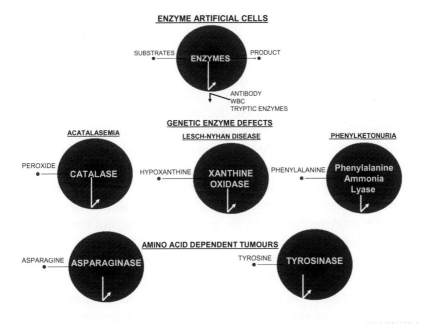

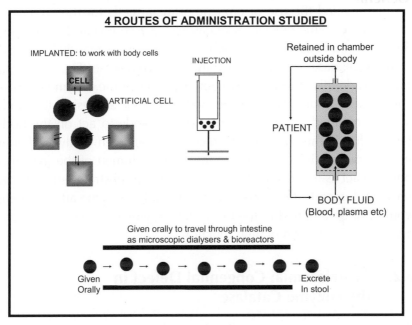

Fig. 6.1. *Upper*: Examples of enzyme artificial cells. *Lower*: Routes of administration.

chapter will concentrate on the type where the substrates accumulate throughout the body fluids.

Enzyme replacement therapy

Much study has been carried out on the use of enzyme therapy in treating defective enzymes. However, as discussed in Chapter 2, the use of enzymes directly has a number of problems: a) the enzyme must be able to be located at the site of action; b) the enzyme has to be available in a highly purified and non-toxic form; c) the heterogeneous enzyme may result in immunological and hypersensitivity reactions; and d) the enzyme has to act for a sufficient length of time. Some of these problems might be avoided by using artificial cells containing the *defective* enzyme (Chang, 1964, 1965; Chang *et al.*, 1966; Chang & Poznznsky, 1968) (Fig. 6.1). In this way, the enclosed enzyme does not leak out and become involved in immunological or hypersensitivity reactions, but can act on the permeant substrates. We investigated the use of enzyme artificial cells for three types of congenital enzyme defects: Acatalasemia, Lesch-Nyhan Disease and phenylketonuria using respectively, artificial cells containing catalase, xanthine oxidase and phenylalanine ammonia lyase (Fig. 6.1). Four routes of administration were studied (Fig. 6.1) i.e. the enzyme artificial cells are: 1) implanted into the body to work alongside the body cells; 2) retained in chambers outside the body to act on the body fluid perfusing the chamber; 3) injected; or 4) given orally so that as the enzyme artificial cells move through the lumen of the intestine, they will act as microscopic bioreactors on the substrates entering the intestinal lumen from the body. In this way, the enzyme artificial cells would not be retained in the body as they would be excreted in the stool after completion of their function.

6.2.   Acatalasemia: Congenital Defect in the Enzyme Catalase

The first study is to use the simplest animal model to determine if enzyme therapy using artificial cells is feasible, effective and free from immunological effects. Acatalasemia is a congenital defect in the enzyme catalase leading to a deficiency in the removal of its substrate peroxide. This is a rare condition in humans, but there is a convenient strain of mice with congenital defect in the enzyme catalase (Feinstein *et al.*, 1966). These mice have a blood catalase activity of only 1–2% of normal mice and a total body catalase activity of 20% of normal mice. Furthermore, the enzyme catalase can be obtained easily in a highly purified form. Our initial study shows that injection of catalase artificial cells into acatalasemic mice can effectively remove peroxides (Chang & Poznansky, *Nature*, 1968b). We therefore carried out further studies to determine the *in vivo* kinetics, effectiveness and immunological and hypersensitivity effects (Chang, 1971b, 1972f; Poznansky & Chang, 1974). The details are as given in the next section.

6.3. Preliminary Study on the Use of Catalase Artificial Cells to Replace Defective Catalase in Acatalasemia

The first study is to answer the question "Can implanted enzyme artificial cells work alongside the body cells to replace their enzyme defects?" (Fig. 6.1).

Implantation of catalase artificial cells

Collodion membrane catalase artificial cells were prepared as described in the Appendix under "Methods of Preparation of Artificial Cells." In acatalasemic mice which are injected with sodium perborate without the protection of a previously injected catalase, the red color of their pupils turned brown and the animals became increasingly immobile, so that by 20 min they became flaccid and were in respiratory distress. Acatalasemic mice protected by an intraperitoneal injection of catalase artificial cells were slightly immobile 5 min after the injection of perborate, and the color of their pupils darkened. However, in 10 min, the mice became more active and their pupils returned to their normal red color, and except for one animal, they were moving about freely at 20 min. Acatalasemic mice injected with a catalase solution and the normal control mice recovered even faster.

Figure 6.2 shows that within 20 min after injection of perborate, normal mice could remove most of the perborate, leaving only 2.5% ± 2.5% S.D. in the body. Acatalasemic mice receiving no catalase injection could only remove a small amount of the injected perborate, leaving 70% ± 8.2% S.D. after 20 min. Acatalasemic mice receiving either catalase artificial cells or catalase solution, could more efficiently remove the injected perborate, leaving respectively, 16% ± 3.5% S.D. and 7% ± 3.6% S.D. 20 min after injection. Thus, intraperitoneally injected liquid catalase and catalase artificial cells both acted efficiently in replacing the enzyme defect. Catalase in solution is expected to be more efficient since there is no membrane barrier for the diffusion of the substrate. On the other hand, catalase in solution may give rise to immunological and hypersensitivity reactions and these will be studied in a later section. In the case of catalase

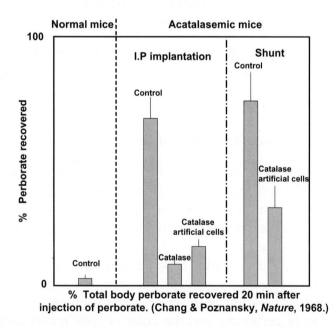

Fig. 6.2. Implanted catalase artificial cells remove peroxide from the body. They are also effective in removing peroxide from the body when retained in a chamber outside the body and perfused with peritoneal fluid. (Modified from Chang & Poznansky, *Nature,* 1968b.)

artificial cells, there was no increase in blood catalase level 20 min after injection, showing that there is no leakage of catalase into the systemic circulation.

Catalase artificial cells retained in a chamber for body fluid perfusion

In the second set of experiment, to avoid the introduction and accumulation of catalase artificial cells, especially in long-term use, we carried out the following study. We retained catalase artificial cells in a chamber outside the body and perfused these cells by recirculating peritoneal fluid from the animal (Fig. 6.1). Here the animals were anesthetized with Nembutal. Each animal received an intraperitoneal injection of 4.5 ml of inperinol fluid (Abbott Laboratory). This fluid was continuously recirculated through a smaller modified version of the extracorporeal shunt chamber (Chang, 1966) for 20 min after injection of perborate. In one group of animals, the shunt contained 4 ml of a 50% suspension of catalase artificial cells. In the control group, no catalase artificial cells were present in the shunt. Figure 6.2 shows that in the control group, the perborate remaining in the body was $76.1\% \pm 13.4\%$ S.D. after 20 min. In the chamber containing catalase artificial cells, only $32\% \pm 9.8\%$ S.D. of the injected perborate was recovered 20 min after injection.

6.4. *In Vivo* Kinetics of Catalase Artificial Cells for Acatalasemic Mice

Summary

The enzyme kinetics and immunological properties of collodion artificial cells containing catalase (H_2O_2: H_2O_2 oxidoreductase, EC 1.11.1.6) were studied and compared with catalase in free solution. *In vitro* studies showed that the K_m for both forms of catalase was 0.55 M. The Vm of the catalase artificial cells was 0.5 mM/min, while that of catalase in free solution was 2.5 mM/min. *In vivo* studies on acatalasemic mice showed that catalase artificial cells were as effective as the same assayed amount of catalase in free solution in reducing total body substrates. After injection, the catalase activity of the artificial cells in the body had a half-time of 4.4 days. This is more stable than catalase in free solution which had a half-life of 2.0 days after injection. *In vitro* studies showed that antibody to catalase would not enter the artificial cells. Immunizing doses of catalase artificial cells did not produce antibodies. Immunizing doses of catalase solution resulted in the production of antibodies to catalase. Injection of catalase solution into these immunized mice resulted in anaphylactic reactions. On the other hand, catalase artificial cells injected into these immunized mice continued to act without causing any adverse reactions.

Methods of preparation of catalase artificial cells and experimental procedures

These are described in detail in Appendix II.

Characteristics of acatalasemic mice

Experiments showed that the acatalasemic mice bred in this laboratory from the original Feinstein strain were the same as the original strain. Thus, the blood and total body catalase levels were respectively, 1–2% and 15%, of the normal mice.

Kinetics of recovery of injected perborate

Figure 6.3 represents the time course of removal of injected perborate from acatalasemic mice and from normal mice. Normal mice removed injected perborate rapidly, whereas acatalasemic mice could only remove it at a very slow rate. While the normal mice showed no outward symptoms after the injection of perborate, the acatalasemic mice were in some distress. The eyes of these acatalasemic mice changed from red to dark brown due to methemoglobin formation, and they became increasingly immobile and by 20 min most of them were flaccid and in respiratory distress. Quantitative analysis showed that injection of either catalase solution or catalase artificial cells could significantly increase the rate of removal of the injected perborate (Fig. 6.3). Acatalasemic mice given prior intraperitoneal injections of

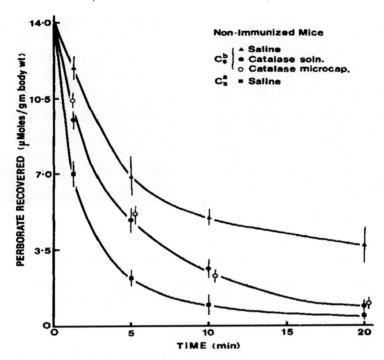

Fig. 6.3. Rate of removal of peroxide in nonimmunized rats using PAL artificial cells as compared to catalase solution and control saline. (Modified from Poznansky & Chang, 1974.)

either catalase in free solution or catalase artificial cells did not suffer any distressful symptoms and were able to remove the peroxide in a manner similar to the normal mice. Thus, both catalase solution and catalase artificial cells were effective and safe after the first implantation.

6.5. Immunological Studies on Catalase Artificial Cells in Acatalasemic Mice

Effects of immunizing doses of catalase solution and catalase artificial cells on antibody production

The next step is to study whether there is a difference in the immunological response when catalase solution and catalase artificial cells, respectively were administered. Acatalasemic mice and rabbits

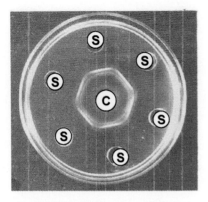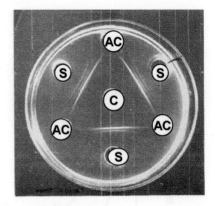

C: catalase solution (the antigen)

S: serum from mice that received immunizing doses of catalase solution contains antibody to catalase as shown by white line of precipitation

AC: serum from mice that received immunizing doses of catalase artificial cells does not contain antibody to catalase; no white line of precipitation

(Modified from Poznansky and Chang, 1974, BBA.)

Fig. 6.4. Ouchterlony double-diffusion technique showed that antibodies to catalase were produced in the acatalasemic mice and rabbits immunized with catalase in free solution. When catalase artificial cells were used, the same test showed no detectable antibodies.

were immunized over a period of 90 days. Each immunizing dose consisted of Freunds' adjuvant with catalase solution or catalase artificial cells (0.1 ml containing 4.8 perborate units of catalase). Ouchterlony double-diffusion technique showed that antibody titers were produced in the acatalasemic mice and rabbits immunized with catalase in free solution (Fig. 6.4). The same test showed that no detectable antibody titers were formed in those animals receiving immunizing doses of catalase artificial cells (Fig. 6.4)

Permeability of membrane of collodion artificial cells to catalase antibody

Catalase antibodies were incubated with catalase artificial cells or with control artificial cells containing no catalase. No antibodies were detected inside the artificial cells when tests were carried out using either the radioactive labeled antibody technique or the micro-complement fixation technique. This suggests that the membrane of artificial cells is not permeable to catalase antibodies.

Response of immunized acatalasemic mice to injections of catalase artificial cells or catalase solution

Intraperitoneal injection of catalase solution into the immunized acatalasemic mice caused them to go into anaphylactic shock and 80% of them died within the first 10 min (Fig. 6.4). Death was accompanied by violent gasping, convulsions and collapse similar to that observed in anaphylactic shock in mice. In the mice that survived, the injected catalase solution was not as effective in removing the peroxide (Fig. 6.5) Injection of catalase artificial cells did not produce

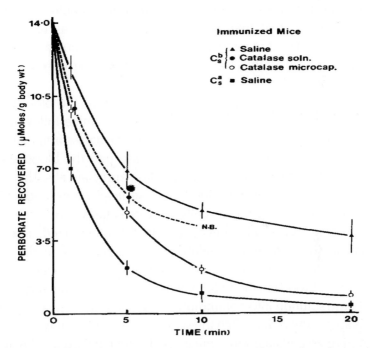

Fig. 6.5. Acatalasemic rats receiving immunizing doses of catalase solution to produce antibody to catalase. This was then followed by the implantation of either catalase solution or catalase artificial cells. (Modified from Poznansky & Chang, 1974.)

any adverse reactions. In addition, Fig. 6.5 shows that catalase artificial cells removed peroxide at the same rate as when the cells were injected into the nonimmunized acatalasemic mice (Fig. 6.3).

Appearance of catalase activity in the blood after injection of catalase solution or catalase artificial cells

Intraperitoneally injected catalase solution is expected to move quickly into the blood stream. However, similarly injected catalase artificial cells should stay inside the peritoneal cavity where they act on the perborate diffusing into the artificial cells. Thus, the enzyme should not leak out of the artificial cells into the circulating blood. To analyze this, the amount of catalase in free solution or of catalase artificial cells was reduced to one-third. This reduces the mortality rate from 80 to 20% for the immunized mice receiving the catalase solution. In this way, the kinetics of removal of the perborate could be studied. The result shows that free catalase solution is not as effective as catalase artificial cells in removing the peroxide in immunized mice. To analyze the reason for this, the following study was carried out in these animals.

Figure 6.6 shows the appearance of catalase in the bloodstream of immunized and nonimmunized acatalasemic mice following the intraperitoneal injection of catalase solution or catalase artificial cells. There was no detectable leakage of catalase into the blood stream after the intraperitoneal injection of catalase artificial cells. There was no significant change in the blood catalase level when followed for up to 3 months. The artificial cells with its catalase are retained in the peritoneal cavity after injection. On the other hand, injection of catalase solution into acatalasemic mice resulted in the rapid appearance of the enzyme in the blood stream. The peak catalase activity in the blood reached 182.8 perborate units 8 h after injection of the catalase solution into nonimmunized acatalasemic mice (Fig. 6.6). What is surprising is that when the same catalase solution was injected into immunized acatalasemic mice, the blood catalase level reached a peak of only 19.6 perborate units 4 h after the injection. This is most likely due to the formation of antigen-antibody complexes that are quickly removed.

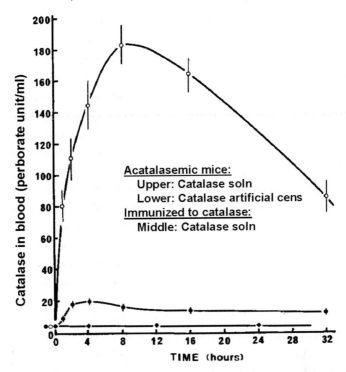

Fig. 6.6. Appearance of catalase activity in the circulating blood after intraperitoneal injection of catalase solution or catalase artificial cells. After intraperitoneal injection, catalase in solution moves rapidly into the circulation catalase artificial cells remains inside the artificial cells and does not leak into the circulation (From Poznansky & Chang, 1974.)

6.6. Conclusion from Results of Basic Study Using the Acatalasemic Mice Model

On a theoretical level, this basic study supports the proposal of using enzyme artificial cells to replace the defective enzyme (Fig. 6.1). Thus, artificial cells containing an enzyme catalase can effectively replace the defective catalase enzyme in this animal model. The result also shows that catalase inside artificial cells is prevented from giving rise to immunological reactions. Furthermore, the membrane of artificial cells is not permeable to catalase antibodies in mice previously immunized with catalase solution. Therefore, the catalase artificial cells can continue to function in those acatalasemic mice previously immunized against catalase. Therefore, in theory, enzyme artificial cells should allow for enzyme therapy in congenital enzyme defects where the accumulation of substrate is in the body fluid.

However, at that time, enzymes for the most common genetic enzyme defect, phenylketonuria, could only be extracted from the liver. These liver enzymes are very complex and unstable and not suitable for use in artificial cells for enzyme therapy. With the later availability of simple and stable enzymes from microorganisms, enzyme therapy becomes much more feasible and practical. However, the major barrier is that enzymes are not stable at body temperature and thus can only function in the body after injection for a few days. In congenital enzyme defects, the duration of treatment is for the life of the patients. Repeated injection of enzyme artificial cells over the many years of treatment would result in the accumulation of much foreign material in the body. To solve this problem, catalase artificial cells retained in small chambers outside the body and perfused by body fluid can also remove perborate. However, a more convenient route of administration would be better for pediatric patients. We, therefore, looked at giving enzyme artificial cells orally (Fig. 6.1). This is based on our earlier finding of the safety and effectiveness of oral administration of urease artificial cells in rats to remove systemic urea (Chang, 1972a). The safety of oral urease artificial cells has later been demonstrated in clinical trials in kidney failure patients (Kjellstrand *et al.*, 1981).

The next two studies will therefore, look at the use of oral administration. This way, the enzyme artificial cells move down the intestinal lumen where they act as combined microscopic diaysers and enzyme bioreactors to remove unwanted substrates moving into the lumen from the body fluid. After completion of their functions, they are excreted in the stool without being accumulated in the body (Fig. 6.1). The enclosed enzymes are protected from intestinal tryptic enzymes.

6.7. Oral Xanthine Oxidase Artificial Cells in a Patient with Lesch-Nyhan Disease

Lesch-Nyhan disease is caused by a genetic defect in the enzyme hypoxanthine phosphoribosyltransferase (HPRT) (Wilson *et al.*, 1983).

This leads to overproduction of purine and accumulation of oxypurine intermediates and uric acid. It has been suggested that damage to the brain in Lesch-Nyhan disease may be secondary to the accumulation of oxypurines such as hypoxanthine (Lloyd *et al.*, 1981). The enzyme HPRT is very complex and difficult to extract on a large scale. However, there is an enzyme, xanthine oxidase, produced by fermentation and can be easily purchased. We, therefore, look into giving xanthine oxidase (XOD) artificial cells orally (Chang, 1989a; Palmour, Chang *et al.*, 1989). As they move through the intestine, the XOD artificial cells act as a microscopic combined dialyser-enzyme reactor (Fig. 6.1). The enclosed XOD is protected from tryptic enzymes, but the ultrathin selectively permeable membranes allow passive entry of small substrate molecules like hypoxanthine. Hypoxanthine is highly lipid soluble, and can therefore equilibrate rapidly between the body fluid and the intestine. In this way, hypoxanthine and other oxypurines from the body fluid entering the intestinal lumen could be converted into uric acid. The uric acid in the intestine would be excreted in the feces directly or as allantoin. This may result in the depletion of the oxypurine body pools. To test this idea, we administered XOD artificial cells to a patient with Lesch-Nyhan disease.

Collodion membrane artificial cells containing XOD were prepared using the updated method as described in Appendix II under "Preparation of Enzyme Artificial Cells." XOD was obtained from Boehringer, Montreal (grade III from buttermilk, chromatographically purified, 50 U at 1–2 U/mg protein). 1 ml artificial cells contained 5 U XOD. The XOD artificial cells were stored as a 50% suspension in saline solution at 4°C for 0–3 days before use. Since the artificial cells are destroyed by pH of less than 3.5, it is important to administer the cells together with a buffer.

Volume ratio of artificial cells/substrate solution

In order to analyze the amount of XOD artificial cells needed for oral administration, the following *in vitro* study was carried out. Fresh substrate solutions were prepared containing 40 μM hypoxanthine in phposhate buffer (0.1 M), pH 7.5. Four different volume ratios of artificials/substrate solution were studied, i.e. 1/10, 1/100, 1/500 and

1/1000. Hypoxanthine was converted by xanthine oxidase into uric acid and the reaction was followed by the rate of formation of the product, uric acid.

Effects of pH on enzyme activity

For oral administration, it is important to analyze the relationship between pH and enzyme activity. It is also important to analyze the stability of the collodion artificial cells and their contents at different pH environments. The pH's of the substrate solutions were adjusted using citric-phposphoric acid-borate-HCl buffer. The following pH's were studied: pH 2.5, 3.5, 4.5, 5.5, 6.5, 8.5, 9.5 and 10.5, there was leakage of all the contents of the artificial cells due to destruction of the membrane. At pH 3.5 and above, there was no leakage. This aspect is important in the clinical use of the artificial cells by oral administration. The artificial cells have to be buffered in order to prevent their destruction in the very acidic stomach environment. The enzyme activity of artificial cells containing xanthine oxidase is optimal in the pH range of 8.5 to 9.8 that is normally present in the intestinal lumen.

Patient with Lesch-Nyhan disease

The patient, a 29-month-old boy with severe Lesch-Nyhan disease (HPRT activity less than 3 pmol inositic acid/mg protein/h) was withdrawn from allopurinol therapy over 5 days. Blood and urine were obtained at 0830 h and 1630 h. After baseline studies for 3 days, the patient was treated with 100 U of XOD artificial cells via a nasogastric tube twice a day (0900 and 1200 h) for 10 days. A sample of his cerebrospinal fluid was obtained at 1630 h before treatment and on the final day of therapy.

Purines were measured by high-performance liquid chromatography and uric acid and creatinine were measured enzymatically and colorimetrically, respectively.

After withdrawal of the allopurinol, uric acid excretion rose rapidly; baseline uric acid/creatinine ratios at 0900 h ranged between 3.5

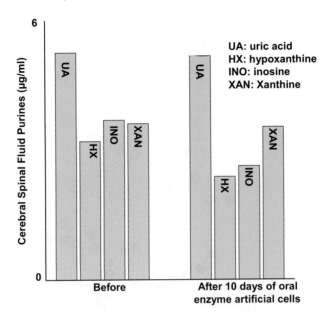

Fig. 6.7. Effect of oral XOD artificial cells on cerebral spinal fluid concentrations of uric acid, hypoxanthine, inosine and xanthine (modified from Palmour *et al.*, 1989).

and 4.1. With the introduction of XOD artificial cells, the morning and afternoon ratios fell to normal within 3 days. During a febrile episode of otitis media the ratios rose transiently. Urinary hypoxanthine excretion also responded promptly to the XOD treatment, the pattern of transient escape during the febrile illness being repeated. The CSF hypoxanthine levels fell by 25% after 10 days of XOD artificial cell administration (Fig. 6.7) and the CSF inosine levels fell by 32% (380 to 260 ng/ml). The levels of CSF xanthine and uric acid, both normal, remained unchanged.

The effect of XOD on plasma inosine was rapid (Fig. 6.8) and within 3 days, the plasma inosine levels had fallen to below 5 μg/ml. There was a reduction in plasma xanthine levels after day 6. The hypoxanthine levels manifested wide diurnal fluctuations but a regression line indicated a gradual decrease.

Artificial Cells

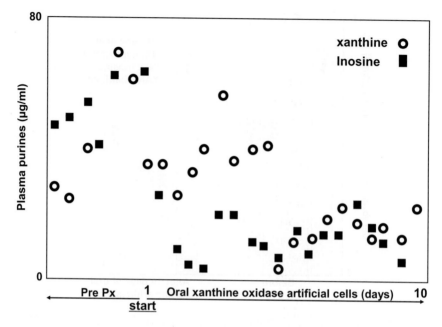

Fig. 6.8. Plasma xanthine and inosine levels before and after administration of oral XOD artificial cells on day 1 shown as "start" in the figure. Hypoxanthine levels, not shown here, manifest wide diurnal fluctuations but a regression line indicates a gradual decrease (modified from Palmour *et al.*; Chang, 1989).

6.8. Conclusion Based on Study

The purpose of our short trial of ten days is only to determine if administration of oral XOD artificial cells can change the systemic levels of the different substrates. As such, the very preliminary result suggests the feasibility of pursuing the study further. Much longer clinical trials would be needed to observe the neurological changes or behavioral changes. This is an extremely rare inborn error of metabolism, and thus we could not find any commercial interest in manufacturing this product for use in long term clinical trials. However, the very preliminary demonstration of the safety and effectiveness of oral enzyme artificial cells in a patient has led us to develop this approach at treating one of the most common genetic enzyme defects, phenylketonuria.

6.9.   Phenylketonuria: Genetic Defect in Enzyme Phenylalanine Hydroxylase

Phenylketonuria

Phenylketonuria (PKU) is caused by a genetic defect in the liver enzyme, phenylalanine hydroxylase. The resulting increase in the level of systemic phenylalanine in the first few years of life can lead to severe mental retardation (Nylan, 1974; Scriver, 1989). A low phenylalanine diet is, at present, the only effective treatment for phenylketonuria (Scriver, 1989). This diet must begin very early after birth. However, this diet is difficult to follow and does not prevent elevation of blood phenylalanine during episodes of fever and infection. Also, it is difficult to use the diet to control the level phenylalanine during pregnancy and elevation can lead to a higher incidence of birth defects. Furthermore, a low phenylalanine diet itself can result in abnormalities in the fetus.

Enzyme replacement therapy

An alternative to this dietary treatment is the use of enzyme therapy to replace the defective enzyme. However, this liver enzyme is complex and requires cofactors. It is also not stable and there is major problem related to large scale isolation and purification. Fortunately, there is a nonhuman enzyme, phenylalanine ammonia-lyase (PAL) that can be readily obtained from *Rhodotorula glutinis* (Hodgins, 1971). It does not require any cofactors and converts phenylalanine into trans-cinnamic acid. The liver converts the product, trans-cinnamic acid into benzoic acid that is excreted via the urine as hippurate. Trans-cinnamate has very low toxicity and has no adverse effects on the fetus of laboratory animals (Hoskins & Gray, 1982). Earlier attempts to use oral PAL did not produce any conclusive results (Hoskins *et al.*, 1980), most likely because the enzyme was destroyed by the tryptic enzymes in the intestine.

Our research on enzyme artificial cells suggests that placing PAL inside artificial cells would prevent the enzyme from coming into contact with intestinal tryptic enzymes (Fig. 6.1). At the same time, phenylalanine can enter the artificial cells to be converted by PAL. This, together with our basic research on the use of enzyme artificial cells for acatalasemia and Lesch-Nyhan disease described earlier, is an encouragement to us to look into the possible use of oral PAL artificial cells for phenylketonuria. However, one problem immediately comes to mind. The intestinal contents have amino acid concentrations that are many times higher than those of the plasma or body fluid. How, then, are we going to extract the amino acid phenylalanine from the body fluid into the intestine for removal by the PAL artificial cells?

Oral PAL artificial cells for PKU based on novel theory of enterorecirculation of amino acids

Our research has shown the presence of an extensive enterore-circulation of amino acids (Chang *et al.,* 1989b). Pancreatic and other glandular secretions into the intestine contain large amounts of proteins, enzymes and polypeptides. Tryptic digestion converts these into amino acids which are then reabsorbed back into the body as they pass down the intestine. They are converted by the liver into proteins, enzymes and polypeptides, some of which are again secreted into the intestine. This forms a large enterorecirculation of amino acids between the body and intestine. Our study shows that the dietary protein source of amino acids is negligible when compared with this recirculation of amino acids. Based on this theory, it would be possible to administer, orally, artificial cells containing one specific enzyme (e.g. phenylalanine ammonia lyase) to lower one specific amino acid (phenylalanine). PAL artificial cells given orally to PKU rats is more effective than a phenylalanine-free diet in lowering the level phenylalanine in the intestine, plasma and cerebrospinal fluid. Whereas PKU rats on PHE-free diet lost weight, those on oral PAL artificial cells continued to grow and gained weight during the study. The following sections describe the above studies in detail.

6.10. Research Leading to Proposal of Enterorecirculation of Amino Acids

Classical theory of the main source of amino acids in the intestinal lumen

It is commonly thought that the tryptic digestion of proteins from ingested food constitutes the major source of amino acids in the intestinal lumen. These intestinal amino acids are then absorbed as they pass through the intestinal tract. The pancreatic and other glandular secretions are mainly to function in supplying the required digestive enzymes. The contributions of these secretions as a source of intestinal amino acids are not usually thought to be of major significance.

Research to test the classical theory

We give the animal nothing by mouth except sucrose and water *ad libitum* for 24 h. If dietary protein is indeed the major source of intestinal amino acids, then there should be a significant decrease in the amino acids in the intestine compared to the control group on standard diet.

Animals on 24 h of protein-free diet

Male Sprague Dawley rats (275–330 grams) were purchased from Charles River Co., St. Constant, Quebec. For 24 h, the control groups received Purina Rat Chow (Ralston Purina, Montreal, Quebec) plus tap water *ad libitum*. The protein-free diet groups received a diet of sugar cubes (Redpath Sugar Co., Montreal) and 10 g/dl glucose in tap water. Both groups of rats stayed in metabolic cages to prevent their access to the feces.

Results of study

Figure 6.10 shows the results obtained. In the duodenum there is surprisingly no significant difference in the amount of amino acids between the control group on normal diet and the group on protein-free diet. The same results are observed for the jejunum and ileum. The concentration of amino acids decreases down the intestinal tract as amino acids are absorbed. However, there is no significant difference between the two groups throughout all segments.

There is an extremely high concentration of intestinal amino acids as compared to the plasma concentration — more than 1000-fold in the duodenum. Since the main source of amino acids in the intestine does not appear to be the proteins from ingested food, what then is the source of the intestinal amino acids in the protein-free diet group, and indeed in the normal control group?

Proposed theory of extensive enterorecirculation of amino acids

I thought that based on the above results, the most likely major source of intestinal amino acids could be the gastric, pancreatic, intestinal and other intestinal secretions (Fig. 6.9). Tryptic digestion converts the large amount of proteins, enzymes, polypeptides and peptides in the secretions into amino acids. These amino acids are then reabsorbed by the body as they pass through the intestine. Continuing pancreatic and intestinal secretions into the intestine constitutes the source of amino acids. This forms a large enterorecirculation of amino acids between the body and intestine. Since the protein-free diet did not alter the intestinal amino acid concentration, the dietary source of amino acids was negligible when compared to that from the pancreatic and intestinal secretions.

In enterorecirculation, the reabsorption of amino acids from the intestine into the plasma could be based on standard amino acid transport mechanisms. The transport of amino acids from the gastrointestinal lumen into the plasma has been extensively investigated (Stevens *et al.*, 1984). Amino acids are generally transported across membranes by passive or active carrier transport mechanisms (Berteloot *et al.*, 1982). Stevens *et al.*, (1984) added to this classification a non-saturable diffusion PHE transport system that contributes significantly to the transport of phenylalanine in both the brush border and the basolateral membranes.

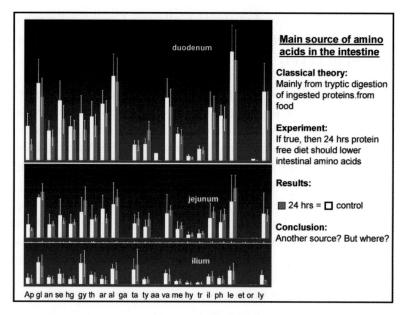

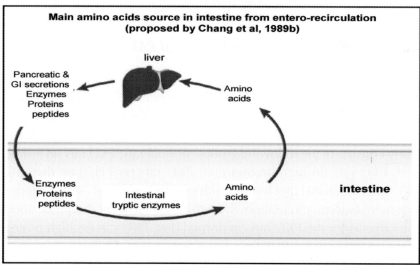

Fig. 6.9. *Upper:* Most of amino acids in the intestine do not come from ingested food. *Lower:* Enzymes, proteins and peptides in pancreatic and GI secretions, are digested by tryptic enzymes in the intestine into amino acids that are reabsorbed and converted by the liver into enzymes, proteins and peptides that are again secreted into the intestine — enterorecirculation.

6.11. Oral Enzyme Artificial Cells for Genetic Enzyme Defects that Result in Elevated Systemic Amino Acid Levels

As discussed above, very surprisingly and unexpectedly, our result shows that the major source of amino acids in the intestine appears to be the extensive enterorecirculation of amino acids. Immediately, this raises the possibility of using this for selective depletion of specific body amino acids. This is somewhat along the line of using binders to remove the enterorecirculation of bile acid, thereby lowering the systemic cholesterol level. However, unlike bile acid, there are 26 amino acids in the body and usually only one amino acid needs to be selectively depleted. This is where the specificity of artificial cells microencapsulated enzyme plays a part. Oral administration of artificial cells containing one specific enzyme (e.g. phenylalanine ammonia lyase) should be able to remove only one specific amino acid (e.g. phenylalanine) (Fig. 6.9).

General design for study of use of oral PAL artificial cells in PKU rats

In this study, we will determine whether oral PAL artificial cells can deplete a specific amino acid, phenylalanine, from PKU rats. This study will also serve as a further test of the enterorecirculation theory. This study consists of the following groups: normal rats; PKU rats on normal diet; PKU rats on high protein diet; PKU rats on PHE-free diet; and PKU rats on normal diet treated with oral PAL artificial cells. If there is extensive enterorecirculation, the intestinal PHE level should be about the same as for the PKU rats on normal diet, PKU rats on high protein diet and PKU rats on PHE-free diet. Furthermore, PKU rats on normal diet treated with PAL artificial cells should have very low phenylalanine in the intestine. In addition, if there is an extensive enterorecirculation of amino acids, then the treatment using oral PAL artificial cells should be more effective than that using PHE-free diet alone in lowering the systemic PHE levels (Bourget and Chang, 1985, 1986). The results to be discussed in detail later is summarized in Fig. 6.10.

Application of entero-recirculation for oral enzyme therapy (Chang et al, 1989b)

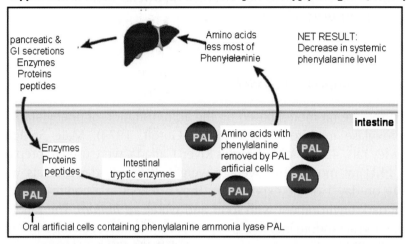

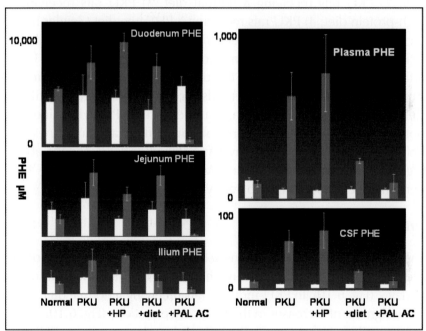

Fig. 6.10. *Upper*: Oral administration of enzyme artificial cells to remove one specific amino acid. *Lower* (left column, day 1 and right column, day 7): PHE: Phenylalanine levels; PKU phenylketonuria rats; HP high protein diet; + diet low PHE diet: PAL AC phenylalanine ammonia lyase artificial cells.

Methods of preparation of catalase artificial cells and experiment procedures

These are described in detail in Appendix II.

Phenylketonuria rat model

We used the rat model based on the administration of phenylalanine hydroxylase inhibitors (Anderson & Guroff, 1974). Male Sprague Dawley rats, 42-day-old, (Charles River, Canada) were used to prepare the PKU rat model. The average weight of the 42-day-old rats was 150 g on day 1 of the experiment. They are kept in a controlled 12 h light/dark environment with food and water *ad libitum*. The rats are separated at random into five groups: 1) Normal rats receiving a normal diet; 2) PKU rats receiving a normal diet; 3) PKU rats receiving a high-protein diet; 4) PKU rats receiving a PHE-free diet (Zeigler Bros., Inc. (USA)) and 5) PKU rats receiving a normal diet plus oral PAL artificial cells — dose of 0.8 ml containing 10 unit PAL/ml. These artificial cells are prepared by the updated method described in the Appendix under "Updated Methods for the Preparation of Artificial Cells" using collodion membrane artificial cells. Phenylalanine was obtained from the plasma, cerebral spinal fluid and intestines just before PKU induction and 7 days after induction following the above regimes. The samples were analyzed for amino acids using high performance liquid chromatography.

Intestinal phenylalanine

The intestinal PHE concentrations in the normal rats receiving a normal diet did not show any significant changes in the intestine from day 1 to day 7 (Fig. 6.10). In PKU-induced rats receiving a normal diet, there were significant increases in the PHE concentrations (Fig. 6.10). What is most striking is that even in the PKU-induced rats receiving PHE-free diets, there were also significant increases in the PHE concentrations in the duodenum, jejunum and ileum (Fig. 6.10). This is so even though no PHE was given by the oral route. This serves to further support the theory of enterorecirculation of amino acids from pancreatic and other glandular secretions. PKU-induced rats receiving a normal diet and oral PAL artificial cells showed a dramatic decrease in the intestinal PHE especially in the duodenum. This seems to show that PAL artificial cells enzymatically degraded most of the PHE in the intestine (Fig. 6.10). This lends further support to the proposal that PAL artificial cells remove PHE from the enterorecirculation.

Plasma PHE levels in PKU rats

Normal rats receiving a normal diet showed no significant changes in plasma PHE levels during the 7 days (Fig. 6.10). PKU rats receiving a normal diet or high-protein diet showed an elevated PHE level which is more than 10-fold compared to normal rats (Fig. 6.10). By the 7th day, the plasma levels of PKU rats receiving PHE-free diet was still significantly higher than those at the normal control rats (Fig. 6.10). On the other hand, by the 7th day, the plasma PHE levels in PAL artificial cells-treated PKU rats were not significantly different from those of the normal control rats (Fig. 6.10).

CSF cerebrospinal fluid in PKU rats

The normal rats showed no significant changes in the CSF PHE level from day 1 to day 7 (Fig. 6.10). The PKU rats on normal diet or high-protein diet showed a significantly elevated level of CSF PHE (Fig. 6.10). On day 7, PKU rats receiving PHE-free diet showed a significantly higher level of CSF PHE when compared to day 1 or the control (Fig. 10). However, the level is significantly lower than that in the PKU rats receiving a normal diet or high-protein diet. With PAL artificial cells given to PKU rats on normal diet, the PHE levels in the CSF were not significantly different from those of the normal rats (Fig. 6.10).

Growth as shown by body weight changes

Normal rats receiving a normal diet grew normally as shown by a significant increase of body weight after 7 days (Fig. 6.11). PKU

Artificial Cells

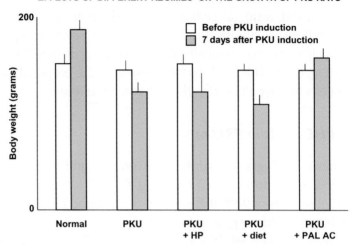

Fig. 6.11. Effects on the growth of rats followed for 7 days. From left to right: (1) normal rats; (2) PKU (phenylketonuria) rats on normal diet; (3) PKU rats on (HP) high-protein diet; (4) PKU rats on low PHE diet (diet); and (5) PKU rats receiving (PAL AC) phenylalanine ammonia lyase artificial cells.

rats receiving a normal diet or high-protein diet showed a significant decrease in body weight. Furthermore, there was a decrease in muscle tone over the same period of time. PKU rats receiving a PHE-free diet lost significantly more weight than PKU rats on normal diet. They also showed a loss of muscle tone. PKU rats on a normal diet treated with oral PAL artificial cells had a small but significant increase in body weight. Abnormal signs such as piloerection, aggression and hyperactivity, observed in PKU-induced rats, were no longer present after 5 days of treatment with PAL artificial cells.

We then looked at the daily changes in body weight, comparing normal rats, with PKU rats on PHE-free diet and PKU rats receiving different amounts of PAL artificial cells.

As shown in Fig. 6.12, normal rats receiving a normal diet showed a steady and significant increase in body weight throughout the 7-day period. PKU rats receiving a normal diet but no PAL artificial cells, showed a significant decrease in body weight, and in addition, a decrease in muscle tone over the same period. PKU rats on a normal

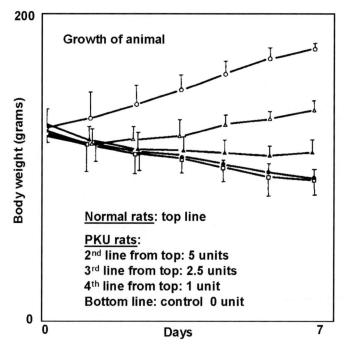

Fig. 6.12. Growth of PKU rats receiving different amounts of PAL artificial cells compared to normal rats. (modified from Bourget and Chang, 1986).

diet treated with 5 units of oral PAL artificial cells lost weight initially. However, they started to gain weight in the next 4 days at the same rate as the normal control rats. This corresponded to the time it took for the elevated plasma PHE level to return to normal. Abnormal signs such as piloerection, aggression and hyperactivity, observed in PKU-induced rats, were no longer present after 5 days of treatment with PAL artificial cells. PKU rats on a normal diet receiving lower doses (1.0 or 2.5 units) of PAL in artificial cells lost weight at the same rate as the PKU rats on a normal diet. Those receiving 2.5 units was only slight better off.

Congenital PKU mice model

We tried to use a congenital mouse PKU model (Shedlovsky *et al.*, 1993) for the above studies. However, the small size of the very young mice and the tendency for "hairy balls" to accumulate in the stomach of these mice, has made it difficult in our hands for the long term daily gastric tube administration of 50 μ PAL artificial cells. XOD artificial cells also tended to jam the very small diameter gastric tubes used for oral administration.

Summary discussion

The above results contribute to the scientific basis for the potential therapeutic uses of PAL artificial cells in PKU, and also serve to compare the use of PAL artificial cells with a PHE-free diet. PAL in artificial cells acts on PHE diffusing into the artificial cells and convert the substrate into transcinnamic acid. Transcinnamic acid is not toxic and is converted to hippuric acid by the liver and excreted in the urine. The artificial cells carry out their function as they move down the intestinal lumen and finally leave the body in the feces. There is, therefore, no accumulation of artificial cells in the body. However, it is important to administer enzyme artificial cells together with a buffer to protect them during their passage through the highly acidic stomach. At a pH of 3.5 or lower, there will be breakage of the membranes of the artificial cells.

The safety of oral administration of enzyme artificial cells has been shown. Thus, we have given oral artificial cells containing xanthine oxidase to lower the systemic hypoxanthine in experimental therapy in Lesch-Nyhan disease (Chang, 1989a; Palmer & Chang, 1989). Artificial cells containing urease and ammonium adsorbent have also been given to animals (Chang, 1972a) and patients in clinical trials with no adverse effect reported (Kjellstrand, 1981).

Before using this in human phenylketonuria, a key challenge has to be solved — phenylalanine ammonia lyase is extremely expensive. There are two groups developing recombinant phenylalanine ammonia lyase that is less costly (Sarkissian *et al.*, 1999; Liu *et al.*, 2002).

Oral enzyme artificial cells to deplete other amino acids

The proposal of oral enzyme artificial cells based on the new finding of enterorecirculation of amino acids could also be studied for other genetic enzyme defects. These include the removal of tyrosine in tyrosinemia, histidine in histidinemia, and others. In addition, the basic results obtained here can be useful for analyzing the feasibility for use in other conditions. One example is the use of oral administration of artificial cells containing asparaginase to lower asparagine in leukemia. Asparaginase is used as an adjunct treatment in chemotherapy, but it has side effects when injected intravenously on a repeated basis. Oral use of asparaginase artificial cells if effective, would avoid the adverse effects related to intravenous injection. Removal of glutamine is another example. The use of oral tyrosinase artificial cells can effectively lower the systemic tyrosinase level. As described in the next chapter, we are studying this for potential application in melanoma, a skin cancer that depends on tyrosine for growth.

CHAPTER 7

Enzyme Artificial Cells in Substrate-dependent Tumors and Activation of Prodrug

7.1. Introduction

Our basic laboratory research described in Chapter 6, shows that after parenteral injection, enzymes in artificial cells can carry out its function for a much longer period of time when compared with free enzymes. Furthermore, enzyme artificial cells can effectively lower systemic amino acids and other unwanted molecules and at the same time prevent the enclosed enzymes from giving rise to immunological reactions. This principle can also be applied to the use of enzyme artificial cells in substrate-dependent tumors and also for activation of prodrugs (Fig. 7.1A). The first study carried out was on the effect of parenteral injection of asparaginase artificial cells on 6C3HED lymphosarcoma in mice (Chang, 1971a). The promising result obtained from this study has led to further detailed studies in this laboratory followed by development of this principle by other laboratories in the form of different configurations of the artificial cell idea (Fig. 7.1B). One successful extension is the use of a polymer, polyethylene glycol, to surround a single asparaginase molecule, PEG-asparaginase, to prolong *in vivo* action and prevent immunological problems (Park *et al.*, 1981). Different configurations of enzyme artificial cells have also been investigated for tyrosine-dependent melanoma (Yu and Chang, 2004) and for the activation of a prodrug into its active form for pancreatic CA (Lorh *et al.*, 2001).

160

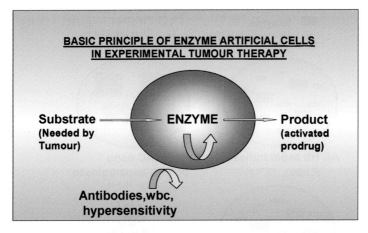

Fig. 7.1. **(A)** Basic principle of enzyme artificial cells in experimental tumor therapy.

7.2. Asparaginase in Lymphosarcoma and Lymphoblastic Leukemia

Guinea pig serum suppresses the growth of mice lymphosarcoma (Kidd, 1953) because its asparaginase can deplete the supply of asparagine to asparagine-dependent tumors (Broome, 1961). Asparaginase can be more readily obtained from the bacteria *Escherichia coli or Erwinia* for use in large-scale studies (Mashburn and Wriston, 1964). Early studies (Broome, 1968; Adamson and Fabro, 1968; Haskell *et al.*, 1969; Whitecar *et al.*, 1970) show that parenterally injected E. *coli* asparaginase is removed rapidly as a foreign protein and the antitumor activity of asparaginase is related to the time it can stay in the circulation. More recent studies show that asparaginase is effective in the treatment of some types of leukemia by removing systemic asparagine (Hawkin *et al.*, 2004). However, being a foreign protein, there are frequent allergic and immunological reactions involved. These problems severely limit the usefulness of asparaginase even though asparaginase therapy can be an important component in the treatment of children with acute lymphoblastic leukemia (Hawkin *et al.*, 2004).

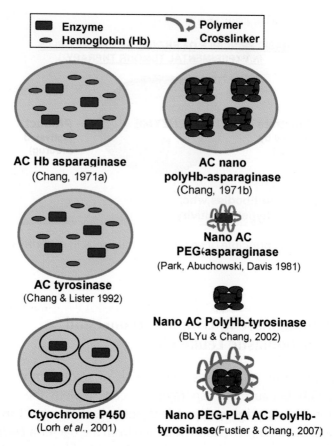

Fig. 7.1. **(B)** After the demonstration of effectiveness of asparaginase artificial cells in a mice model of lymphosarcoma (Chang, 1971a), different extension of this principle have been developed.

7.3.   Asparaginase Artificial Cells for Lymphosarcoma

Asparaginase artificial cells on 6C3HED lymphosarcoma in mice (Fig. 7.2A)

In the *in vivo* experiments (Chang, 1971a), each mouse received a subcutaneous injection of 500,000 6C3HED lymphosarcoma cells in the groin. Immediately after implantation, each mouse was given

ASPARAGINASE ARTIFICIAL CELLS

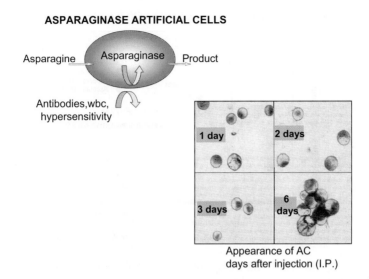

Appearance of AC
days after injection (I.P.)

Fig. 7.2. **(A)** *Top:* Basic principle of the use of artificial cells to prevent immunological and hypersensitive reactions and at the same time remove asparaginase needed by the tumor.
Lower: Appearance of asparaginase artificial cells recovered at different times after intraperitoneal injection.

one of the following intraperitoneal injections: 0.05 ml/g body weight of saline; 0.05 ml/g of a 50% suspension of control artificial cells containing no asparaginase; 0.05 ml/g of a solution containing 3.5 IU of asparaginase; or 0.05 ml/g of a 50% suspension of artificial cells containing 3.5 IU of asparaginase. Figure 7.2B shows the time when the tumors first appeared. The time when the tumor first appeared was 9.0+ 0.9 days (mean ± SD) for those which had received saline injections; 8.7 + 1.6 days for those which had received control artificial cells injections; and 14.0 + 4.5 days for those which had received a single intraperitoneal injection of asparaginase solution. In the group that had received a single intraperitoneal injection of asparaginase-loaded artificial cells, no tumor appeared in 50 percent of the implanted sites after 120 days. Thus, when compared with the asparaginase solution, the microencapsulated form was much more effective in suppressing the growth of implanted mouse lymphosarcoma (Fig. 7.2B). This

Effects of 1 injection on % of mice with lymphosarcoma

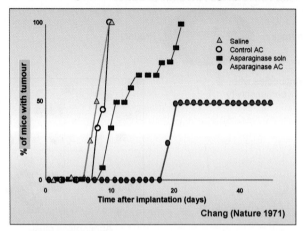

Chang (Nature 1971)

Fig. 7.2. (B) Effects of one intraperitoneal injection on the first appearance of lymphosarcoma in (i) saline: control group receiving saline; (ii) control AC: control group receiving artificial cells containing no asparaginase; (iii) asparaginase soln: group receiving asparaginase solution; (iv) asparaginase AC: group receiving asparaginase artificial cells. (Graph prepared from results in Chang 1971a, *Nature.*)

promising preliminary result led to further detailed research as described below.

Effects on plasma asparagine level

Four groups of mice were used (Chang, 1973). Each mouse in the two control groups received either an intraperitoneal injection of 1 ml of saline or 1ml suspension of control artificial cells containing no asparaginase. Each mouse in the third group received an intraperitoneal injection of 1 ml of saline containing 5 IU of asparaginase. Each mouse in the fourth group received an intraperitoneal injection of asparaginase artificial cells with an assayed activity of 5 IU. Figure 7.3 shows that intraperitoneal injection of control saline and control artificial cells did not significantly change the plasma asparagines levels. A single intraperitoneal injection of 5 IU of asparaginase in saline caused a lowering of the plasma

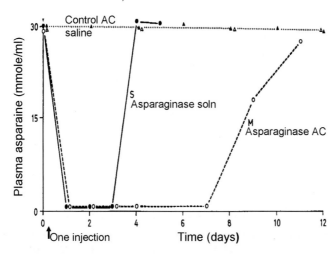

Fig. 7.3. Effects on plasma tyrosine level from one intraperitoneal injection of (a) saline: control group receiving saline; (b) control AC: control group receiving artificial cells containing no asparaginase; (c) asparaginase soln: group receiving asparaginase solution; or (d) asparaginase AC: group receiving asparaginase artificial cells. (From Chang, 1973.)

asparagines level to zero and maintained it at zero for 3 days, after which the plasma asparagines returned to the pre-injection level. The same assayed amount of asparaginase in artificial cells retained a zero plasma asparagine level for up to 7 days, the level returning to normal only after 11 days (Chang, 1973).

Asparaginase activities in the body

Each mouse in the two control group received an intraperitoneal injection of 1 ml saline or 1 ml 50% suspension of control artificial cells containing no asparaginase. The third group each received 0.5 IU/g body weight of asparaginase solution. The fourth group each received artificial cells containing asparaginase (0.5 IU/g body weight). Total body asparaginase were measured at the time intervals shown in Fig. 7.4 for a total of 16 days (Sui-chong and Chang, 1974). As shown in Fig. 7.4, there were no significant detectable asparaginase in the two control groups. In those groups that received one intraperitoneal injection of asparaginase solution, the total

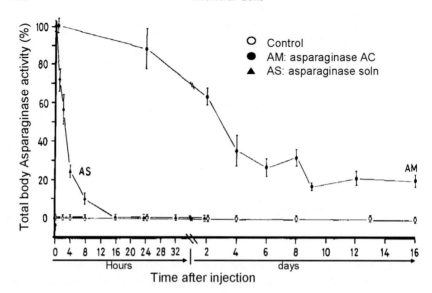

Fig. 7.4. Effects on total body asparaginase levels from one intraperitoneal injection of (a) control AC: control group receiving artificial cells containing no asparaginase; (b) asparaginase soln: group receiving asparaginase solution; or (c) asparaginase AC: group receiving asparaginase artificial cells. (From Sui-Chong and Chang, 1974.)

body enzyme activity decreased to less that 50% by 4 h and became undetectable 16 h after injection. On the other hand, in those mice receiving one intraperitoneal injection of asparaginase artificial cells, it took 60 to 72 h for the asparaginase activity to decrease to 50% of the initial acitivity. By 16 days, 20% of the asparaginase activity can still be found (Siu and Chang, 1974). Thus, the use of asparaginase artificial cells can very significantly increase the duration of asparaginase activity in the body after one parenteral injection.

Asparaginase does not leak out of artificial cells after injection

In order to avoid hypersensitivity and immunological reactions, the asparaginase should, after injection, stay inside the artificial cell to act on asparagines diffusing into the artificial cells. Our study shows that there was no detectable asparaginase leakage from the artificial cells into the plasma for the entire 16 days period of this study (Sui-chong and Chang, 1974). Thus, no detectable asparaginase leaked out of the artificial cells into the circulating blood. On the other hand, after intraperitoneal injected of asparaginase solution, the enzyme rapidly leaked out of the implanted site into the plasma, reaching a peak after 4 h. It was then cleared rapidly from the circulation.

Further studies shows that after intraperitoneal injection, artificial cells could be found intact and dispersed in the peritoneal cavity after 1, 2 and 3 days. On the 6th day, they still remained intact and dispersed, but tended to form small aggregates (Fig. 7.2A). In separate experiment using 80 μm mean diameter nylon, artificial cells containing ^{51}Cr-labelled hemoglobin were injected intraperitoneally. After the first week, the artificial cells were well dispersed over the whole peritoneal cavity. After the second week, they were found in larger numbers in the upper parts of the peritoneal cavities. No significant radioactivity was detected in the lung, liver, spleen, lymph nodes, blood or particle- free peritoneal washings. Recovered radioactivity was associated with the artificial cells in the peritoneal cavity for 4 weeks and the ^{51}Cr-labelled hemoglobin did not leak out of the artificial cells. These results further support the observation that after parenteral injection, asparaginase remained inside the artificial cells and continue to act on asparagine diffusing into the cells. In the case of free asparaginase, it enters the circulation and is rapidly removed as foreign proteins.

Discussions

The above results explain why one intraperitoneal injection of as-paraginase artificial cells is effective in significantly delaying the growth of lymphosarcoma when compared with one intraperitoneal injection of asparaginase solution (Fig. 7.2B). Thus, asparaginase inside artificial cells remained active in the body for a much longer length of time when compared with the asparaginase solution. Furthermore, one intraperitoneal injection of asparaginase artificial cells can maintain a low system asparagine level for a significantly longer period of time when compared with the enzyme in free solution.

What is perhaps even more important is that, unlike asparaginase solution, asparaginase inside artificial cells does not leak out into the circulation after injection. This would avoid the hypersensitivity and immunological problem related to asparaginase solution. Indeed, the basic study carried out using catalase artificial cells described in Chapter 6, shows that an enzyme inside artificial cells is prevented from causing immunological reactions after injection. Thus, our research supports the feasibility of using the principle of artificial cells to prevent problems related to the use of free asparaginase. This has led to further development to extend this approach.

7.4. Crosslinking Asparaginase to Hemoglobin (PolyHb-Asparaginase)

Our study shows that we can increase the enzyme stability of urease, catalase and asparaginase by crosslinking them with an excess of hemoglobin molecules (Chang, 1971b). We carried out, further, a study on the effect of crosslinking asparaginase into polyHb-asparaginase (Chang, 1973d) (Fig. 7.5A). This way, asparaginase is surrounded by an excess of hemoglobin molecules, thus preventing the enzyme from interacting with the external environment. At the same time, small molecules like asparagine can diffuse into the enzyme for conversion. Crosslinking to macromolecules also stabilizes the conformation of asparaginase. When kept at a body temperature of 37°C, PolyHb-asparaginase inside artificial cells retained 25% of the initial activity after 110 days (Fig. 7.5B). On the otherhand, without crosslinking, asparaginase with hemoglobin inside artificial cells retained 25% of the initial asparaginase activity after less than 15 days at 37°C (Fig. 7.5B). Both are stable at a storage temperature of 4°C for 110 days without significant loss of enzyme activity (Chang, 1973d). The *in vitro* stability of asparaginase artificial cells at 37°C is comparable to the that after injection into the animal (Fig. 7.4). The much higher stability of asparaginase crosslinked to hemoglobin in the form of polyHb-asparaginase (Fig. 7.5A) should allow for a significantly longer period of action after one injection.

NANO POLYHB-ASPARAGINASE IN AC
(Chang, 1971a)

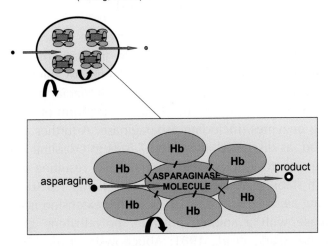

Fig. 7.5. (A) Schematic representation: glutaraldehyde crosslinking asparaginase with an excess of hemoglobin molecule to form polyHb-asparaginase inside artificial cells. Each molecule of asparaginase thus stabilizes and separates the enzyme from the external large molecules while allowing asparagine to enter for conversion.

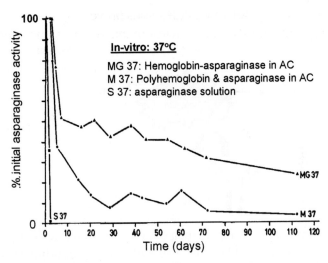

Fig. 7.5. (B) Comparison of the stability of asparaginase stored at body temperature of 37°C: S37 — asparaginase in solution; M37 — asparaginase inside artificial cells; MG37 — polyHb-asparaginase.

7.5. Crosslinking of Asparaginase to Polymer (Polyethylene Glycol — PEG)

The research discussed above showing the use of artificial cells to prevent immunological reactions and to increase the duration of enzyme activity, was based on artificial cells of micron diameter (Chang, 1971a). This is achieved by using interfacial polymerization to form an ultrathin polymer membrane in the form of artificial cells containing enzymes, including asparaginase. A further extension of this method, as discussed in Section 7.5, is to crosslink hemoglobin to asparaginase to form polyHb-asparaginase, resulting in a marked increase in stability at body temperature (Chang, 1973d). Instead of crosslinking asparaginase with Hb, a further extension is to crosslink asparaginase with a soluble polymer, polyethylene glycol (PEG-asparaginase) (Park, *et al.*, 1981; Abuchowski *et al.*, 1984; Asselin *et al.*, 1993, Duncan, 2003) (Figs. 7.1B, 7.6). This also uses the principle of artificial cells (Fig. 7.1A) in that the surrounding polymer molecules prevents the direct contact of the enzyme with the external

NANO PEG-ASPARAGINASE

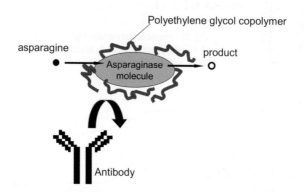

Fig. 7.6. Instead of using hemoglobin to crosslink with asparaginase to form PolyHb-asparaginase, an extension is to use synthetic polymer, polyethylene glycol (PEG), to form PEG-asparaginase (Park *et al.*, 1981). The surrounding PEG similarly stabilize and separate asparaginase from external large molecules while allowing asparaginase to enter for conversion.

environment, thus preventing immunological reactions and at the same time rapid removal of asparagine is achieved (Figs. 7.1A, 7.6). Substrates like asparagine, can diffuse through the surrounding polymer to interact with the enzyme (Fig. 7.1A, 7.6)

Studies show that PEG-asparaginase indeed has reduced immunogenicity as well as a 5-fold increase in half-life when compared with native *E. coli asparaginase* (Park *et al.*, 1981; Abuchowski *et al.*,1984; Asselin *et al.*, 1993). A randomized clinical study was carried out in children with untreated acute lymphoblastic leukemia comparing the native asparaginase from *E. coli* to PEG-asparaginase (Avramis *et al.*, 2002). When compared with the native asparaginase, PEG-asparaginase has a longer duration of *in vivo* enzyme activity and also a lower incidence of high titer antibodies to asparaginase. In children with relapsed acute lymphoblastic leukemia, weekly PEG-asparaginase injection resulted in a higher remission induction rate of 97% as compared to 82% with biweekly injection of native asparaginase (Abshire *et al.*, 2000). With the longer duration of action, fortnightly injection of PEG-asparaginase is as effective as injection 2–3 times/week of native asparaginase (Duncan 2003; Harris and Chess, 2003).

Hawkins *et al.* (2004), reported their study on the combined use of PEG-asparaginase with chemotherapy for relapsed pediatric acute lymphoblastic leukemia. This includes 28 pediatric patients with relapsed medullary ($n = 16$) and extramedullary ($n = 11$) acute lymphoblastic leukemia. Patients received induction therapy that included PEG-asparaginase (2500 IU/m^2 intramuscularly, weekly) and intensification therapy that included PEG-asparaginase (2500 IU/m^2 intramuscularly, once on day 7). Serum and CSF asparagine levels were lowered and maintained during this period in the majority of samples. PEG-asparaginase antibody was detected in only three of the 28 patients. Of the 28 patients, 24 showed remissions; two patients died from infection during induction, and two failed to achieve a remission. They concluded that intensive PEG-asparaginase therapy in relapsed acute lymphoblastic leukemia results in lowering of the asparagine, and the use of PEG-asparaginase in further trials for recurrent acute lymphoblastic leukemia is warranted.

7.6.   PolyHb-Tyrosinase Artificial Cells for Melanoma

Introduction

Melanoma, a fatal skin cancer, is a common tumor which accounts for 10% of all malignancies. The incidence of melanoma has risen dramatically in the last century, doubling every 10 years in many countries. The incidence is now approximately 10 per 100,000 per annum in Europe, giving an approximate lifetime risk of 1 in 200 (Katsambas, Nicolaidou, 1996). Melanoma is most commonly found on the skin, with 10% in the eyes. At least 20% of people diagnosed with melanoma progress to advanced disease and die within five years of diagnosis (Borden, 2002). At present, despite the use of adjuvant therapy, chemotherapy, immunotherapy and radiation therapy, the median survival of patients with metastatic melanoma is about six months.

Tyrosine is important in the metabolic cycle of melanoma and malignant melanomas require higher concentrations of tyrosine than other cells for growth. Research, especially those from Meadows' group, has shown from in animal studies, that lowering of systemic tyrosine using tyrosine and phenylalanine restricted diet can inhibit the growth of melanoma *in vitro* and *in vivo* (Meadow and Oeser, 1983; Fu *et al.*, 1997). However, low tyrosine diets are not well tolerated by human, resulting in weight loss and other adverse effects in patients (Meadow *et al.*, 1982). Furthermore, restricted diet can only lower the systemic tyrosine level to about 67% of the normal levels (Meadow *et al.*, 1982).

Polyhemoglobin-tyrosinase for melanoma

We therefore introduce a new approach to chemically crosslinking tyrosinase with an excess of hemoglobin to form a soluble polyhemoglobin-tyrosinase (PolyHb-tyrosinase) complex (Figs. 7.1B, 7.7A). This is similar to the PolyHb-asparaginase as described in Section 7.5.

PolyHb-tyrosinase carries out two functions: 1) Tyrosinase can remove tyrosine needed for growth by melanoma. With a molar

NANO POLYHb-TYROSINASE
(BL Yu & Chang 2004)

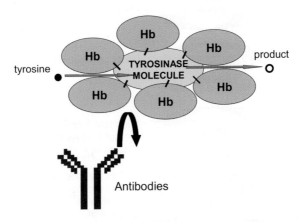

Antibodies

Fig. 7.7. (A) Melanoma needs a higher concentration of tyrosine for growth. Similar to PolyHb-asparaginase, crosslinking of an excess of hemoglobin to tyrosinase also stabilizes and separates the tyrosinase from the external large molecules, while allowing tyrosine to enter for conversion.

ratio of 100 Hb: 2 tyorinase, tyrosinase is covered and protected by the polyhemoglobin (PolyHb) from immunological and other adverse effects and from rapid removal from the circulation after intravenous injection; 2) The polyHb component can supply oxygen during the period of action of tyroinase. It is well known that disturbances of the microcirculations of tumor result in poor perfusion by red blood cells and thus low oxygen tension (Fig. 7.7B). Adequate oxygen tension is needed to facilitates radiation and chemotherapy, and PolyHb by itself, has been shown to improve the effectiveness of chemotherapy in animals with glioma (Pearce *et al.*, 1998)(Fig. 7.7C). Thus, the polyHb component of PolyHb-tyrosinase can supply the needed oxygen. In the form of polyHb-tyrosinase, both components circulate for the same length of time, thus complementing each other (Fig. 7.7B).

Our study shows that intravenous injection of PolyHb-tyrosinase retards the growth of melanoma in a mice model (Yu and Chang, 2004a). We inoculated B16F10 cells subcutaneously into mice and when the tumor volume reaches an average of $125\,mm^3$ on day 9,

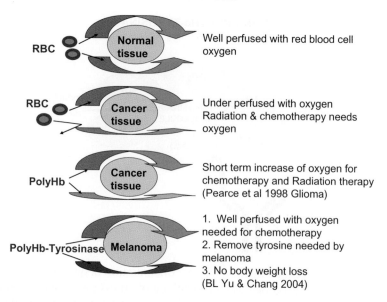

Fig. 7.7. **(B)** PolyHb-tyrosinase being a soluble macromolecule of nanodimension, it can better perfuse the imperfect vasculature of tumor to supply oxygen needed for chemotherapy and radiation therapy. At the same time, the tyrosinase component can remove tyrosine needed by melanoma for growth.

we started the following studies: 1) Sham control group receiving no intravenous injections; 2) Saline control group receiving daily intravenous injection of 0.1 ml saline; 3) Test group receiving 0.1 ml of PolyHb-tyrosinase solution. The end point of this study is based on McGill Faculty of Medicine Animal Care Committee's regulation that tumor burden should not exceed 10% of the animal's normal body weight. Figure 7.7C shows that there is no significant difference in the tumor size between the control group and the sham control group. On the other hand, 4 days after the daily intravenous injection of PolyHb-tyrosinase, the tumor volume is significantly lower than in the control group (Figs. 7.7C and 7.8A). Six days after the daily intravenously injection, the tumor volume in the PolyHb-tyrosinase group is only $53.49 \pm 13.91\%$ of that in the control group (Figs. 7.7C and 7.8A). Nineteen days after the inoculation of B16F10 melanoma cells, the tumor volume of the control group has reached the maximal of 10%

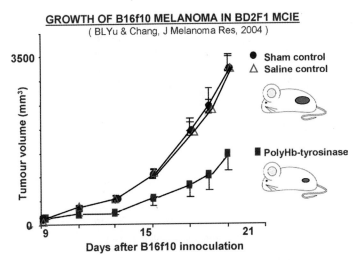

Fig. 7.7. **(C)** Effects of daily intravenous injection of PolyHb-tyrosinase in mice on tumor growth of B16F10 melanoma in mice. (i) sham control: no intravenous injection; (ii) saline control: 0.1 ml intravenous saline daily; (iii) PolyHb-tyrosinase group: 0.l ml intravenous PolyHb-tyrosinase daily. End point of study is when tumor in any one group reaches 10% of normal body weight All values are represented as mean ± SEM. (From Yu and Chang, 2004a).

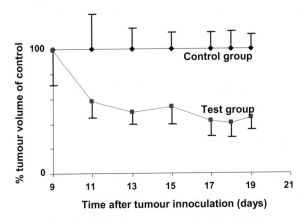

Fig. 7.8. **(A)** Analysis of effect of daily intravenous PolyHb-tyrosinase in test group on tumor size, represented as % of the tumor size of the control group.

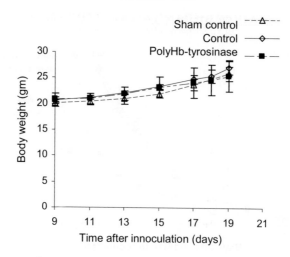

Fig. 7.8. **(B)** Body weight of mice in Fig. 7.6C after tumor implantation in: (i) sham control; (ii) control group (saline); and (iii) test group (PolyHb-tyrosinase). (From Yu and Chang, 2004a.)

of body weight allowed by the Animal Care Committee and we have to terminate the study. At this time, the tumor size in the test group was only $45.28 \pm 10.09\%$ of the control group (Figs. 7.7C and 7.8A). What is also very important is that there is no loss in body weight (Fig. 7.8B). This shows that the use of polyHb-tyrosinase prevents the major problem related to tyrosinase restricted diet that causes nausea, vomiting and weight loss.

Studies on optimization, in vitro and in vivo characterization

Before the above study in the melanoma mice model, we have carried out detailed studies to characterize the *in vitro* and *in vivo* properties of PolyHb-tyrosinase (Yu and Chang, 2002, 2004a, 2004c, 2004e and 2006). This includes optimization of the method of preparation and *in vitro* and *in vivo* characterization of PolyHb-tyrosinase (Yu and Chang, 2004e). These studies showed that PolyHb-tyroinase can lower plasma tyrosinase in mice without causing vomiting and weight loss. Furthermore, polyHb-tyroinase retains its ability to inhibit the growth of melanoma cell culture and retains its ability to carry oxygen. This leads to the above study on the actual use of this preparation in a melanoma mice model. After optimization and *in vitro* and *in vivo* characterization of the preparation, we carried out the following studies.

We carried out a study of the effects of intravenous injection of PolyHb-tyrosinase in normal mice. Daily intravenous injection of PolyHb-tyrosinase in the test group, reduces plasma tyrosine levels rapidly to 0.19 ± 0.09 mg/dl on day 2, compared with 1.44 ± 0.16 mg/dl in the control group (Fig. 7.9A). This low level was maintained by daily intravenous injection of polyHb-tyrosinase. Since one of the adverse effects of the use of tyrosine restricted diet is nausea, vomiting and weight loss, we observed the mice for the occurrence of the side effect. Daily measurements of body weight showed no difference between the control group and the test group receiving daily intravenous injection. Furthermore, we have not observed any

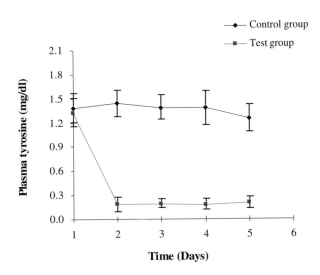

Fig. 7.9. PolyHb-tyrosinase fulfills the duo functions of removing tyrosine and transporting oxygen. **(A)** Effects of i.v. injection of PolyHb-tyrosinase into normal mice: Plasma tyrosine concentration (mg/dl) after daily i.v. injection of 0.1 ml saline (control group: 5 mice) and 0.1 ml PolyHb-tyrosinase (test group: 5 mice), as compared with control group. (From Yu and Chang, 2004a.)

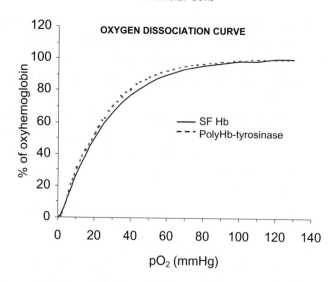

Fig. 7.9. **(B)** After crosslinking with tyrosinase, the hemoglobin component of polyHb-tyrosinase continues to maintain its ability to bind and release oxygen.

vomiting in these mice. The PolyHb component of PolyHb-tyrosinase also retained its ability to carry and release oxygen when as compared to hemoglobin (Fig. 7.9B).

Next, we analyzed the effects of PolyHb-tyrosinase on the growth of melanoma cell culture. We cultured B16F10 melanoma cells in DMEM and added one of the following four solutions to the culture medium: 1) saline solution (0.9 g/dl NaCl); 2) free tyrosinase solution; 3) PolyHb solution; and 4) PolyHb-tyrosinase solution. Figure 7.9B shows that PolyHb by itself does not have any effects on the growth of the B16F10 cells as compared to saline. When compared to native tyrosinase, PolyHb-tyrosinase retains its ability to inhibit the growth of B16F10 melanoma cells *in vitro*.

Summary

In summary, PolyHb-tyrosinase can retard the growth of melanoma in mice without having the adverse effects of weight loss from the

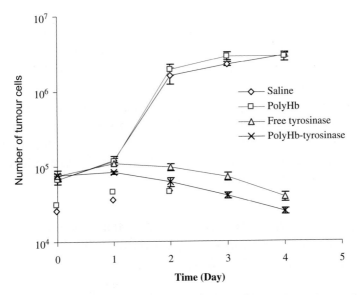

Fig. 7.10. *in vitro* growth curves of B16F10 cell lines after the addition of saline, PolyHb, free tyrosinase or PolyHb-tyrosinase to the medium. Results shown are the mean ± SD of three separate experiments. *Statistical significance $P < 0.05$ compared with saline and PolyHb. (From Yu and Chang, 2004a.)

tyrosine restricted diet. Furthermore, Polyhb-tyrosinase can lower the plasma tyrosine level to 15% of the original level as compared to 67% for tyrosine restricted diet. These results encourage further studies to further optimization and to investigate the combined use of this approach with other methods like radiation and chemotherapy for treating malignant melanoma.

7.7. Tyrosinase Artificial Cells by Oral Route

Introduction

As described in Chapter 6, our earlier basic research showed that there is an extensive enterorecirculation of amino acids between the intestine and the body (Fig. 6.9). This allows us to use oral administration of enzyme artificial cells to effectively lower the amount of a specific amino acid, phenylalanine, in a phenylketonuria rat model. The amount of any specific amino acid in the body can be by selecting artificial cells with the corresponding enzyme (Fig. 6.9). Furthermore, our earlier preliminary results showed that oral tyrosinase artificial cells can lower the level of tyrosine in the small intestine, ileum (Chang *et al.*, 1988/1989). We therefore investigated the use of artificial cells containing tyrosinase for the removal of systemic tyrosine (Yu and Chang, 2004b) (Fig. 7.11A).

Effects of Long-term Oral Administration of Polymeric Microcapsules Containing Tyrosinase on Maintaining Decreased Systemic Tyrosine Levels in Rats

The concentration of tyrosine in plasma, as for all other amino acids, tends to fluctuate even though the time of feeding and blood collection

Oral tyrosinase AC in the intestine

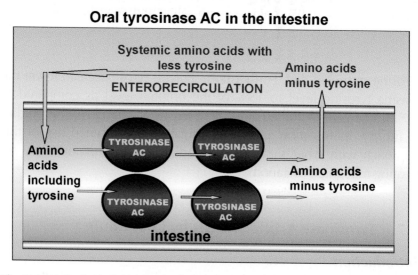

Fig. 7.11. Oral administration of micron-dimension tyrosinase artificial cells for the removal of tyrosine. **(A)** Schematic representation of the removal of tyrosine from the enterorecirculation, thus lowering the system tyrosine level.

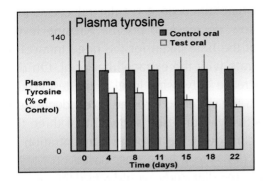
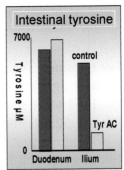

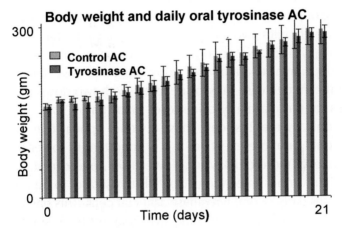

Fig. 7.11. **(B)** Tyrosine in the small intestine, ileum, is significantly lowered. Plasma tyrosine is also lowered, but because of the large intracellular pool of tyrosine, it takes time for the plasma tyrosine level to decrease. Unlike the use of tyrosine restricted diet, this approach does not result in weight loss since there is no significant differences between the animals' weight in the control and the group receiving asparaginase artificial cells.

are the same. This fluctuation is a normal phenomenon in all rat studies. Thus, we take the tyrosine level in the control group as 100%; the other data are expressed as a percentage of the original activity. Each group received 3 times daily either control artificial cells containing no tyrosinase (control group) or tyrosinase artificial cells (test group).

Both groups of rats were maintained on a regular rat chow. In the test group, there was a slow decrease of plasma tyrosine level in the first 3 days, because of the time taken for the large intracellular pool of tyrosine to equilibrate with the extracellular pool. On day 4 there was a significant decrease in tyrosine level in the test group to 68.8%. By days 18 and 22, the systemic tyrosine levels decreased to 56.8% and 52.6%, respectively. Our result showed that three times daily oral administration can lower the tyrosine concentration in a rat's plasma, starting from day 4 to a level that, based on Meadow *et al.* 's (1982) tyrosine restricted diet study, would retard the growth of melanoma. However, unlike the group on the tyrosine restricted diet, the test group on oral treatment did not lose any weight. Instead, they gained weight with a weight gain curve identical to that of the control group (Fig. 7.11B). No abnormal effect or behavior was observed in both groups. In a separate set of study using one dose daily, there was no significant decrease in tyrosine level. With two doses daily, it took 14 days for the plasma tyrosine level to fall to 69.9%, and reaching 55.8% only by the 21st day. Thus, it would appear that three doses daily would be more effective. If it is important to lower the plasma tyrosine level more quickly to a lower level, one can start with one or two intravenous injection of PolyHb-tyrosinase, to be followed by three doses daily of oral tyrosinase artificial cells. This has been tested (Fig. 7.12) and the plasma tyrosine level is shown to be rapidly lowered after 1 day. It may be necessary to give another intravenous injection of PolyHb-tyrosinase on day 2, since the plasma tyrosine level rises on day 3 (Fig. 7.12).

Enzyme kinetics

Before carrying out the above study, we have carried out a number of optimization and characterization studies (Yu and Chang, 2004b). This includes analysis of the enzyme kinetics showing the following. For free tyrosinase, V_{max} is 49.02 ± 7.35 mg/dl.min and K_m is 4.65×10^{-4} M. For tyrosinase artificial cells, apparent V_{max} is 114.94 ± 2.01 mg/dl.min and apparent K_m is 4.65×10^{-4} M. Since tyrosinase artificial cells are to be given orally, we also studied the effects of pH on

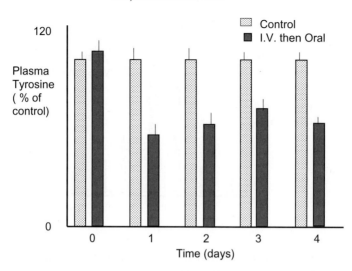

Fig. 7.12. In order to increase the rate of decrease of tyrosine, 1 or 2 intravenous PolyHb-tyrosinase can be given at the beginning followed by oral asparaginase artificial cells to maintain a low asparagine level.

enzyme kinetics and the stability of the enzyme. Figure 7.13A shows that both the free enzyme, and tyrosinase artificial cells have similar enzyme activity at pH of 6 and 7. However, at lower or higher pH, tyrosinase artificial cells have a higher enzyme activity. This is because artificial cells contain a high concentration of hemoglobin (10g dl) that acts as a buffer and the internal pH is different from the external pH. Figure 7.13B shows the stability of tyrosinase in artificial cells after incubation for 1 h at 37°C. Despite the high concentration of Hb, at pH 2 and 4, there is a significant decrease in enzyme activity after 1 h at the body temperature of 37°C. As a result, for oral administration, we have to include a buffer in the suspending medium of tyrosinase artificial cells.

We have also analyzed the ability of the tyrosinase artificial cells to lower the tyrosine level in intestinal juice before using the artificial cells for the study on oral administration. Tyrosinase artificial cells are incubated at 37°C in a shaker with fresh intestinal juice obtained from anesthetized rat. Figure 7.14 shows that in both the groups, there is an initial decrease in tyrosine level due to the dilution of the

Artificial Cells

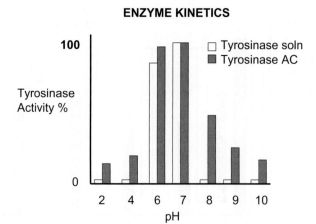

Fig. 7.13. **(A)** *In vivo* characterization of asparaginase artificial cells. pH kinetic of tyrosinase and tyrosinase artificial cells.

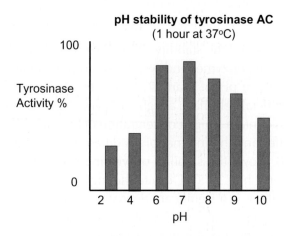

Fig. 7.13. **(B)** Tyrosinase stability when stored at different pHs for 1 h at 37°C.

intestinal juice by the artificial cell suspension. In the case of control artificial cells containing no tyrosinase, the tyrosine level decreased by about 50% from the initial level of 177.58 ± 29.92 mg/dl. This increased to 219.76 ± 15.21 mg/dl at 33 min. As discussed under

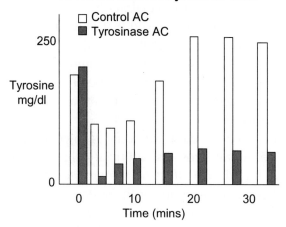

Fig. 7.14. Tyrosinase artificial cells and control artificial cells containing no tyrosinase are each suspended in fresh intestinal juice at 37°C. The initial decrease in tyrosine is due to the dilution effect of adding the suspensions. After this, tyrosine in the control group keeps increasing as protein in the intestine is digested by the intestinal tryptic enzyme. However, tyrosinase artificial cells lowers and maintains a low tyrosine level.

enterorecirculation in Chapter 6 (Fig. 6.1), glandular and intestinal secretions result in intestinal juice containing high concentrations of proteins, enzymes, polypeptides and peptides. Intestinal tryptic enzymes break these down into amino acids including tyrosine, thus, the increase in tyrosine level. For the tyrosinase artificial cell group, the tyrosine concentration in the intestinal fluid fell from the original concentration of 200.25 ± 10.16 mg/dl to 73.34 ± 14.72 mg/dl at 33 min. We used this *in vitro* model to analyze the optimal dose needed and arrived at the final dose for the study in rats described earlier.

7.8.   Discussions on PolyHb-Tyrosinase and Tyrosinase Artificial Cells

Meadow's group show that tyrosine and phenylalanine restricted diets reduce the growth of melanoma and increase the survival of B16 melanoma-bearing mice (Meadow and Oeser, 1983; Fu *et al.*, 1997).

One of the main problems with these low amino acids diets is that they cause nausea, vomiting and malnutrition as well as weight loss in the severely ill patients. We showed that daily intravenous injection of PolyHb-tyrosinase can retard the growth of melanoma in mice without having the adverse effects of vomiting and weight loss of tyrosine restricted diet. Furthermore, the use of PolyHb-tyrosinase can reduce the plasma asparagines level to 15% as compared with 62% for tyrosine restricted diet.

To avoid the need for daily intravenous injections, we have also studied the use of oral tyrosinase artificial cells. Oral administration does not have an adverse effect on the growth and body weight of the animal. It can effectively lower the systemic tyrosine in animals, but takes a longer time to reach a low level. By first giving one or two intravenous injections of nano PolyHb-tyrosinase artificial cells, followed by daily oral tyrosinase artificial cells, the plasma tyrosine level can be quickly lowered and maintained. These results encourage further studies to achieve futher optimization and to investigate the combined use of this approach with other methods like radiation and chemotherapy for treating malignant melanoma.

The same adverse effects of nausea, vomiting and weight loss are also encountered when using diet restriction in attempts to lower the systemic tyrosine in hypertyrosinemia, an inborn error of metabolism (Russo, Mitchell and Tanguay, 2001). Oral tyrosinase artificial cells could be tested for the lowering of systemic tyrosine in hypertyrosinemia.

7.9. Enzyme Artificial Cells for Activating Prodrugs

Introduction

Pancreatic carcinoma ranks fifth in cancer-related death (Günzburg and Salmons, 2001). Few of these can be treated surgically, furthermore, radiotherapy and chemotherapy are not effective. Ifosfamide is converted by cytochrome P450 2B1 (CYP2B1) into 4-hydroxyifosfamide that becomes phosphoramide mustard and acrolein that are potentially effective chemotherapeutic agents for

Artificial cells containing
genetic engineered cells with cytochrome 450
(Lohr et al, 2001)

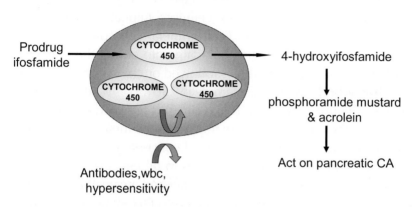

Fig. 7.15. Artificial cells containing genetically modified cells expressing PRO-DRUG-activating cytochrome P450 have been administered intra-artierially to lodged in pancreatic carcinoma (Lorh *et al.*, 2001). Here they activate the prodrug ifosfamide into its active form locally, thus reducing systemic toxicity.

pancreatic carcinoma. Unfortunately, they have a short plasma half-life and the high dose of ifosfamide needed is not well tolerated (Günzburg and Salmons, 2001). Artificial cells containing genetically modified cells expressing PRO-DRUG-activating enzymes, such as cytochrome P450, have been studied as a potential treatment for inoperable pancreatic carcinoma in patients (Lorh *et al.*, 2001) (Fig. 7.15). These artificial cells are administered intra-arterially, to accumulate in the pancreatic carcinoma, where they activate the prodrug, ifosfamide. As a result, the local concentration of the activated ifosfamide is much higher than in the rest of the body, thus reducing systemic toxicity and increasing local efficacy.

Clinical trial results

Lohr *et al.* (2001) encapsulated human cells transfected with a CYP2B1-expression contruct inside artificial cells to prevent immunorejection (Fig. 7.15). The prodrug, ifosfamide, can diffuse rapidly into the artificial cells to be converted into the active form. They carried out a phase I/II study in 14 patients with inoperable pancreatic CA (stages III–IV 86% Stage IV). The typical dose they gave was 250–300 artificial cells (cellulose sulphate membrane) with a mean diameter of 0·8 mm. Each artificial cell contains 10^4 cells. These are administered angiographically into one of the arteries that supply the pancreatic CA. This is followed on the second day by low-dose (1 g/m^2 body surface) ifosfamide in 250 ml 0·9% sodium chloride given on 3 consecutive days. This regimen is repeated on days 23–25. In this way, the artificial cells are trapped in the tumor and can locally activate ifosfamide to give a high local concentration of the activated drug in the tumor. This method proves to be effective in treating the tumor and the low dose of ifosfamide needed means that there is less systemic toxic effect in the rest of the body. Follow-up angiography of the tumor vessels after the infusion of the artificial cells showed no or only minor changes and there was no arterial occlusion. Of the 12 patients who remained in the study, after 10 months two had more than 50% reduction in tumor volume and two had 25–50% reduction. The tumor size of the other eight patients did not increase and remained stable at 75–125% of the initial size. Median survival was doubled in the treatment group as compared with historic controls, and one-year survival rate was three times better. They concluded that further studies of this approach combined with chemotherapy for inoperable pancreatic cancer are warranted.

7.10.   Materials and Methods

Asparaginase artificial cells

Nylon membrane artificial cells containing asparaginase were prepared as follows: 1260 IU of the enzyme is dissolved in 1.5 ml of a 10 gm% hemoglobin (hemoglobin substrate, Worthington Co.) solution. To the 1.5 ml enzyme solution in a 100 ml beaker is added an equal volume of a freshly prepared alkaline 1,6-hexamethylenediamine solution (an aqueous solution containing

4.4 gm% 1,6-hexamethylenediamine (Eastman), 1.6 gm% sodium bicarbonate, and 6.6 gm% sodium carbonate). Immediately upon mixing the two solutions in a 100 ml beaker, the subsequent steps for the preparation of 80 μ mean diameter nylon artificial cells as described in Appendix II are carried out. Only properly prepared artificial cells which do not show any leakage of enzymes are used in these experiments. Control artificial cells are prepared in exactly the same way, except that asparaginase has not been added to the hemoglobin solution.

Collodion membrane artificial cells containing asparaginase were prepared as follows: 1260 units of the enzyme was dissolved in 3 ml of a 10 gm% hemoglobin (hemoglobin substrate, Worthington Co.) solution. The remaining steps for the preparation of collodion membrane artificial cells as described in Appendix II were carried out.

PolyHb-tyrosinase in melanoma

Preparation of PolyHb-tyrosinase

Purified bovine hemoglobin is purchased from Biopure Corporation (Boston, MA, USA). Glutaraldehyde (25%) is obtained from Polysciences (Warrington, PA, USA). L-lysine (monohydrochloride, SigmaUltra > 99%), L-tyrosine (98% TLC), tyrosinase from mushroom (EC 1.14.18.1, 3000 units/mg stated activity) are purchased from Sigma-Aldrich (Ontario, Canada). All other reagents are of analytical grade.

For the preparation of PolyHb-tyrosinase, reaction mixtures are prepared containing hemoglobin (10 g/dl) and tyrosinase (6000 U/ml) in 0.1 M potassium phosphate buffer, pH 7.6. To prepare PolyHb, an equivalent volume of buffer is used instead of the enzyme. Prior to the start of crosslinking, 1.3 M lysine is added at a molar ratio of 7:1 lysine/hemoglobin. Crosslinking reaction is started with the addition of glutaraldehyde (5%) at molar ratio of 16:1 glutaraldehyde/hemoglobin. Glutaraldehyde is added in four equal aliquots over a period of 15 min. After 3.5 to 48 h at 4°C under aerobic conditions with constant stirring, the reaction is stopped by adding 2.0 M lysine at a molar ratio of 200:1 lysine/hemoglobin. The solutions are dialyzed in physiological saline solution and passed through sterile 0.45 μM filter. Aliquots (500 μl) of the 16:1 crosslinked preparation are concentrated using 100 KD microconcentrators (Amicon, Beverly, MA, USA). Samples are centrifuged at 2500 g for 55 min at 23°C. Then, the retentate is collected. Hemoglobin concentration is determined by cyanomethemoglobin at 540 nm.

Determination of tyrosinase activity

Tyrosinase activity is assessed by measuring the formation of enzymatic products at 300 nm. The adsorbance at 300 nm is monitored continuously for 8–14 min using a Perkin Elmer Lambda 4B spectrophotometer, and changes in O.D./min are used to analyze the activity of the enzyme.

Effects of intravenous injection of PolyHb-tyrosinase in rats

Fasted male Sprague-Dawley rats (245–260 g) are obtained from Charles River Canada (St-Constant, Quebec, Canada). The animals are anaesthetized with intraperitoneal injection of 65 mg/kg pentobarbital (Somnotol, Decton Dickinson, NJ, USA). The rats' body temperature is maintained using a warming blanket. Incisions are carefully performed in one side of the hindlimbs below the inguinal ligament, and the femoral vessels are carefully isolated. Polyethylene cannulae are inserted and secured distal to the superficial epigastric branches in the femoral veins (PE-10, PE-50 Clay Adams). Proper vessel access is tested with a small volume injection of heparinized saline (50 IU/ml). Blood samples are taken from each group at the beginning, then the samples are injected through the femoral vein. The femoral artery cannulae is connected to the venous cannulae for blood to circulate thoroughly for a short interval. Then blood samples are taken from the femoral artery at different time intervals. The plasma in each blood sample is separated from the blood and placed in a 1.5 ml plastic tube, then stored at −80°C until analyzed. The tyrosine concentration in plasma is analyzed by fluorometric method using Perkin Elmer Luminescence Spectrometer LS50B.

B16-F10 melanoma cells and culture conditions

B16-F10 murine melanoma cells are obtained from American Type Tissue Collection (Manassas, VA, USA). The tumor cells are cultured in DMEM (Life Technologies, Invitrogen Canada) supplemented with 10% FBS. Cells are passaged every 2–3 days. For the experiment, melanoma cells are cultured in complete DMEM until they become 30–40% confluent. Then, appropriate aliquots of different test solutions (0.57 ml sample per 10 ml medium) are added to the medium. The cell viability is followed up to 4 days thereafter. Tumor cells are monitored by phase microscopy. Cell counts are obtained daily using a hemacytometer. Cell viability is determined by trypan blue exclusion.

Intravenous injection of PolyHb-tyrosinase in normal mice

BD2F1 female mice (C57BL/6 × DBA/2F1) at age 57–63 days are purchased from Charles River Canada (St-Constant, Quebec, Canada). The mice are kept at a 12-h light interval and fed conventional food and water *ad libitum*. All the animals are housed and cared for according to the regulations of McGill University on Animal Care. All the mice are acclimatized for at least 7 days prior to use. Two groups of mice (5 mice per group) are studied. Intravenous injection of saline at 0.1 ml for the control group is carried out every day; 0.1 ml of PolyHb-tyrosinase solution is injected into the mice in the test group every day, and blood is drawn after each injection from both groups.

B16F10 melanoma bearing mice model

B16F10 melanoma cells prepared at 1×10^6 in 0.1 ml of HBSS are injected subcutaneously into shaved lateral flank of the mice. The sizes of primary tumors are measured every 2 days using calipers. Tumor volume is calculated using the formula $V = (A \times B^2)/2$, where V is volume (mm^3), A is long diameter (mm) and B is short diameter (mm). We start intravenous injection at day 9 after tumor implantation when the tumors reach an average size of 125 mm^3. The end point of study is when the tumor in any group of animal reaches 10% of the body weight. This is based on the regulations of the ethics committee of the Faculty of Medicine Animal Care Committee of McGill University.

Statistical analysis is performed using Student's *t*-test within ANOVA and considered significant at $P < 0.05$.

Oral tyrosinase artificial cells

Materials

Tyrosinase from mushroom (EC. 1.14.18.1, 5350 units/mg stated activity), and hemoglobin from bovine (lyophilized powder) are purchased from Sigma Company (Oakville, Ontario, Canada). Collodion is purchased from Fisher Scientific Company (Whitby, Ontario, Canada). All other reagents are of analytical grade.

Control artificial cells

These are prepared using the standard published method as described in Appendix II for collodion membrane artificial cells. Briefly, 1 g hemoglobin and 200 mg Tris are added to 10 ml double distilled deionized water and stirred until dissolved. Gravity filter the solution through a Waterman #42 filter into an Erlenmeyer flask. Take 2.5 ml of this 10 g/dl hemoglobin solution and was microencapsulated within spherical, ultrathin, cellulose nitrate membrane artificial cells as described under Appendix II. All the control artificial cells are prepared daily and stored in 1% v/v Tween 20 solution at 4°C until use.

Tyrosinase artificial cells

1.907 mg of 5350 units/mg tyrosinase is dissolved in 5 ml 10% hemoglobin solution; then the methods described above to immobilize tyrosinase in collodion membrane artificial cells as a 50% suspension is followed.

Before oral administration, the control artificial cells and tyrosinase artificial cells are washed and resuspended in 0.1 M Tris · HCl buffer (pH 8.5). The total volume of artificial cells for feeding is 1 ml, containing tyrosinase 1020 U/ml. This is suspended in 1.5 ml Tris · HCl buffer. This buffer is needed to protect the microencapsulated tyrosinase during its their passage through the stomach with its acidic medium. For the *in vitro* studies, three types of tyrosinase microcapsules are prepared, with tyrosinase concentration of 450 U/ml, 670 U/ml, and 900 U/ml.

In vitro studies using rat intestinal juice

Fasted male Sprague-Dawley rats (245–260 g) purchased from Charles River (St. Constant, Quebec, Canada) are anesthetized with intraperitoneal injection of pentobarbital (Somnotol, 65 mg/kg). The whole length of the small intestine is removed intact with a surgical scissor. The contents are gently expressed into polypropylene centrifuge tubes stored on ice. These tubes are centrifuged at 1500 g for 10 min at 4°C, and the supernatant is recovered for the following experiment. Control artificial cells or tyrosinase artificial cells at different concentration of 450 U/ml, 670 U/ml, and 900 U/ml are incubated with the rats' intestine juice at 37°C, and then the samples are taken at different time intervals. Then, 10% trichloroacetic acid (TCA) is added to stop the reaction. After centrifuging the sample, the tyrosine level in the supernatant is measured by the fluorometric method using Perkin Elmer Luminescence Spectrometer LS50B.

Animal studies

Fasted male Sprague-Dawley rats (130–150 g) are used in this study. They are kept in a controlled 12-h light/dark environment with food and water *ad libitum*. All the rats are acclimatized for at least 3 days prior to use. All the animal experiments are performed according to the regulations of the McGill University on Animal Care. Two groups (5 rats in each group) are studied: 1) control group: each rat receiving oral administration of artificial cells containing no enzyme; 2) test group: each rat receiving oral administration of artificial cells loaded with tyrosinase.

In the following experiments, a blood sample is taken at 4:00 pm on day 0. No artificial cells are administrated on that day. From that day on, for daily dosing, artificial cells are given at 10:00 am using stomach feeding tubes. Blood samples are taken six hours after oral administration of the artificial cells every two days. The procedures for blood collection from the lateral saphenous vein are as follows. First, the animal is placed in a restrainer. Then, the hind leg is extended and the limb is fixed by holding fold of skin between tail and thigh. The hair on outer surface of lower hind leg is shaved to visualize the vein. Vaseline is applied on skin prior to nick to prevent wicking of blood onto skin and to ease collection. The vein is nicked perpendicularly with a 25-gauge needle and the blood sample (about 1 ml) is collected by microhematocrit (capillary) tube. The plasma in each blood sample is separated from the blood and placed in a 1.5 ml plastic tube, then stored at $-80°C$ until analyzed for tyrosine concentration. For twice daily dosing, artificial cells are given at 10:00 am and 4:00 pm, and blood samples are taken every week (day 7, 14 and 21) just before the second feeding for 21 days. For three times daily dosing, artificial cells are administrated at 10:00 am, 2:00 pm, and 6:00 pm for 22 days. Blood samples are taken on day 4, 8, 11, 15, 18, and 22 just after the second feeding. Data are expressed as mean $\pm$ S.D. The differences of tyrosine concentration in the rats' plasma between the control group and the test group at the same time point are determined by using Student's *t*-test within ANOVA and Bonferroni correction, and considered significant at $p < 0.05$.

CHAPTER 8

Artificial Cells for Cell Encapsulation

8.1. Introduction

Different attempts have been made, with the use of membrane systems, to prevent rejection of transplanted cells by the immune system. These include the use of dialysers, ultrafiltrators, membrane sacs, membrane disks and polymeric devices. However, in these configurations, one of the major problems is low cell viability due to the low permeability to oxygen and nutrients. As early as 1964 (Chang, 1964) Chang successfully used artificial cells to bioencapsulate cells via a drop method. He suggested that in addition to protecting the enclosed cells from immunorejection, these artificial cells also provide a high surface-to-volume relationship, which allows for good mass transfer of oxygen and nutrients (Chang, 1964, 1965, 1972; Chang et al., 1966) (Fig. 8.1).

The following quotations come from a 1965 publication by Chang:

"..... microencapsulation of intact cells or tissue fragments... the enclosed material might be protected from destruction and from participation in immunological processes, while the enclosing membrane would be permeable to small molecules of specific cellular product which could then enter the general extracellular compartment of the recipient. For instance, artificial cell bioencapsulated endocrine cells might survive and maintain an effective supply of hormone. The situation would then be comparable to that of a graft placed in an immunologically favorable site." ... "There would be the further advantage that implantation could be accomplished by a simple injection procedure rather than by a surgical operation."

CELLS IN ARTIFICIAL CELLS (AC)
(Chang, 1964 *Science*; Chang, 1965; Chang *et al.*, 1966)

(1) Cells inside AC protected from immuno-rejection (antibody, wbc).
(2) Oxygen & nutrients equilibrate rapidly into artificial cells.
(3) Secretion (e.g. insulin) controllable by permeant material (e.g. glucose).
(4) Conversion of waste metabolites and toxins (e.g. hepatocytes).

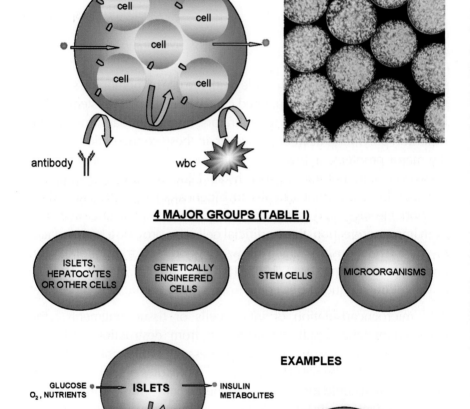

Fig. 8.1. Basic principles of artificial cells containing cells, tissues, genetically engineered cells, stem cells and microorganisms for therapeutic uses.

The method is based on the extension of the drop technique described in Chapter 1 (Chang, 1957). For the encapsulation of cells inside artificial cells, a hemoglobin solution containing cells is added dropwise to a silicone liquid containing a diacid, sebacyl chloride. The hydrophobic diacid crosslinks hemoglobin only at the interface to form a membrane by interfacial polymerization and the encapsulated cells in the resulting artificial cells thereby remain intact (Fig. 2.5) (Chang, 1964, 1965, 1972a; Chang *et al.*, 1966). "Microencapsulation of intact cells... erythrocytes were suspended in hemolysate...; and a silicone oil was substituted for the stock organic liquid. Microencapsulation was then carried out... based on the principle of interfacial polymerization for membranes of cross-linked proteins... A large number of human erythrocytes suspended in hemolysate within a microcapsule of about 500μ diameter was prepared by the syringe [drop] method."

However, there was not much interest in biotechnology in the 1960s and it is only with the more recent interest in biotechnology that many groups around the world have extended the author's approach of cell encapsulation. Thus, when Chang first approached the Connaught Laboratories in Canada to ask them to develop his microencapsulation method for use in diabetic patients there was not much interest. Conaught Laboratories later assigned the project to Dr. Sun and they have since extended Chang's original drop-method by using milder physical crosslinking to prepare alginate-polylysine-alginate [APA] microcapsules containing cells (Lim and Sun, 1980). They showed that after implantation, the islets inside artificial cells remain viable and continued to secrete insulin to control the glucose levels in diabetic rats.

Recent reviews show that many groups are now developing artificial cells to encapsulate biological cells for cell therapy (Kulitreibez *et al.*, 1999; Orive *et al.*, 2003; Chang, 2005). These include bio-encapsulation of endocrine tissues, hepatocytes and other cells for cell therapy. More recently, there have been much interests in the use of artificial cells to bioencapsulate genetically engineered cells for potential applications in amyotrophic lateral sclerosis, Dwarfism, pain treatment, IgG_1 plasmacytosis, hemophilia B, Parkinsonism and axotomized septal cholinergic neurons. Even more recently, we have been studying the use of artificial cells to encapsulate stem cells for use in liver regeneration. A 2003 consensus paper in *Nature Medicine* from several leading groups around the world (Orive *et al.*, 2003) shows that there has been increasing realization of the potential of cell encapsulation with polymeric artificial cells in recent years. Indeed, cell-based products, including cells with synthetic biomaterials, are now being recognized as a new category of therapeutic products by the US Pharmacopeia and National Formulation (US Pharmacopeia and National Formulary, 2002). Despite all the promising results, it should be emphasized that further developments will be needed before this approach is ready for actual routine clinical applications.

As summarized in Table 8.1, artificial cells for bioencapsulation of biological cells is a very large area of research that includes artificial cells containing 1) islets, liver cells, endocrine cells and others; 2) genetically engineered cells; 3) stem cells; and 4) microorganisms (Fig. 8.1). A detailed description of this area will require at least one large multi-author book and cannot be done in this monograph alone. This chapter contains an overview of the therapeutic uses of artificial cell encapsulated cells, including discussions of the potential application as well as problems. Examples of the detailed procedures for bioencapsulation of biological cells inside artificial cells are included at the end of this chapter.

8.2. Artificial Cells Containing Islets, Liver Cells, Endocrine Cells and Others

Artificial cells containing islets

The largest effort in this area is the use of artificial cells to encapsulate islets for the treatment of diabetes mellitus. As mentioned above, we first put forward the idea of using polymeric cells to encapsulate cells for transplantation (in particular, islet cells) (Chang, 1964, 1965, 1972; Chang *et al.*, 1966) (Figs. 2.5, 8.1) and thereby to protect them from immune rejection. The author approached the Conaught Laboratories (of insulin fame) to develop his idea of encapsulation of islets for the treatment of diabetes. However, it was not until 1980 that this

Table 8.1 Therapeutic Applications of Artificial Cells Encapsulated Cells

Items	Comments	First Proposed
I. ARTIFICIAL CELLS CONTAINING CELLS	Basic method of preparation Application in cell therapy proposed: 1) prevent immunorejection. 2) rapid mass transfer	Chang, 1964, 1965, 1972, Chang *et al.*, 1966
II. CELLS and TISSUES		
Pancreatic islets	Feedback controlled secretion of insulin for diabetes mellitus. Tested in phase I trial. Development needed for further clinical trials	Chang, 1965, 1972a, Chang *et al.*, 1966 Lim and Sun, 1980
Hepatocytes	To support liver function in liver failure	Chang, 1965, 1972a, Wong and Chang, 1986
Kidney cells Parathyroid cells	Erythropoeitin for anemia Parathyroid hormone for hypoparathyroidism	Koo and Chang, 1993 Hasse *et al.*, 1997
II. GENETICALLY-ENGINEERED CELLS		
Mouse myoblast (mGH gene)	Secrete mGH for dwarfism	Al-Hendy *et al.*, 1996
SK2 hybridoma cells	Secrete anti-hIL-6 monoclonal antibodies for IgG1 plasmacytosis	Okada *et al.*, 1997
Mouse myoblasts (human factor IX gene)	Secretes human factor IX for hemophilia B	Basic *et al.*, 1996
Hamster kidney cells (CNTF gene)	Secrete ciliary neurotrophic factor for amyotrophic Lateral sclerosis	Aebischer *et al.*, 1996
Neuro2A cells (propio-melanocortin gene)	Secrete endorphin for pain treatment	Saitoh *et al.*, 1995
Hamster kidneyfibroblasts (hNGF gene)	Secrete human nerve growth factor (hNFG) for Parkinsonism	Basic *et al.*, 1996

(Continued)

Table 8.1 (*Continued*)

Items	Comments	First Proposed
Senogeneic BHK cells (hNGF gene)	Secrete human nerve growth factor (hNGF) for axotomized septal cholinergic neurons	Winn *et al.*, 1994
Cells producing CNTF	Alleviate Huntington's disease To phase I clinical trial in patients	Aebischer *et al.*, 1996 Bloch *et al.*, 2004
iNOS-expressing cells	Tumor suppression: Clinical trials in patients with inoperable pancreatic cancinoma	Lorh *et al.*, 2001
IL-2 secreting C2C12 cells	Tumor suppression: Effective in animal study	Xu *et al.*, 2002
III. STEM CELLS		
BM Stem cells plus hepatocytes	Increased duration of viability and function of hepatocytes *in vitro* and *in vivo*	Liu and Chang, 2000
BM Stem cells	Increase liver regneration and improve survival rates of 90% hepatectomized rats	Liu and Chang, 2006
IV. MICROORGANISMS		
Pseudomonas pictorum	To remove cholesterol: *in vitro* studies	Garofalo and Chang, 1991
Erwinia herbicola	To convert ammonia and phenol to L-lysine and L-DOPA: *in vitro* studies	Lyold-George and Chang, 1955
CDH5 *E. coli*	Urea removal in rat studies	Prakash and Chang, 1996
Lactobacillus delbrueckii	Metabolic induction to remove urea	Chow *et al.*, 2003
Lactobacillus plantarum 80 (pCBH1)	For bile acid deconjugation to remove cholesterol	Jone *et al.*, 2004

idea was successfully implemented by them (Lim and Sun, 1980). When implanted into diabetic rats, artificial cell encapsulated islets corrected the diabetic state for several weeks (Lim and Sun, 1980). In their experiments, Lim and colleagues used a modification of Chang's drop technique and instead of chemically crosslinking protein to form the membrane, they used a milder electrostatic crosslinking of alginate and polylysine.

Since 1980, many groups have been investigating the approach or extensions of the approach for islets. The results in animals seem promising and the implanted islets function for up to one year (Sun *et al.*, 1996; Calafiore *et al.*, 1999; De Vos *et al.*, 2002; Duvivier-Kali *et al.*, 2001; Sakai *et al.*, 2001; Cruise *et al.*, 1999). A typical result of animal studies is shown in Fig. 8.2. A clinical trial in a patient with type 1 diabetes by Soon-Shiong's group has shown that the procedure is safe and without side effects (Soon-Shiong *et al.*, 1994). However, efficacy studies using larger amounts of artificial cell encapsulated islets have not yet been carried out. This is because there are a number

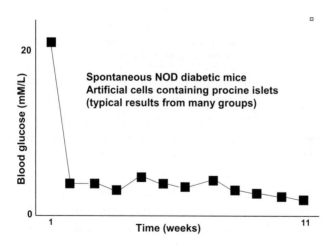

Fig. 8.2. Typical results obtained by a number of groups showing that heterologous islets inside artificial cells are not immuno-rejected. They continue to function for various lengths of time in maintaining a normal blood glucose level. However, the results are not always reproducible especially in the duration of function after implantation.

of challenges that need to be addressed before the initial clinical trial can be explored more extensively. These challenges are so important that they will be discussed in detail in a later section.

Artificial cells containing hepatocytes

Artificial cells containing hepatocytes have been used as a model for cell and gene therapy (Wong and Chang, 1986, 1988; Kashani and Chang, 1991; Bruni and Chang, 1989, 1991, 1995; Chang, 2001). Implantation of artificial cell encapsulated hepatocytes increases the survival of rats with fulminant hepatic failure (Wong and Chang, 1986). Artificial cell encapsulated rat hepatocytes, after implantation into mice are not rejected; instead, there is an increase in viability after intraperitoneal implantation (Wong and Chang, 1986), owing to the retention of hepato-stimulating factors inside the microcapsules as they are secreted by the hepatocytes (Kashani and Chang, 1991).

The Gunn rat is the model for the Crigler-Najjar syndrome in humans, the disease being is due to defects in the liver enzyme UDP-glucuronosyltransferase (UDPGT). Intraperitoneal implantation of artificial cells containing hepatocytes lowered the high systemic bilirubin levels that are characteristic of this condition (Bruni and Chang, 1989, 1991). UDPGT of the hepatocytes in artificial cells converts bilirubin into conjugated bilirubin that can be excreted in the urine, as in normal animals (Bruni and Chang, 1995).

The most promising use of artificial cell encapsulated hepatocytes is for short-term "bridging" in acute liver failure, as a liver support to allow the patient's own liver to regenerate and carry out its function. This will be described in the next chapter under regenerative medicine. However, this approach will probably have to be combined with hemoperfusion to first remove the large amount of toxins and products released by the breakdown of the acutely damaged liver. Hemoperfusion has already been successfully used for removing these materials in fulminant hepatic failure, resulting in the recovery of consciousness of patients from hepatic coma (Chang, 1972a; Gazzard *et al.*, 1974). It is only after the removal of hepatic toxins that the implanted artificial cell encapsulated hepatocytes can carry out their function to keep the patient alive, in the hope that the liver can regenerate sufficiently to assume normal function (Legallais *et al.*, 2001; Allen *et al.*, 2001).

Other areas

Other areas of research have focused on the microencapsulation of erythropoietin (EPO)-secreting renal cells to treat anemia (Koo and Chang, 1993), and parathyroid cells for secreting parathyroid hormone to treat hypoparathyroidism (Hasse *et al.*, 1997). However, in these two cases, one would want to compare these approaches with standard hormone replacement therapy with respect to cost-effective.

8.3. Challenges of Artificial Cell Encapsulated Cells in Therapy

Availability of cells

A key challenge for the approaches described above is the availability of human cells (allogeneic cells) for use in encapsulation. Some have proposed the use of xenogeneic cells from nonhuman sources, as the polymeric membrane can exclude leukocytes and antibodies, resulting in immune protection (Chang, 1965, 1972a; Chang *et al.*, 1966; Sun *et al.*, 1996; Lacy *et al.*, 1991). In allografts using human cells, immune mechanisms are mainly mediated by T lymphocyte sub-populations (i.e., CD8$^+$) with donor cells. So, the physical isolation provided by the capsules should prevent cell-to-cell contact between the artificial cell encapsulated tissue and the host's immune system. However, one group has carried out research that indicates that both xenografts and allografts might provoke an inflammatory cell response (de Vos *et al.*, 2003). They proposed that cells in capsules produce proteins, which can provoke inflammatory responses in immune cells, and that this even occurs with isografts. Another possibility is that the immune response is due to the use of the standard method for cell encapsulation, resulting in the occasional cells protruding on to the surface of the artificial cells

(Wong and Chang, 1991a); the use of the two-step method to prevent cell extrusion seems to prevent this problem (Wong and Chang, 1991b; Liu and Chang, 2002, 2003). This will be discussed in a later section.

In xenografts using non-human cells, there are other factors involved, including antibodies, complement fractions and cytokines (de Vos *et al.*, 2003; Hunkeler *et al.*, 2001). Proteins and other therapeutic factors secreted by xenografts, unlike those secreted by allografts, might result in immunological problems in the long term. If this should prove to be the case, then it would mean that only human cells can be used. The method involving xenogenic cells may require the use of a membrane with very limited permeability (Schuldt and Hunkeler, 2000; Uludag *et al.*, 2000; Dionne *et al.*, 1996). In this regard, our group uses chromatography to study variations in microcapsule membrane composition on the diffusion of larger molecules (Coromili and Chang, 1993). Other groups (Omer *et al.*, 2003; Binette *et al.*, 2001) have shown that xenogeneic neonatal porcine islets, which are far less immunogeneic than the adult counterpart, could be used to lessen the problems related to xenografts. Moreover, the support mechanisms consisting of either cells, or pharmacological agents coencapsulated with xeno cells, might help to attenuate the xenoimmune problem. However, there are also other sources of cells, such as genetically engineered cells and stem cells as described below.

Mass transfer, biocompatibility and stability

Many groups are looking into optimizing the biocompatibility, mass transfer, stability, and reproducibility of polymeric artificial cells in cell encapsulation (Kulitreibez *et al.*, 1999; Orive *et al.*, 2003; Hunkleler *et al.*, 2003; Gill and Ballesteros, 2000). Long-term biocompatibility is very important, as foreign-body reaction with resulting fibrous reaction will decrease the mass transfer of oxygen, nutrients and metabolites and result in the death of the artificial cell encapsulated cells. In addition, much work is being done to investigate the structural–functional relationship and other aspects in an attempt to increase the long-term stability of the polymeric artificial cells, and to ensure reproducibility of the cells (Calafiore *et al.*, 1999; Duvivier-Kali, 2001; Sakai, 2001; Uludag *et al.*, 2000). There are a number of commercial machines for automatic production, resulting in much improved reproducibility and narrower size distribution, as seen in the artificial cells that we have prepared using one such machine (Fig. 8.3).

Cell protrusion or entrapment in artificial cell membrane

However, improving the reproducibility, biocompatibility and strength of the polymeric membrane might not be enough. The standard alginate-polylysine-alginate membrane artificial cells containing cells are based on the following. Alginates are heteropolymer carboxylic acids, coupled by glycosidic bonds of β-D-mannuronic (M) and α-L-gluronic acid unit (G). Alkali and magnesium alginate are soluble in water, whereas alginic acids and the salts of polyvalent metal cations are insoluble. Thus, when a drop of sodium alginate solution containing a cell suspension enters a calcium chloride solution, rigid spherical gels are formed by ionotrotpic gelation (Fig. 8.3). The cells are dispersed randomly throughout the gel spheres, including some near or at the surface. When polylysine is added, it physically crosslinked alginate on the surface to form a polylysine-alginate membrane. However, during this process those cells near to or on the surface may protrude out of the membrane as it is formed. In the final step, sodium citrate is added and the gel sphere is liquefied, and the entire artificial cell is now supported only by the membrane. When using the standard method to encapsulate high concentrations of cells, even with automatic production, some cells are exposed on the surface or entrapped in the membrane matrix (Fig. 8.3) (Wong and Chang, 1991a). Entrapment of cells in the membranes weakens the membrane, resulting in breakage with time after implantation, thus releasing the contents of the artificial cells. In those cases when cells are exposed on the surface, they would not be immunoisolated and the rejection process would involve not only the exposed cells, but the whole artificial cells (Fig. 8.3).

Artificial Cells

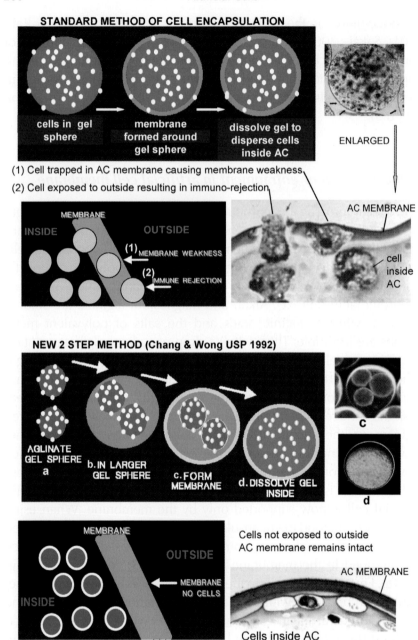

Fig. 8.3. Standard drop method of preparing artificial cells to encapsulate cells results in weakening of the membrane and exposure of cells, resulting in immuno-rejection. New 2-step method solves this problem.

Method to prevent cell protrusion or entrapment in artificial cell membrane

For encapsulation of high concentrations of small cells, we have designed a two-step method (Fig. 8.3) (Wong and Chang, 1991b), which should be particularly useful for the encapsulation of hepatocytes and genetically engineered cells. This method involves first forming very small alginate gel microspheres containing cells (Fig. 8.3). These small microspheres are then enclosed in larger gel microspheres. In this way, there will be no cells on the surface of the larger gel spheres. As a result of this, no cells will be trapped in the alginate-polylysine-alginate membrane when it is formed around each larger gel microspheres. The small alginate gel microspheres as well as the larger gel microspheres are then dissolved to release the cells, allowing them to be freely suspended in the artificial cells. Thus, the cells are dispersed freely inside the artificial cells, and so an extra diffusion barrier's avoided (Fig. 8.3). The exact procedure can be found at the end of this chapter under "Method". Detailed studies comparing this approach to the standard methods show that with this approach, protrusion of cells as well as local invasion of inflammatory cells are avoided, and the viability of the implanted cells is prolonged (Fig. 8.3) (Wong and Chang, 1991b; Liu and Chang, 2002; 2003). Others have attempted to use our earlier technology of enclosing smaller microcapsules in larger microcapsules (Fig. 2.4) (Chang, 1965, 1972; Chang *et al.*, 1966) or to use a double coating for the membrane. However, these approaches add further diffusion barriers.

Other improvements to microencapsulation method

One group has developed a promising approach of coating islets directly with a polymer membrane (Kulitreibez *et al.*, 1999). This would result in an artificial cell with a minimal diameter, thus allowing for optimal diffusion. Many other extensions and modifications of the original method have also been studied, including the use of different types of membrane materials (Kulitreibez *et al.*, 1999). Further technologies include the use of electrostatic devices (Halle *et al.*,

1994); emulsification/thermal gelation (Raymond *et al.*, 2004); and JetCutter technology (Schwinger *et al.*, 2002). The use of high voltage electrostatic pulses can produce much smaller alginate microcapsules (less than 300 microns). The smaller microcapsules would allow for better mass transfer than the larger ones (Halle *et al.*, 1994). The. emulsification/thermal gelation method (Raymond *et al.*, 2004) can be used to prepare large batches of microcapsules. This helps to solve the problem faced by most of the present cell encapsulation methods that involve variations of the drop approach that is very laborious when producing large batches of microcapsules. The JetCutter technology (Schwinger *et al.*, 2002) is another approach to solving the problem of the standard drop technique and it can produce larger batches with greater ease.

Capillary fiber cell encapsulation

Implantation of artificial cell encapsulated cells for extended periods is a long-term aim, but several groups are looking into other configurations for more immediate clinical applications. For example, Aebischer's ingenious use of capillary fibers to encapsulate cells has allowed his group to insert these subcutaneously into the cerebral spinal fluid on a short-term basis, and to remove and renew the fibers (Aebischer *et al.*, 1996). This would prevent the problem related to permanent retention of genetically engineered cells and other effects of long-term implantation. However, the fiber encapsulation and insertion approach is not applicable to conditions that require large amounts of artificial cell encapsulated cells, as is the case for islets and hepatocytes.

Oral administration

Another approach that does not require implantation or insertion is oral administration of artificial cells (Chang, 1972a, 1974f, 1997l, 2005l; Chang and Prakash, 1996). However, this is only for use in certain conditions, e.g., in the removal of unwanted metabolites from the body as in the removal of phenylalanine in phenylketonuria, removal of hypoxanthine in Lesch-Nyhan Disease, removal of tyrosine in melanoma and removal of uremic waste metabolites in kidney failure. For oral administration, very strong polymeric membranes that will not break down in the presence of powerful digestive enzymes in the intestine are needed and are being developed (Chen *et al.*, 2006). Furthermore, a buffer should be added to the oral formulation to protect the enclosed cells during their passage through the acidic environment of the stomach. There are other requirements needed for the most optimal use of oral administration and this is being studied (Chen *et al.*, 2006) As with all medical therapies, different conditions will require different types of approaches for the use of artificial cell encapsulated cells — either implantation, insertion or oral administration, as appropriate.

8.4. Artificial Cells Containing Genetically-engineered Cells

General

Cells can be genetically engineered outside the body to fulfill very specialized and specific functions. However, use of cells from the same person (autologous cells) is too individualized and thus not suitable for scale-up production. The use of cells from other humans (allogeneic cells) is more suitable for scale-up production, but this requires the use of drugs to suppress immunological rejection. As a result, there has been much ongoing research interest in the use of artificial cells containing genetically engineered cells (Table 8.1). Thus, the genetically engineered cells are immunoisolated, but oxygen and nutrients can enter, and the therapeutic agents from the cells can diffuse out to carry out their function in the body (Fig. 8.1). Examples of the use of artificial cells containing genetically engineered cells are: 1) for conditions requiring growth hormone, monoclonal antibodies, factor XI, or erythropoietin; 2) some neurological disorders; and 3) tumors.

Growth hormone, monoclonal antibodies, factor XI, erythropoietin

Artificial cell encapsulated mouse fibroblasts with a human growth hormone (hGH) fusion gene continue to secrete and deliver human growth hormone (hGH) when implanted into animals in an animal model of dwarfism (Al-Hendy *et al.*, 1996). Implantation of microencapsulated SK2 hybridoma cells that secreted anti-hIL-6 monoclonal antibodies (SK2 mAb) into transgenic mice (hIL-6 Tgm) suppressed IgG1 plasmacytosis, resulting in significant increases in survival time of these mice (Okada *et al.*, 1997). Artificial cell encapsulated genetically engineered mouse myoblasts that secrete human factor XI have also been studied for potential use in hemophilia B (Basic *et al.*, 1996). Artificial cell encapsulated myoblasts have been used to deliver erythropoietin in mouse thalasemia (Dalle *et al.*, 1999).

Neurological disorders

Neurological disorders represent a potentially exciting area for the therapeutic application of polymeric artificial cells. For example, artificial cell encapsulated neuro2A cells containing the propio-melanocortin gene that released endorphin, when injected into the cerebral spinal fluid of rats, decreased sensitivity to pain (Saitoh *et al.*, 1995; Hagihara *et al.*, 1997). In another example, artificial cell encapsulated cells that secreted human nerve growth factor promoted the survival of axotomized septal cholinergic neurons (Winn *et al.*, 1994). An extension of the principle of polymeric artificial cells involves the preparation of polymeric hollow fibers instead of spherical polymeric membrane artificial cells. An example of this approach is the use of artificial cell encapsulated baby hamster kidney (BHK) cells that contain the gene for human or mouse ciliary neurotrophic factor (CNTF), for the treatment of the neurodegenerative disease, amyotrophic lateral sclerosis (ALS) (Aebischer *et al.*, 1996). Other examples (Emerich *et al.*, 1997; Block *et al.*, 2004; Bachoud-Levi *et al.*, 2000) are highlighted in Table 8.1.

In the case of Huntington's disease, a phase I clinical trial was carried out using a semipermeable membrane encapsulating a BHK cell line engineered to synthesize and release CNTF (Bloch *et al.*, 2004). The membrane was implanted into the right lateral ventricle of six subjects with stage 1 or 2 Huntington's disease. It was retrieved and exchanged for a new one every 6 months, over a period of two years. Improvements in electrophysiological results were observed, and the results were correlated with the amount of CNTF released. The group reported that this phase I study showed the safety, feasibility, and tolerability of this gene therapy procedure. They also stressed the need to improve on the technique as although all retrieved capsules were intact, they contained variable numbers of surviving cells, and CNTF release was low in 13 of 24 implants.

Tumors

There has also been much recent interest in the potential use of polymeric artificial cells containing genetically modified cells for the treatment of tumors (Read *et al.*, 2001; Joki *et al.*, 2001; Xu *et al.*, 2002; Cirone *et al.*, 2002; Lorh *et al.*, 2001). For example, artificial cell encapsulated genetically modified cells expressing pro-drug-activating enzymes, such as cytochrome P450, have been studied as a potential treatment for inoperable pancreatic carcinoma in patients (Lorh *et al.*, 2001). The artificial cell encapsulated cells are given intra-arterially so that they are located in the pancreatic carcinoma where they activate the drug ifosfamide. Therefore the local concentration of the activated ifosfamide is much higher than in the rest of the body, thus minimizing systemic toxicity and increasing local efficacy. Artificial cell encapsulated transfected cells that secrete the anti-angiogenic factor endostatin have also been found to show promise in animal models of malignant brain tumors.

Another approach involves on polymeric artificial cells containing genetically modified myoblast cells that secrete interleukin-2 linked to the Fv region of a humanized antibody with affinity to tumors overexpressing the oncogene ERBB2 (Xu *et al.*, 2002). Regulatory agencies are more likely to approve clinical trials for the use of genetically engineered cells for conditions that have no other effective treatments. Malignant tumor that has no available effective therapy is one typical example. Thus, as discussed above, clinical trials in inoperable pancreatic carcinoma in patients have been carried out (Lorh *et al.*, 2001). Another example is the use of artificial cell encapsulated transfected cells that secrete endostatin for malignant brain tumors (Xu *et al.*, 2002). This is another case that would be more likely approved for clinical trials in patients.

Discussions

The research described above has shown that polymeric artificial cells containing genetically modified cells have potential use in treatment of tumors. However, further research is needed to study the long-term safety and effects in order to address regulatory concerns regarding the use of any products containing genetically modified cells. Until then, regulatory agencies would be relunctant to approve clinical trials of genetically engineered cells in humans unless there is no other treatment available. Malignant tumors that have no available treatment are therefore better candidates for this approach.

8.5. Artificial Cells Containing Stem Cells

More recently, there has been much interest in the use of stem cells for therapy (Alison *et al.*, 2002; McNeisch, 2004) as they can differentiate into different types of cells. Allogeneic stem cells still require immunoprotection inside polymeric artificial cells, but would avoid problems related to xenografts. However, at present, the amount of stem cells that can be harvested for large-scale use is still limited. There has also been considerable controversy regarding the use of fetal stem cells. We have therefore investigated the use of adult bone marrow cells.

For instance, in order to increase the duration of the viability and function of artificial cell encapsulated hepatocytes after implantation, we have co-encapsulated hepatocytes with bone marrow stem cells (Liu and Chang, 2000, 2002, 2003, 2006a, 2006b). This resulted in increased viability of the hepatocytes both *in vitro* and *in vivo* (Liu and Chang, 2000, 2002). This has also a significantly longer effect on lowering of the high systemic bilirubin levels in congenital Gunn rats (Liu and Chang, 2003). The exact reason for this effect has to be further investigated. However, it is known that bone marrow stems cells in the presence of hepatocytes can differentiate into hepatocytes (Alison *et al.*, 2000). Furthermore, it has been proposed that bone marrow stem cells secrete a factor that can help maintain the viability of the coencapsulated hepatocytes (Liu and Chang, 2000, 2006b). Even more recently we studied the use of artificial cells containing only bone marrow stems and no hepatocytes (Liu and Chang, 2005, 2006a, 2006c). In hepatectomized rats with 90% of their liver resected, one intraperitoneal injection of these artificial cells resulted in recovery of the animal and regeneration of the liver (Liu and Chang, 2005, 2006a, 2006c). However, in the control groups, most of the animals did not recover. In those receiving one injection of free bone marrow stem cells, the results were not significantly different from the group that did not receive any injection. Our analysis showed that the stem cells encapsulated within artificial cells stayed in the peritoneal cavity where they secrete a hapato-growth factor that drain into the portal circulation to stimulate the regeneration of the resected livers in the rats. On the other hand, the free bone marrow stem cells were shown to disappear quickly from the peritoneal cavity. The next chapter contains a detailed description of the use of artificial cells containing bone marrow stem cells in the treatment of liver failure.

8.6. Artificial Cells Containing Microorganisms

Another area is the study of artificial cells containing nonpathogenic microorganisms, including genetically engineered microorganisms. We have encapsulated in artificial cells a NADH *oxidizing bacterium* for the repeated recycling of NAD to NADH (Fig. 8.4) (Ergan and Chang, 1984). Another study involved the encapsulation of *pseudomonas-pictorum*, a microorganism that removes cholesterol (Fig. 8.4) (Garofalo and Chang, 1991). Artificial cells containing *erwinia herbicola bacterium* can convert the substrates to L-DOPA

Artificial Cells

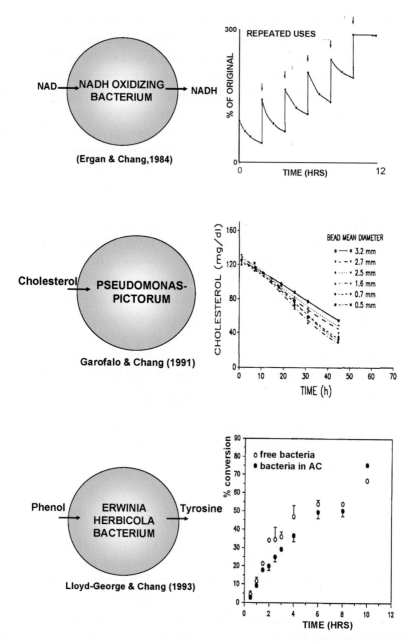

Fig. 8.4. Artificial cells containing NADH oxidizing bacterium to repeatedly recycle NAD to NADH; artificial cells containing *Pseudomonas pictorum* to remove cholesterol; and artificial cells containing *Erwinia herbicola* to convert the substrates into tyrosine and L-DOPA.

and phenol to tyrosine *in vitro* (Fig. 8.4) (Lyold-George and Chang, 1995).

Artificial cells containing microorganisms to remove urea have also been studied. For example, fermentation induction of *Escherichia coli* with *Klebsiella aerogenes* gene increases its capacity for urea removal. This can then be placed inside artificial cells to remove urea (Fig. 8.5) (Prakash and Chang, 1996a). However, microorganisms cannot be injected. An earlier study here on enzyme therapy using artificial cells showed that artificial cells can be given orally (Chang, 1972a, 1974f, 1997l, 2005). We therefore administer these orally to uremic rats. Our study showed that oral administration of polymeric artificial cells containing *Escherichia coli* with *Klebsiella aerogenes* gene can lower the elevated systemic urea level in uremic rats (Prakash and Chang, 1996a). Artificial cells containing another genetically

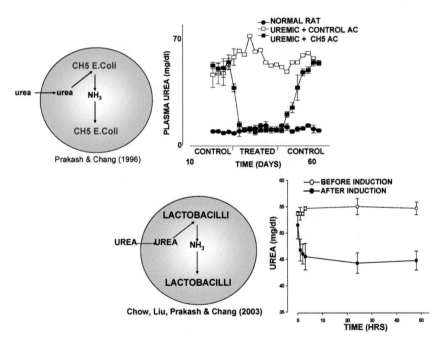

Fig. 8.5. Artificial cells containing a genetically engineered CH5 *E. coli,* that after fermentation induction, can remove systemic urea. Artificial cells containing fermentation induced *Lactobacillus delbrueckii* that can remove urea when tested *in vitro.* This is not a genetically engineered cells and is used in yougart, being much safer than CH5 *E. coli.*

engineered microorganism,*bactobacillus plantarum* 80 (pCBH1), have
been studied *in vitro* for bile acid deconjugation (Jones *et al.*, 2004).
However, it is important to ensure that any genetically engineered
nonpathogenic microorganisms would have no adverse effects after
oral administration. Until then, regulatory agencies seem to require
that there should be absolutely no leakage of genetically engineered
microorganisms from orally administered polymeric microcapsules,
even though they are classified as nonpathogenic. In order to avoid
the use of genetically engineered cells, we used metabolic induction
to increase the urea removal activities of *Lactobacillus delbrueckii*,
and also to encapsulate this in artificial cells to remove urea (Fig. 8.5)
(Chow *et al.*, 2003). *Lactobacillus delbrueckii* has been used widely in
regulatory agency approved food products like yougart for many years.

8.7.   Concluding Remarks

More complicated applications of artificial cells, such as cell
encapsulation, have considerable therapeutic potential. As discussed
in a recent consensus publication (Orive *et al.*, 2003), there has
been significant progress toward the successful transplantation of
artificial cell encapsulated cells in experimental animal models,
which has resulted in promising potential therapies for a number of
conditions. However, there are major challenges to be resolved before
meaningful, large-scale clinical trials can be carried out. The key
requirements include reliable and safe sources of cells, biocompatible
and stable membranes with suitable molecular cut-off to prevent
immune rejection, reproducibility of the product and long-term
survival of the graft. In addition, if xenografts or genetically engineered
cells are used, there are also ethical, political and regulatory questions
that need to be resolved. To promote development towards routine
clinical applications, it is important to change our attitudes towards
polymeric artificial cells. There has been a mistaken impression that
all polymeric artificial cells involve very simple technologies that can
be easily developed for clinical use. We now know that this is not the
case, and we have to approach their use in a highly interdisciplinary
manner, integrating basic and applied research and development

involving polymer chemistry, cell physiology, molecular biology, biotechnology and other fields. This interdisciplinary approach will accelerate the clinical realization of the many exciting potential applications, ushering us into a new and exciting era in which development of polymeric artificial cells can tap into the advances in biotechnology and molecular biology.

8.8. Method I: Standard Alginate-Polylysine-Alginate Method

Materials

Sodium alginate is usually prepared as a 2.0% solution in 0.90 gm % sodium chloride solution. Sodium alginate, from Kelco Gel℗ low viscosity alginate (Keltone LV), MW 12,000–80,000, is obtained from Kelco, Division of Merck and Co. (Clark, NJ). Prior to its use, it is either filter sterilized or heat sterilized for 5 min.

The reagents poly-l-lysine, MW 15,000–30,000; type IV collagenase; and type I-S trypsin inhibitor are purchased from Sigma Chemical Co. (St. Louis, MO). HEPES (4-(2-hydroxyethyl)-1-piperazine ethane sulphonic acid) buffer is purchased from Boehringer Mannheim (Montreal, PQ). Analytical grade reagents, including sodium chloride, sodium hydroxide, calcium chloride di-hydrate, tri-sodium citrate di-hydrate, and D-fructose. William's E medium (Gibco; Burlington, Ont), supplemented with streptomycin and penicillin (Gibco Laboratories; Burlington, On), were used for hepatocyte.

All the solutions are kept in an ice-cold bath before use as well as during the process of bioencapsulation. The pH of the solutions is kept at 7.40 by buffering with N-2-hydroxylethyl piperazine-N-2 enthanesulfonic acid (HEPES). Except for sodium alginate, the solutions are sterilized by filtering through a sterile 0.2 μm Millipore filter.

Droplet generators

We have used two differently sized droplet generators prepared by ourselves in the laboratory as well as a more elaborate commercially available one.

The first laboratory prepared droplet generator consists of two co-axially arranged jets: the central jet consisted of a 26 G stainless steel needle (Perfektum; Popper and Sons, Inc., New Hyde Park, NY), and a 16 G surrounding air jet, through which the sample and air was respectively passed. To prevent the extruding sample from occluding the outlet of the surrounding air jet, the tip of the sample jet was constructed such that the tip projected 0.5 mm beyond the end of the air jet.

The second laboratory prepared droplet generator was a larger and a slightly modified variation of the first droplet generator described above. This generator consists of a 13 G sample jet, and a 8 G surrounding air jet. The ends of the jets are cut flush to each other. A 1.7 × 1.1 mm PTFE capillary tube (Pharmacia P-L Biochemicals; Montreal, PQ) is inserted into the sample jet until it protruded approximately 15 mm from the outlet of the sample jet. The end of the capillary tubing is tapered to facilitate shearing resulting from the flow of passing air from the air jet. The capillary tubing is approximately 3.2 m in length, and has the capacity to be filled with microspheres suspended in 2.5 ml of sodium alginate.

Commercial generator: For the preparation of larger samples, we used one of the commercial generators.

Procedure

1. Hepatocytes, hepatocytes and bone marrow cells, or bacterial cells are suspended in an autoclaved sodium alginate in ice-cold 0.90% sodium chloride solution.

2. The viscous alginate suspension is pressed through a 23 G stainless steel needle using a syringe pump (Compact infusion pump model-975 Harvard App. Co. Inc., MA). Sterile compressed air, through a 16 G coaxial stainless steel needle, is used to shear the droplets coming out of the tip of the 23 G needle.

3. The droplets are allowed to gel for 15 min in a gently stirred heat sterilized and ice-cold solution of calcium chloride (1.4%, pH 7.2). Upon contact with the calcium chloride buffer, alginate gelation is immediate.

4. After gelation in the calcium chloride solution, alginate gel beads are allowed to react with polylysine (PLL), MW 16,100 (0.05% in HEPES buffer saline, pH 7.2) for 10 min. The positively charged PLL forms a complex of semipermeable membrane.

5. The beads are then washed with HEPES (pH 7.2) and coated with an alginate solution (0.1%) for 4 min.

6. The alginate-poly-L-lysine-alginate capsules so formed are then washed in a 3.00% citrate bath (3.00% in 1:1 HEPES-buffer saline, pH 7.2) to liquefy the gel in the microcapsules.

7. The APA microcapsules formed, which contain entrapped hepatocytes or bacterial cells. coli, are stored at 4°C to be used for experiments. The conditions are kept sterile during the process of microencapsulation.

8.9. Method II: 2-Step Method

The standard method described above is not optimal for encapsulating high concentrations of cells or microorganisms. As shown schematically in Fig. 8.3, cells or microorganisms may be trapped in the membrane matrix. This can weaken the membrane. If cells are exposed to the surface, the result may be loss of immunoisolation and rejection. Therefore, the two-step method has been developed with the aim to prevent this problem (Fig. 8.3) (Wong and Chang, 1991b).

1. First, the hepatocytes or hepatocytes and bone marrow stem cells or other cells suspended in sodium alginate are entrapped within solid calcium alginate microspheres. This is done by filling a 5 ml syringe (Becton Dickinson and Co.; Rutherford, NJ) with the cell suspension, and extruding the sample with a syringe infusion pump (Harvard Apparatus; Mill, MA) through the sample jet of the first droplet generator. The droplets formed at the end of the sample jet are allowed to fall drop-wise into a PYREX dish (125.65 mm) containing 300 ml 100 mM $CaCl_2$ (100 mM $CaCl_2$, 10 mM HEPES, 20 mM D-fructose, pH 7.4). Every 5 min the cells in the syringe are resuspended by gentle inversion of the syringe to minimize the effect of cells sedimenting in

the alginate solution. The air flow and infusion rate through the droplet generator are 2.0–3.0 l/min. and 0.28–0.39 ml/min, respectively; the clearance height between the end of the sample jet and the surface of the calcium solution is set at approximately 20 cm. A strainer cup is fitted inside the dish to collect the droplets, and to facilitate the removal of the formed microspheres.

2. The microspheres are allowed to cure for approximately 15 min, after which they are removed and temporarily stored in Hank's Balanced Salt Solution (Gibco Laboratories; Burlington, Ont) supplemented with 10%, 100 mM $CaCl_2$.

3. In the second step, 1.0 ml of formed microspheres is collected and washed three times with buffered saline (0.85% NaCl, 10 mM HEPES, 20 mM D-fructose, pH 7.4).

4. The final saline washing is aspirated and 1 ml of 1.2–1.6% sodium alginate is added to the 1.0 ml of washed microspheres. The sodium alginate is prepared by diluting the 4% stock solution with buffered saline. With a 5 ml syringe, the length of the PTFE capillary tubing is filled with the sodium alginate and suspension of microspheres. The tapered end of the capillary tubing is inserted through the top of the sample jet of the second droplet generator until the tip of the tubing extends approximately 15 mm beyond the end of the sample jet. The air flow and extrusion rate through the modified droplet generator are 7.0–9.0 l/min. and 0.28–0.39 ml/min, respectively. The tip of the capillary tubing is set at approximately 20 cm above the surface of the calcium solution. With the 5 ml syringe still attached to the other end of the tubing, the microsphere suspension in the tubing is extruded using the Harvard infusion pump. Similarly, the drops formed at the end of the sample jet are allowed to fall dropwise into a PYREX dish containing a strainer cup and filled with 300 ml of 100 mM $CaCl_2$.

5. The spheres are allowed to cure in the calcium solution for approximately 15 min, after which they are removed and washed with buffered saline.

6. The alginic acid matrix on the surface of the sphere is stabilized with poly-l-lysine by immersing 5 ml (settled volume) of macrospheres

in 80 ml of 50 mg% poly-l-lysine (50 mg% poly-l-lysine, 0.85% NaCl, 10 mM HEPES, 20 mM D-fructose, pH 7.4) for 10 min.

7. The spheres are then drained, washed with buffered saline, and immersed into 200 ml of 0.2% sodium alginate (0.2% sodium alginate, 0.85% NaCl, 10 mM HEPES, 20 mM D-fructose, pH 7.4) for 10 min to apply an external layer of alginate.

After 10 min, the spheres are collected and immersed in 200 ml 50 mM sodium citrate (50 mM sodium citrate, 0.47% NaCl, 20 mM D-fructose, pH 7.4) to solubilize the intracapsular calcium alginate. This may require up to 30 min with frequent changes of the sodium citrate solution.

8.10.   Method III: Preparation of Cells for Encapsulation

Preparation of rat hepatocytes

1. Each rat is anesthetized with sodium pentobarbital and cannulated via the portal vein.

2. The thoracic vena cava is cut and the liver perfused with a calcium free perfusion buffer (142 mM NaCl, 6.7 mM KCl, 10 mM HEPES, pH 7.4) for 10 min at 40 ml/min.

3. Next, the liver is perfused with a collagenase perfusion buffer (67 mM NaCl, 6.7 mM KCl, 100 mM HEPES, 5 mM $CaCl_2$, 0.05% collagenase, pH 7.5) for an additional 15 min at 25 ml/min.

4. The liver is then excised, placed in William's E medium supplemented with 100 μg/ml streptomycin and penicillin, and gently shaken to free loose liver cells from the liver tissue.

5. The cells are collected, filtered through a 74 μm nylon monofilament mesh (Cistron Corp., Elmford, NY), and centrifuged to remove connective tissue debris, cell clumps, non-parenchymal cells, and damaged cells.

6. Isolated hepatocytes are prepared for encapsulation by first washing and suspending the cells with buffered saline (0.85% NaCl, 10 mM HEPES, 20 mM D-fructose, pH 7.4).

7. The cells are then mixed with a 4% stock solution of sodium alginate (4% sodium alginate, 0.45% NaCl), to make a cell suspension consisting of 20 x 10^6 cells/ml of 2% sodium alginate.

Bone marrow stem cells from rats

Each rat is anesthetized with sodium pentobarbital, and both femur are isolated. Iscove's Modified Dulbecco's Medium (IMDM, GIBCOBRL, Life Technologies, NY) is used to flush out bone marrow cells from the femurs using a 5 ml syringe with a 22-gauge needle. The cell suspension was filtered through a nylon sieve (85 μm). The bone marrow cells are then washed with IMDM and centrifuged at 50 G for 10 min at 4°C, this is repeated three times. After the last wash, the bone marrow cells (nucleated cells) are kept on ice until use.

Genetically-engineered E. coli DH5 cells and microorganism

1. Genetically-engineered *E. coli* DH5 cells are grown in an L.B. medium. The composition of the LB medium is 10 g/l bactotryptone (Difco), 5 g/l bacto yeast extract (Difco), and 10 g/l sodium chloride (Sigma). The pH is adjusted to 7.5 by adding about 1 ml of 1 N NaOH. The medium is sterilized in Castle Labclaves for 30 min at 250°C and autoclaved before use.

2. Log phase bacterial cells are harvested by centrifuging at 10,000 *g* for 20 min at 4°C. The supernatant is discarded.

3. The cell mass is then washed and centrifuged at 10,000 *g* for 10 min at 4°C five times with sterile cold water to remove media components.

4. Bacterial cells are suspended in an autoclaved sodium alginate in ice-cold 0.9% sodium chloride solution.

5. The viscous alginate-bacterial suspension is pressed through a 23-gauge needle using a syringe pump (Compact Infusion Pump Model 975, Harvard App. Co. MA).

6. Compressed air through a 16-gauge needle is used to shear the droplets coming from the tip of the 23-gauge needle.

7. The droplets are allowed to gel for 15 min in a gently stirred ice-cold solution of calcium chloride (1.4%).

8. After gelation in the calcium chloride, the alginate gel beads are coated with poly-l-lysine (0.05% in HEPES buffer saline, pH 7.2) for 10 min.

9. The beads are then washed with HEPES and coated with an alginate solution (0.1%) for 4–8 min.

10. The alginate-poly-L-lysine-alginate capsules are then washed in a 3% citrate bath (3% in 1:1 HEPES-buffer saline, pH 7.2) to liquefy the gel in the microcapsules.

8.11. Method IV: Artificial Cells with Macromolecular Cutoffs

When using cells or microorganisms to act on macromolecules, the above methods cannot be used. For microorganisms to act on cholesterol bound to lipoprotein, the microorganisms have to be encapsulated in macroporous microcapsules (47).

Microorganism

Pseudomonas pictorum (ATCC #23328) was used because of its ability to degrade cholesterol. It is first cultured in nutrient broth (DIFCO) at 25°C, followed by harvesting and resuspension in a cholesterol medium for 15 days at 25°C. The suspension can then be used as an inoculum for biomass production. The culture is grown in bovine calf serum (Sigma) at 37°C for 36 h, and then harvested. This is used to prepare bacterial suspensions for immobilization. The concentration is about 0.4 mg of dry cell/ml.

Inoculum medium: Bovine calf serum is from SIGMA. It is used in all experiments unless otherwise specified. Nutrient agar plates are prepared by dissolving 8 g of nutrient agar (Difco) in 100 ml of water. The solution is autoclaved for 15 min at 121°C, allowed to cool at 50°C and poured into plastic Petri dishes (Fisher Scientific). The plates are stored at 4°C for up to 2 months.

Cholesterol medium: The composition of the medium is: ammonium nitrate (0.1%); potassium phosphate (0.025%); magnesium sulfate (0.025%); ferric sulfate (0.0001%); yeast extract (0.5%); cholesterol (0.1%), all dissolved in water with pH adjusted to 7 and autoclaved for 15 min.

Procedure

1. A 2% agar (Difco) and 2% sodium alginate (Kelco) solution is autoclaved for 15 min and cooled to 45°C to 50°C.

2. *P. pictorum* suspended in 0.4 ml of 0.9% NaCl is added drop by drop to 3.6 ml of agar alginate solution at 45°C, with stirring done vigorously.

3. 3 ml of the mixture obtained is kept at 45°C while it is being extruded through the syringe. The extruded drops are collected into cold (4°C) 2% calcium chloride and allow to harden. These agar-alginate beads are about 2 mm in diameter.

4. After 15 min, the supernatant is discarded and the beads are resuspended in 2% sodium citrate for 15 min.

5. Then, they are washed and stored in 0.9% saline at 4°C.

To test for immobilized bacterial activity, 1 ml of beads/microcapsules are placed in a sterile 50 ml Erlenmeyer. A 5 ml of serum is added and a foam plug is fitted. Samples are withdrawn at specified intervals. When empty beads or microcapsules are being prepared, the bacterial suspension is replaced by saline, and all the other steps are kept the same.

CHAPTER 9

Artificial Cells Containing Hepatocytes and/or Stem Cells in Regenerative Medicine

9.1. Introduction

Artificial cells and regenerative medicine

Chapter 8 describes the potential of artificial cells containing islets, cells, genetically engineered cells, microorganisms and stem cells for treatment of a number of clinical conditions. Many of these clinical conditions, like diabetes and neurological diseases, require long term treatment. Chapter 8 also discusses the further development of cell encapsulation technologies needed for long term use. To cater for more immediate use, one area of research in this laboratory focuses on the use of artificial cells in regenerative medicine. We have been studying the use of artificial cells containing hepatocytes and/or bone marrow stem cells for liver regeneration. Fulminant hepatic failure or extensive liver resection for metastatic cancer or severe injury can result in severe decrease in liver function not compatible with life. On the other hand, the liver is an organ that can regenerate itself, given the required time and condition. Thus, our aim is to study whether artificial cells containing hepatocytes and/or bone marrow stem cells can maintain the experimental animal alive, long enough to allow the liver to regenerate sufficiently for the animal to recover (Fig. 9.1).

Artificial Cells

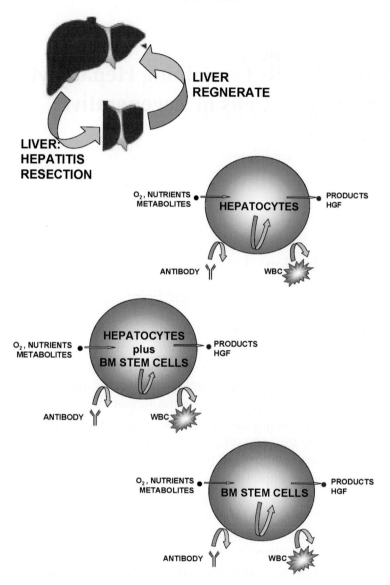

Fig. 9.1. *Upper:* Liver failure severe enough to be not compatible with life. This can be caused by acute hepatitis, massive traumatic injury or extensive cancer resection. Liver has the ability to regenerate itself to its original size if the patients can survive for a sufficient length of time under suitable conditions. *Lower:* We studied the use artificial cells with 3 different contents for liver regeneration: (1) hepatocytes; (2) hepatocytes plus bone marrow stem cells; or (3) bone marrow stem cells alone.

Design of study

We use artificial cells with three different contents (Fig. 9.1): 1) hepatocytes; 2) hepatocytes plus bone marrow stem cells; and 3) bone marrow stem cells alone.

We first carried out a preliminary feasibility study on the use of artificial cells containing hepatocytes in the galactosamine induced fulminant hepatic failure rat model. We then used the Gunn rats, a congenital rat model of hyperbilirubinemia, for more quantitative measurement. These studies show the feasibility of using artificial cells to support liver function. Next, we carried out a study to determine whether artificial cells can prevent immunorejection, for example, in artificial cells containing rat hepatocytes injected into mice. We also developed an improved method for the encapsulation of hepatocytes in artificial cells. Using this improved method of preparing artificial cells containing both hepatocytes and bone marrow stem cells, we showed that the duration of functions after implantation was improved.

Furthermore, we studied the use of artificial cells containing only bone marrow stem cells. The reason for this is that, although the use artificial cells can help prevent immunorejection without the need for immunosuppressant, it is difficult to obtain hepatocytes from humans. If we use hepatocytes from animals, we can similarly prevent immunorejection. However, since smaller protein molecules like albumin from the hepatocytes can diffuse out of the artificial cells, there is the potential of immunological problems. On the other hand, obtaining bone marrow from humans for use in other patients is a routine hematological procedure. Thus, if we can use artificial cells containing only bone marrow stem cells, then it would solve many of the potential problems. As described in this chapter, we show that artificial cells containing bone marrow stem cells results in the regeneration of the liver and recovery of the rats which had 90% of their livers surgically resected.

9.2.   Artificial Cells Containing Hepatocytes

Galactosamine-induced fulminant hepatic failure in rats

This is a preliminary feasibility study (Wong and Chang, 1986). Hepatocytes from 125–135 g Wistar rats were isolated and enclosed in alginate-polylysine-alginate artificial cells as described under "Methods" in Appendix II. Intraperitoneal injection of galactosamine (140 mg/100 g body weight) resulted in acute liver failure in the Wistar rats of 275–285 g. Forty-eight hours after the galactosamine injection, the rats in grade II coma were separated into pairs. One rat in each pair was randomly selected as control and the other for treatment. A total of 14 rats were used. In the control group, 4 ml of alginate artificial cells containing no hepatocytes was injected intraperitoneally. In the treated group, 4 ml of alginate artificial cells containing hepatocytes were similarly injected. Each 300 micron diameter artificial cell contained 120 ± 20 S.D. hepatocytes. The total number of artificial cells was about 62,000 in the 4 ml of artificial cells injected. Thus, each injection consists of artificial cells containing a total of about 7.4×10^6 hepatocytes. These artificial cells are flexible and can be easily injected using syringes. The animals were given food and 5% glucose oral *ad libitum*. Survival and other parameters were monitored.

Galactosamine-induced fulminant hepatic failure rats in the control group died $66.1 \pm$ S.D. 18.6 hours after galactosamine injection. The survival time in the group that received artificial cells containing hepatocytes was $117.3 \pm$ D.S. 52.7 h. Paired analysis showed that this is significantly ($P < 0.025$) higher than that of the control group. The total number of hepatocytes injected was only a small fraction of the total hepatocytes of the rat liver. This preliminary data encourages us to pursue further studies using artificial cells containing higher concentrations of hepatocytes to study the effect on survival rates. However, we need to obtain more basic information before pursuing these studies on survival rates. The galactosamine model is not always reproducible. Thus, we opted to first use a more stable Gunn rat animal model where we can carry out more quantitative measurements (Bruni and Chang, 1989). This would allow us to carry out more meaningful analysis on how to improve the artificial cells system.

Gunn rats with severe elevation of bilirubin, hyperbilirubinemia

The homozygous phenotype of Gunn rats (Gunn, 1938), like their human counterpart in Crigler-Najjar syndrome, lacks the liver enzyme, UDP-glucuronyltransferase. This results in severe elevation of bilirubin, termed hyperbilirubinemia, appearing shortly after birth. In humans, death of the infant usually occur in the first year of life. We used the Gunn rat model to carry out the following study (Bruni and Chang, 1989). Artificial cells containing hepatocytes were prepared (as described in the Methods section of Chapter 8). For this study, 2.2 ml of a 50% suspension of artificial cells were injected intra-peritoneally into the rats. The total number of viable hepatocytes implanted in each rat was 15×10^6.

Figure 9.2 shows the results of the first study on 3.5-month-old female Gunn rats (Bruni and Chang, 1989). During the 16-day control

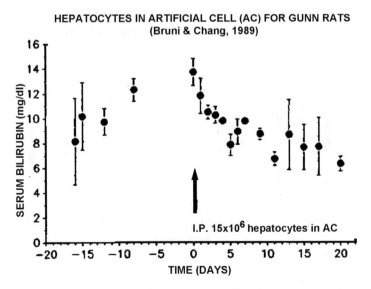

Fig. 9.2. One intraperitoneal injection of artificial cells containing a total of 15×10^6 hepatocytes into Gunn rats. Effects on the lowering of systemic bilirubin (Bruni and Chang, 1989).

period, serum bilirubin increased steadily at a rate of 0.32 ± 0.07 mg/dl per day. Twenty days after the intraperitoneal injection of artificial cells containing hepatocytes from Wistar rats, the serum bilirubin level decreased from the original 14 ± 1 mg/dl to 6.0 ± 1 mg/dl. Ninety days later, the serum bilirubin was still lower than the control period.

This preliminary study was followed by a second study using both control and treated groups.

The effect of artificial cells containing hepatocytes (total hepatocytes in the artificial cells being 15×10^6) was compared to two control groups (Burni and Chang, 1989): 1) Gunn rats received control artificial cells containing no hepatocytes; and 2) Gunn rats receiving free hepatocytes (total of 15×10^6). Figure 9.3 shows that the serum bilirubin levels in the Gunn rats which received control artificial cells containing no hepatocytes are significantly higher than those in the Gunn rats that received artificial cells containing hepatocytes. Figure 9.3 shows that both free hepatocytes and artificial cells containing the same number of viable hepatocytes have the same effect

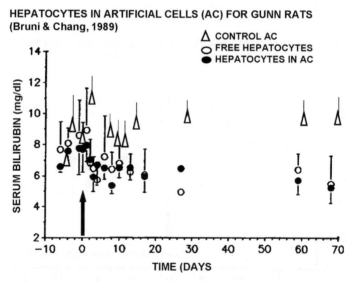

Fig. 9.3. Effect of one intraperitoneal injection of (1) control artificial cells containing no hepatocytes; (2) 15×10^6 free hepatocytes; or (3) artificial cells containing a total of 15×10^6.

in lowering the serum bilirubin levels. This shows that the method of preparing artificial cells does not affect the ability of hepatocytes to conjugate bilirubin. The result also shows that free hepatocytes from the closely related Sprague-Dawley rats are not rejected by the Gunn rats.

Summary

The results up to now show that hepatocytes in artificial cells when injected, can carry out their functions. This is shown by the increase in survival time in a galactosamine fulminant hepatic failure rat model and the lowering of bilirubin in a hyperbilirubinemic Gunn rat. However, before pursuing further how this can help in the regeneration of the liver, it would be important first, to study whether artificial cells can protect hepatocytes from immunorejection.

9.3.   Immunoisolation

Can artificial cells protect hepatocytes from immunorejection?

In order to answer this question, we follow the viability of rat hepatocytes implanted into mice either in the free form or inside the artificial cells (Wong and Chang, 1988). However, with implanted artificial cells, the viability of hepatocytes depends on more than one factor. In addition to immunorejection, there is also fibrous coating and aggregation of artificial cells related to biocompatibility. The result is decreased viability due to lack of oxygen and nutrients. We opted first to study only the factor related to immunorejection. This is done by following only those artificial cells that are not aggregated inside the fibrous mass.

Hepatocytes were obtained from the livers of male Wistar rats using the procedure as described in Appendix II. The viability of the hepatocytes prepared in this way was 86.0% as measured by trypan blue exclusion . The hepatocytes were then encapsulated in alginate-polylysine-alginate artificial cells using the method described in Appendix II. The viability of the hepatocytes using this procedure was 63.4%. The recipients were 20–22 g male CD-1 Swiss mice. A total of 5×10^6 hepatocytes in either the free form or inside artificial cells were implanted into each mouse intraperitoneally. At predetermined intervals, the mice were sacrificed and the peritoneal cavity was washed twice with 2.0 ml of iced Hanks balanced salt solution. The viability of the hepatocytes was determined by trypan blue stain exclusion. As mentioned above, only the free artificial cells were analyzed in this study. The microcapsules were stained for 5 min and then punctured with sharp forceps to release the hepatocyte from inside the artificial cells. The trypan blue stained and unstained cells were counted under a light microscope and the total number of cells was measured using a hemocytometer.

Figure 9.4 shows the results obtained (Wong and Chang, 1988). The viability of free rat hepatocytes injected into mice decreased rapidly.

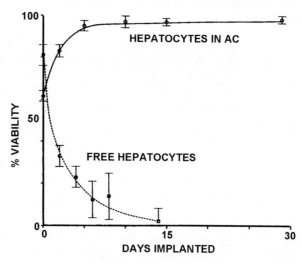

Fig. 9.4. Intraperitoneal injection into each mouse of (1) 15×10^6 rat hepatocytes or (2) artificial cells containing a total of 5×10^6 rat hepatocytes. Trypan blue dye exclusion is the method for analyzing the viability of free hepatocytes and hepatocytes in the artificial cells that are free in the peritoneal cavity. (Wong and Chang, 1988).

No viable free hepatocytes were recovered by the 14th day. On the other hand, the viability of the hepatocytes inside the free artificial cells steadily increased, and by 29 days after implantation, the viability has increased from the original 63.4% to more than 90%. The result of the present study seems to show that rat hepatocytes in the free form are rejected rapidly after being injected intraperitoneally into mice. For the artificial cells that were not aggregated, the rat hepatocytes inside were protected from rejection after the cells were injected into mice. However, the increase in viability of hepatocytes inside the artificial cells was an unexpected observation. We have therefore carried out a separate study to find out the reason.

Why do rat hepatocytes inside artificial cells recover their viability after implantation into mice?

Factors which can stimulate liver regeneration have been isolated from the liver extract and serum of newborn rats or partially hepatectomized rats. One of these factors has a molecular weight of <10,000 and the other one, a molecular weight of >100,000 (Michalopoulos *et al.* 1984). Alginate-polylysine-alginate membrane of artificial cells is only permeable to molecules of up to 64,000 mol wt (Ito and Chang, 1992). Our study (Kashani and Chang, 1988) shows that the higher molecular weight hepatic growth factor of >100,000 m.w. is continuously secreted by the hepatocytes inside the artificial cells. Since this type of artificial cells is not permeable to molecules larger than 64,000, we found that the >100,000 m.w. hepatic growth factor continues to accumulate inside the artificial cells. Trypan blue stain exclusion only shows the increase in permeability and damage of the hepatocytes and not necessarily death of the hepatocytes. Thus, it is likely that the increasing concentration of this hepatic growth factor inside the artificial cells may have played a role in the regeneration of the damaged hepatocytes inside the artificial cells.

Aggregation of artificial cells after implantation

It has been generally accepted that aggregation of artificial cells after implantation is due to the problem of biocompatibility. As mentioned earlier, the aggregation of the artificial cells was due to fibrous coating. We now look at another possible reason. We examined the artificial cells containing rat hepatocytes that have form fibrous aggregates after implantation into mice (Wong and Chang, 1991a). In many such cases, one can find the occasional hepatocytes extruding out of the membrane of the artificial cells after implantation (Fig. 9.5). The most severe tissue reaction seems to occur especially at these sites.

We then carried out another study on newly prepared alginate-polylysine-alginate artificial cells containing hepatocytes that have not been implanted. Using serial sections, we observed that in a significant number of the artificial cells, hepatocytes were extruding out of the membrane or entrapped in the membrane of the artificial cells (Fig. 9.5) (Wong and Chang, 1991a). We hypothesize that this may be one of the causes of the aggregation of the artificial cells containing hepatocytes. Each artificial cell contains more than 100 hepatocytes. Even when one hepatocyte out of these is exposed to the surface, the body will generate an immune response, resulting in the rejection of the whole artificial cell. In those cases where the hepatocytes are entrapped in the membrane but not exposed to the outside, weakening of the membrane will result. With time, erosion can result in the hepatocytes being exposed to the outside, followed by immunorejection.

My analysis of the standard method of preparation shows that with more than 100 hepatocytes in one artificial cell, the exposure of one or two hepatocytes to the surface is likely a random process (Fig. 9.5). I therefore devised a two step method to overcome this problem. The procedure and principle have been described in detail in Chapter 8 and Appendix II (Wong and Chang, 1991b). Thus, extrusion of hepatocytes to the outside and entrapment in the membrane are prevented (Fig. 9.5).

Artificial cells prepared by standard method: cell viability after implantation

Implanted artificial cells were recovered by laparotomy and analyzed (Liu and Chang, 2002). For the artificial cells prepared using the standard method, from week 2, most of the microcapsules were seen attached to the greater momentum, portal tract, pancreas and surface

STANDARD METHOD: ARTIFICIAL CELLS CONTAINING HEPATOCYTES

SERIAL SECTIONS : hepatocytes exposed to outiside or entrapped in membrane

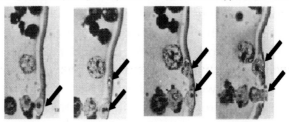

14 DAYS AFTER IMPLANTATION: RESULT OF IMPERFECT AC
Fibrous reaction to extruding hepatocyte   Erosions at sites of imperfection

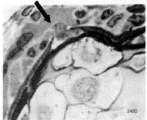 

2 STEP METHOD: ARTIFICIAL CELLS CONTAINING HEPATOCYTES

CROSS SECTION   14 DAYS AFTER IMPLANT

 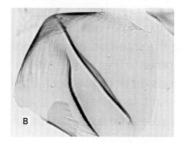

Fig. 9.5. *Upper.* Serial section of artificial cells containing hepatocytes freshly prepared using the standard method. A hepatocyte is seen to extrude to the outside and another one is entrapped in the membrane of the artificial cell. *Middle:* After implantation in mice, fibrous tissue can be seen at site where rat hepatocyte extrudes to the outside. Membrane defects are present 14 days after implantation. *Lower.* Artificial cell containing rat hepatocytes prepared using the new 2-step method show no hepatocytes extrusion or entrapment. No membrane defect was observed 14 days after intraperitoneal injection into mice. (Wong and Chang 1991a, 1991b; Liu and Chang, 2002).

of the liver and the spleen. At week 3, very few intact free artificial cells could be recovered; at week 4, the aggregated microcapsules were enclosed by more connective tissue, and some newly formed blood vessels were observed among the aggregates of microcapsules. Under the microscope, macrophages and lymphocytes were found infiltrating the capsular membrane (Fig. 9.5). No further morphological and viability studies were carried out in these groups after week 4, as little or no intact microcapsules could be recovered.

Artificial cells prepared by two-step method: cell viability after implantation

In contrast to the above results, when artificial cells were prepared using the new two-step method, the following results were obtained (Liu and Chang, 2002). At week 4, more than 90% of transplanted intact microcapsules could be recovered. At 8 weeks, 12 weeks and 16 weeks, the recovery rate was 60%, 50%, 20%, respectively. From week 4 on, some artificial cells were found attached to the greater omentum. Microscopic studies showed little or no inflammatory cells infiltration in the recovered microcapsules membrane (Fig. 9.5). Viability study shows that after the two-step encapsulation, the hepatocytes viability inside the artificial cells was 74%–78%. In the worst example of artificial cells containing hepatocytes prepared using the standard method, one could see membrane holes and cell entrapments (Fig. 9.5). Artificial cells prepared using the two step methods do not show such imperfections after implantation (Fig. 9.5). The result serve to show that biocompatibility is not the only factor that causes the aggregation and fibrous enclosure of artificial cells after implantation.

9.4.   Artificial Cells Containing Hepatocytes or Hepatocytes Plus Stem Cells

Introduction

The next step is to try to further increase the duration of function and viability of the hepatocytes inside artificial cells. For this purpose, we compare artificial cells containing hepatocytes to artificial cells containing hepatocytes and stem cells prepared by the two-step method.

Viability of free cells in culture

We first studied plated culture hepatocytes and compare this to plated co-culture of hepatocytes and bone marrow stroma cells (Liu and Chang, 2000). Free hepatocytes are detached 4 h after plating, and reached the optimal growth stage at day 7. At day 14, most hepatocytes became detached and cell death occurred. In the co-culture of hepatocytes and bone marrow stem cells, the hepatocytes were maintained for a longer period of time (14–21 days) than those of the hepatocyte culture. At day 14, most of the hepatocytes remained attached, and showed normal growth. This result leads us to the next step of coencapsulating hepatocytes and bone marrow stem cells in artificial cells.

In vitro viability of encapsulated cells

We compare the *in vitro* viability of hepatocytes encapsulated alone in artificial cells with that of hepatocytes co-encapsulated with a bone marrow stem cells (Liu and Chang, 2000). They are prepared as described under Methods in Chapter 8. The most striking microscopic observation is that when only hepatocytes are encapsulated, most of the cells remain single after encapsulation and during the culture period (Fig. 9.6). When co-encapsulated with bone marrow stem cells, most of the hepatocytes are attached to one another after encapsulation and during the culture period (Fig. 9.6). Mechanical rupturing of the artificial cells releases the encapsulated cells for viability studies using trypan blue exclusion. After isolation from the intact liver, the viability of the hepatocytes was 85%–90%, and after isolation from the femur, the bone marrow stem cells (nucleated cells) was 95%–100%. Immediately after encapsulation, the hepatocytes viability was 76%–83%. Starting from day 14 in the culture medium, the viability of hepatocytes in co-encapsulation was significantly higher than the

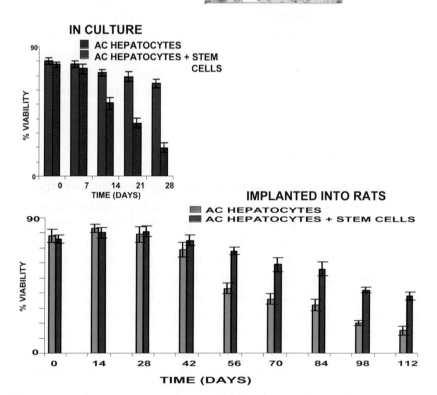

Fig. 9.6. Comparision between artificial cell containing hepatocytes and artificial cell containing both hepatocytes and bone marrow stem cells. After 28 days in culture; *Upper*: appearance of hepatocytes when encapsulated alone (*left*) and in combination with stem cells (*right*); and *Middle*: viability in culture; *Lower*: viability after implantation into rats. (Liu and Chang, 2002).

hepatocytes encapsulated alone ($P < 0.01$) (Fig. 9.6). By the 28th days, the viability of the hepatocytes co-encapsulated with bone marrow cells was maintained at a high level (Fig. 9.6). On the other hand, the viability of the hepatocytes encapsulated alone has fallen to very low levels (Fig. 9.6).

After implantation, the viability or hepatocytes in 1) artificial cells contain hepatocytes alone; and 2) artificial cells containing both hepatocytes and bone marrow stem cells, are as follows. From week 1 to week 6 post-transplantation, there was no significant difference in hepatocytes viability between the 2 groups (Fig. 9.6). However, from week 7 to week 16 post-transplantation, the viability of the hepatocytes co-encapsulated with bone marrow stem cells, was significantly higher than that of the group with only hepatocytes (Fig. 9.6). The average cell number in each artificial cell containing both hepatocytes and bone marrow stem cells remained the same.

Discussion of above results

Artificial cells containing both hepatocytes and bone marrow stem cells make possible a greater hepatocyte viability in culture and after implantation. The exact mechanism by which the bone marrow cells facilitate the maintenance of the hepatocytes is unknown. When co-encapsulated with bone marrow stem cells, the hepatocytes were found to attach to one another, whereas when encapsulated alone, most hepatocytes remained unattached (Liu and Chang, 2002). The above results seem to show that both cell-cell interactions and secretions from the bone marrow stem cells may contribute to this outcome. In addition, other groups have shown that attachment among cells is very important for maintaining hepatocytes viability and liver specific function inside the artificial cells (Theise *et al.*, 2000). Our results show that the co-encapsulation of hepatocytes and bone marrow stem cells improves hepatocytes viability when implanted into normal rats. In addition the new two-step encapsulation method provides for improved viability of implanted hepatocytes. This hepatocyte viability was increased to more than 3 months and therefore approached a level such that they could possibly be used in our study of liver regeneration. First, a quantitative study was carried out to see if this increase in viability corresponds to actual increase in function.

Artificial Cells

Study using Gunn rat model

We compare the effect of implantation of artificial cells prepared using the two-step method and containing 1) no hepatocytes; 2) hepatocytes; and 3) both hepatocytes and bone marrow stem cells (Fig. 9.7) (Liu and Chang, 2003). In the Gunn rats, the blood total bilirubin level was 4.4 ± 0.86 mg/dl before injection of the artificial cells. In the control groups of empty artificial cells, the bilirubin levels increased steadily throughout the 8 weeks and by week 8, the bilirubin level had reached 9mg/dl, about doubled the initial values. For artificial cells containing only hepatocytes, the plasma bilirubin levels decreased significantly, reaching the lowest level (2.7 mg/dl) at week 2 after transplantation. After this, the bilirubin level increased progressively reaching the same levels as the control groups at week 7 post transplantation. For artificial cells containing both hepatocytes and bone marrow stem cells, the bilirubin levels also decreased to the lowest level at week 2 after

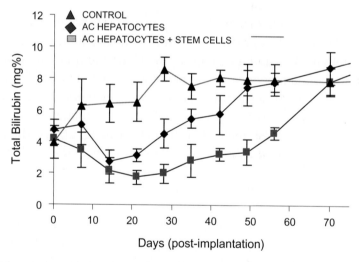

Fig. 9.7. One intraperitoneal injection into each Gunn rat of (1) control artificial cells containing no hepatocytes; (2) artificial cells containing hepatocytes; or (3) artificial cells containing both hepatocytes and bone marrow stem cells. The results on the total systemic bilirubin levels was followed for 70 days. Artificial cells were prepared using the 2-step method. (Liu and Chang, 2003).

transplantation. However, the important difference is that in this case, the level remained significantly lower than that of the control group up to 8 weeks after injection (Fig. 9.7) (Liu and Chang, 2003).

9.5. Artificial Cells Containing Hepatocytes in Rats with 90% of Liver Surgically Removed

Introduction

The above studies show that co-encapsulation of hepatocytes and bone marrow stem cells inside artificial cells allows us to attain 8 weeks of both hepatocyte viability and hepatocyte function. This means that we are ready to proceed to the next step of studying the use of this method for liver regeneration. Our original plan was to study the regeneration of liver in rats with 90% of liver surgically removed using the following: 1) control with no hepatectomy; 2) control with hepatectomy; 3) free hepatocytes; 4) artificial cells containing hepatocytes; 5) artificial cells containing both hepatocytes and bone marrow stem cells; 6) free bone marrow stem cells; 7) artificial cells containing bone marrow stem cells. We started with (1), (2), (3) and (4) and obtained the following results.

Artificial cells prepared using 2-step method on survival of rats with 90% of liver surgically removed

In the 90% hepatectomy group (PH) and the PH groups that received empty control microcapsules, most of the rats died in the first three days after hepatectomy. There was no death in the sham control group of rats that did not undergo liver resection. In the two groups receiving artificial cells containing hepatocytes or free hepatocytes, the survival rate was not significantly different from that of the sham control group (Fig. 9.8) (Liu and Chang, 2006). Free hepatocytes from Wistar rats were not immunorejected by the Gunn rats. Thus, both free hepatocytes and artificial cells containing hepatocytes alone enabled the rat liver to regenerate resulting in the

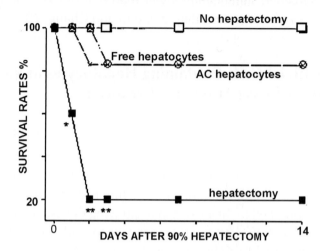

Fig. 9.8. Except for one group with no hepatectomy, the other groups are rats with 90% of their liver removed surgically. One intraperitoneal injection of (1) free hepatocytes; or (2) hepatocytes in artificial cells (AC) results in increased in survival rates that are not significantly different from the normal rats. (Liu and Chang, 2005).

recovery of the rats. However, three important points need to be considered.

Factors to be considered

(1) Rat liver can regenerate much faster than human liver. The results discussed earlier in this chapter show that artificial cells prepared using the new two step method containing hepatocytes alone can remain viable after implantation for up to 14 days and they can also function effectively in maintaining a low bilirubin in the Gunn rats for up to 14 days. In humans, liver regeneration takes much longer and 14 days may not be sufficient and most likely need up to 8 weeks. Thus, one would likely need to consider the use of artificial cells containing both hepatocytes and bone marrow stem cells that have a much longer duration of function and viability after implantation, i.e. up to 8 weeks.

(2) The present rat model with 90% of the liver removed surgically can be applied to clinical situations where liver resection is so extensive that it is not compatible with life. This is especially the case in extensive cancer metastasis to the liver or in extensive traumatic injuries to the liver. However, in the case of fulminant hepatic failure, in addition to inadequate liver function, there is also the other important factor of toxins released from the necrotic liver cells. These toxins can inhibit the regeneration of the remaining liver tissue, in addition to being toxic to the body. The next chapter describes the use of an adsorbent artificial cell hemoperfusion device to remove these toxins. This results in the removal of the toxins and the recovery of consciousness in grade IV hepatic coma patients (Chang, 1972b). However, hemoperfusion only removes the toxins and does not enable the production of hepatic growth factor to stimulate the regeneration of the liver or render to support the inadequate liver functions. A combination of hemoperfusion at the beginning to remove the toxins and use of artificial cells containing hepatocytes and bone marrow stem cell should provide the needed functions.

(3) Although artificial cells containing hepatocytes can prevent immunorejection without the need for immunosuppressant, it is difficult to obtain hepatocytes from humans. The use of hepatocytes from animals can also help prevent immunorejection by using artificial cells. However, since smaller protein molecules like albumin secreted by the enclosed hepatocytes can diffuse out of the artificial cells, there is the potential of immunological problems. On the other hand, obtaining bone marrow from humans for use in other patients is a routine hematological procedure. Thus, if we can use artificial cells containing only bone marrow stem cells, then we would solve many of the potential problems. Thus, we carried out the next study on this possibility.

9.6. Artificial Cells Containing Bone Marrow Stem Cells

In this study, we injected artificial cells containing bone marrow stem cells intraperitoneally into 90% hepatectomized rats (Liu and

SURVIVAL OF RATS WITH 90% OF LIVER SURGICALLY REMOVED
Liu & Chang; *J Liver Transplant* (2006)

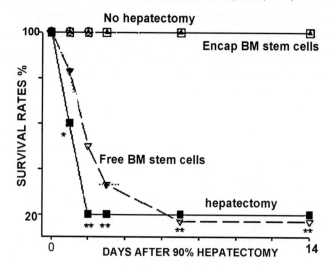

Fig. 9.9. Survival rates of rats with 90% of liver surgically removed, hepatectomy compared to those without removal of liver tissues (no hepatectomy). One peritoneal injection of artificial cells containing bone marrow stem cells resulted in the survival of the 90% hepatectomized rats. On the other hand, free bone marrow stem cells did not significantly increase the survival rates. (Liu and Chang, 2006).

Chang, 2006). We then studied the survival rates of the rats and the transdifferentiation of bone marrow stem cells.

Viability of isolated cells

About 3×10^8 nucleated BMCs could be obtained from the two femurs of each rat. BMCs viability after bioencapsulation was 98%. Hepatocytes viability after encapsulation was 82%.

Survival rates of rats with 90% of liver surgically removed

Figure 9.10 shows the survival rates of the following groups: 90% hepatectomized rats receiving 1) no artificial cells as control; 2) free bone marrow stem cells; and 3) artificial cells; that contain bone

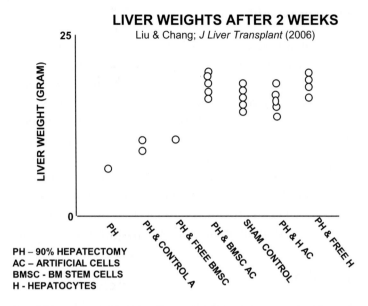

Fig. 9.10. This summarizes the liver weights of the rats that have survived 2 weeks after 90% of liver in each rat was surgically removed (hepatectomy). In the 3 groups where most of the rats died, those few that survived have a much lower liver weight. In those groups where all the rats survived after 2 weeks, the liver weights are not significantly different from the sham control group. This shows that in these groups, the 10% remaining livers have regenerated to the normal size. The rats in the sham control group did not have any of their liver removed. (Liu and Chang, 2006.)

marrow stem cells. The 4th group consisted of sham control rats that have not received hepatectomy. In the 90% hepatectomized control group, and the groups that received free bone marrow stem cells, the survival rates were significantly lower than the sham control group with most of the mortality occurring in the first three days post-surgery. There was no significant difference in the survival rates between the sham control group and the group receiving artificial cells containing bone marrow stem cells. Thus it would appear that in rats, artificial cells containing bone marrow stem cells are as effective as artificial cells containing hepatocytes in improving the survival rates of 90% hepatectomized rats.

Remnant liver weight

For those rats that survived at week 2 post hepatectomy, the remnant livers were removed, and the weights were measured. In the 90% hepatectomized rats that received 1) artificial cells containing bone marrow stem cells; 2) free hepatocytes; or 3) artificial cells containing hepatocytes, the liver wet weights were not significantly different from those of the sham control group (Fig. 9.11) (Liu and Chang, 2006). On the other hand, most of the animals did not survive in the 90% hepatectomized rats that 1) received no injection; 2) received control artificial cells containing no cells; and 3) received free bone marrow stem cells (Fig. 9.11). In those few that survived, the liver weights were much lower, but the small number did not allow for statistical analysis.

Blood chemistry

From blood chemistry (Liu and Chang, 2006), it was shown that except for the sham surgery group, the albumin levels in all the other groups decreased from day 1 post-surgery. After this, in the group that received artificial cells containing hepatocytes, the albumin levels increased and reached the normal level at day 2 post surgery. In the groups that received artificial cells containing bone marrow stem cells, the albumin only started to increase from day 3 post-surgery and reached the pre-surgery level in about one week. Blood levels of ALT and AST increased and reached peak levels at day 1 post-surgery for the control artificial cell group, free bone marrow stem cell group and the control group receiving no treatment. In the groups that received free hepatocytes or artificial cells containing bone marrow stem cells or hepatocytes, the results were different. There were significantly lower increases in the ALT, AST levels as compared to the hepatectomized group that received no injection or injection of control artificial cells containing no cells.

Plasma hepatic growth factor (HGF) levels

In the hepatectomized groups that received artificial cells containing hepatocytes or artificial cells containing bone marrow stem cells, the

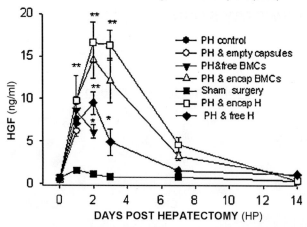

HEPATIC GROWTH FACTOR (HGF) IN PLASMA
Liu & Chang; *J Liver Transplant* (2006)

Fig. 9.11. Plasma hepatic growth factor (HGF) followed for those rats that have survived for 14 days. The results are shown as Mean ± S.D. only for those groups where all the animals have survived. PH = surgical removal of 90% of the liver. BMCs = bone marrow stem cells. H = hepatocytes. Encap = in artificial cells. (Liu and Chang, 2006.)

blood levels of HGF peaked at days 2 and 3 post surgery, and were significantly higher than those in the other groups (Liu and Chang, 2006). After this, the levels decreased and returned to pre-surgery level on day 14. In the free hepatocytes transplantation group, the peaked level was significantly lower at day 2 post surgery, decreasing to pre-surgery level at day 7. In the control hepatectomized groups that received 1) no treatment 2) artificial cells containing no cells group; or 3) free bone marrow stem cells, the blood HGF levels were much lower (Fig. 9.11).

Laparotomy and histology

Two weeks after transplantation, laparotomy showed that in the group that received artificial cells containing bone marrow stem cells, the artificial cells remained in the peritoneal cavity. They were found freely distributed throughout the peritoneal cavity or to be behind the liver or under the spleen had adhered or attached to the large omentum. Histology examination of the bone marrow stem cells in the artificial

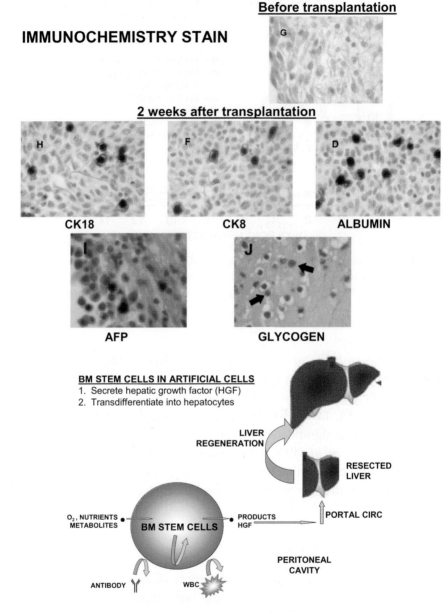

Fig. 9.12. *Upper*: Immunochemistry stain for hepatocyte markers (CK18 and CK8) for albumin production, glycogen production and for AFP. *Lower*: Possible mechanisms for artificial cells containing bone marrow stem cells in liver regeneration. (Liu and Chang, 2006.)

cells showed that before transplantation most cells were polygonal, star like, some with tail shape cytoplasm (Fig. 9.12). When they were retrieved two weeks after transplantation, much of cells morphology was transformed into round or oval shape (Fig. 9.12) (Liu and Chang, 2006).

Immunocytochemistry

Hepatocytes recovered from artificial cells show cluster of cells positively stained with hepatocyte markers CK8 and CK18 and also showing production of albumin ALB (Fig. 9.12), but no cells was found to stain positive with AFP. Before implantation, the bone marrow stem cells recovered from artificial cells did not show positive immunochemistry stain. However, when bone marrow stem cells were recovered from the artificial cells retrieved at week 2 after implantation, it was found that scattered cells were positively stained with hepatocyte markers CK8 and CK18 and also albumin production (Fig. 9.12). There were also cells that stained positively with AFP (Fig. 9.12) (Liu and Chang, 2006).

PAS glycogen stain

PAS staining revealed that hepatocytes recovered from artificial cells retrieved 2 weeks post-transplantation were positive for PAS, indicating these hepatocytes remained capability of glycogen production. In the bone marrow stem cells recovered from artificial cells retrieved two weeks post-transplantation, there were also cells that stained positive for PAS, indicating that the cells are capable of producing of glycogen (Fig. 9.12).

Possible mechanisms responsible for recovery of 90% hepatectomized rat model

What are the mechanisms responsible for the recovery of 90% hepatectomized rat model in our study? Most likely two combined mechanisms are involved 1) The bone marrow stem cells in the artificial cells release the hepatic growth factors HGFs that stimulate the regeneration of the remaining 10% liver in the rats that had 90% of the livers surgically removed; and 2) transdifferentiation of the bone marrow stem cells in the artificial cells into hepatocytes (Fig. 9.12).

(1) Transdifferentiation into hepatocytes. Our immunochemistry stain studies show that some of the bone marrow stem cells recovered from the artificial cells transdifferentiated into hepatocyte-like cells that expressed ALB, CK8, CK 18 and AFP, typical markers of hepatocytes. They also produced albumin and glycogen. Could these transdifferentiated hepatocytes contribute to the recovery of the 90% hepatectomized rats? Transdifferentiation into hepatocyte-like cells is not immediate (Lagasse *et al.*, 2000). Thus, the number of transdifferentiated cells was limited at the beginning and this alone could not have provided all the liver support observed in our study.

(2) Another possible mechanism involves the hepatic growth factors (HGFs) an important factor in liver regeneration (Rokstad *et al.*, 2002) as well as in stimulation of the transdifferentiation of BMCs into hepatocytes (Spangrude *et al.*, 1988). The level of HGF increases in acute or chronic liver failure (Uchida and Weissman, 1992). As discussed earlier, there are two subgroups of HGF, one being of higher molecular weight of >100,000 and the other of smaller molecular weight of <10,000 (Michalopoulos *et al.*, 1984). Alginate-polylysine-alginate membrane artificial cells allow the passage of molecules of 64,000 or less and retain those >64,000 (Ito and Chang, 1992). Our earlier study showed that HGF of >100,000 m.w. secreted by hepatocytes were retained and they accumulated in the artificial cells, thus helping to increase the regeneration of hepatocytes in the artificial cells (Kashani and Chang, 1988). The smaller molecule weight HGF of <10,000 can diffuse out of the artificial cells to stimulate the regeneration of the remaining 10% liver mass in the 90% hepatectomized rats. Indeed, Rokstad *et al.* (2002) showed that certain type of HGF was able to pass through the APA membrane and these most likely are those in the <10,000 m.w. range. In this regard, our study showed

that in the 90% hepatectomized rats that received artificial cells containing bone marrow stem cells, the blood HGF levels were significantly higher than those in the other groups, including the 90% hepatectomized group (Fig. 9.11). The HGF-like factors were most likely released from the bone marrow stem cell inside the artificial cells. As a result, the HGF stimulated the remnant liver to regenerate faster in the first days post transplantation before there was enough transdifferentiation of stem cells inside the artificial cells into hepatocytes (Fig. 9.12). Further support for this explanation is that artificial cells containing bone marrow stem cells stayed in the intraperitoneal cavity throughout the 14 days of the study. Thus, the HGF secreted can be drained into the portal circulation to reach the 10% remaining liver mass to stimulate its regeneration. On the other hand, intraperitoneal injection of free bone marrow stem cells did not increase the survival rates. This is most likely because the free bone marrow stem cells are rapidly removed from the peritoneal cavity through the lymphatic drainage and any HGF released does not drain into the portal circulation and stimulate the regeneration of the remnant liver.

9.7. Artificial Cells Containing Stem Cells in Regeneration Medicine

It would appear from the above study that implantation of artificial cells containing bone marrow stem cells results in the regeneration of the 90% hepatectomized liver and the survival of the animal. This is as effective as the injection of free hepatocytes or artificial cells containing hepatocytes. It is likely that hepatic growth factor HGF plays an important initial role followed by transdifferentiation into hepatocytes. These observations could stimulate further investigation of the potential for an alternative to hepatocytes transplantation for the treatment of acute liver failure or extensive liver resection. The use of artificial cells containing stem cells could also be investigated in other areas of regenerative medicine.

CHAPTER 10

Hemoperfusion in Poisoning, Kidney Failure, Liver Failure, and Immunology

10.1. Introduction

Artificial kidney machine

Kolff (1944) invented the artificial kidney, hemodialyzer, that has for many years supported the lives of thousands of kidney failure patients and also saved many lives from fatal accidental or suicidal poisoning. Unfortunately, only a small fraction of the world's patients can benefit from this treatment due to the high costs of the machine and treatment. In accidental or suicidal poisoning, in addition to the costs of the machine there are additional factors preventing its more common uses. For example, in many centers, especially the smaller ones, a hemodialyzer is not readily available. Furthermore, there are problems related to the length of treatment required, the possible fluctuation in extracorporeal blood volume and the specialized personnel required.

Hemoperfusion based on artificial cells

While working on the biotechnological aspect of artificial cells, I did some analysis and found that artificial cells can be the basis of a very simple and inexpensive detoxifier that can more efficiently carry out the detoxifying function of the hemodialyzer (Chang, 1966). Being a physician, my first aim is to develop something for more immediate use in patients. As a result, while continuing with basic research on the biotechnology of artificial cells, I spent a major part of my earlier 10 years of research career in developing this idea for clinical use in patients. This resulted in artificial cells containing adsorbents forming a miniaturized detoxifier that is much simpler, less costly and more efficient than the hemodialyzer (Chang, 1966, 1969a, 1971d, 1973a, 1974c, 1974h, 1975i, 1976g, 1977b, 1980e) (Fig. 10.1). The principle will be described in detail later, but briefly it is based on thousands of 90 micron diameter adsorbent artificial cells retained inside a small container by screens at either end (Fig. 10.1). Blood from patients containing toxins or drug perfuses through the screen to come in contact with the artificial cells. Toxins or drugs diffusing into the artificial cells are removed by the adsorbents inside the artificial cells. This cleanses the blood that returns to the patient. The membrane of the artificial cells prevents the adsorbent from being released into the body and also prevents the adsorbent from damaging the blood cells (Fig. 10.1).

10.2. Development and Clinical Trials

When I presented this concept and the prototype at the annual meeting of the ASAIO (Chang, 1966), Professor Kolff was most supportive and encouraging. He wrote a letter to the Medical Research Council of Canada giving his strong support. This letter played an important role in my successful application for an MRC special project grant for me to develop and test the concept in my laboratory for clinical trials. At that time this was a very new idea and physicians were not ready to use it on their patients in clinical trials. Professor Kolff again offered his support by writing a letter stating that he would invite me to go to his center to carry out clinical trials on the invention if no one else was prepared to do so. With this strong support, McGill's human ethics committee having reviewed in details all the preclinical safety and efficacy animal studies (Chang, 1969a, 1972a, Chang & Malave, 1970, Chang *et al.*, 1971a) approved the use of the invention in clinical trials at McGill's teaching hospitals. The arrangement was for me to prepare the hemoperfusion device in my laboratory which could then be used at the hospitals for clinical trials. It was agreed that the procedure as described in the next section be followed. The physicians at the

Artificial Cells

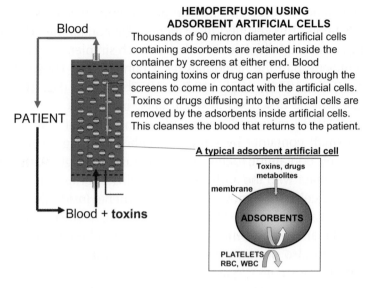

HEMOPERFUSION USING ADSORBENT ARTIFICIAL CELLS

Blood

Thousands of 90 micron diameter artificial cells containing adsorbents are retained inside the container by screens at either end. Blood containing toxins or drug can perfuse through the screens to come in contact with the artificial cells. Toxins or drugs diffusing into the artificial cells are removed by the adsorbents inside artificial cells. This cleanses the blood that returns to the patient.

PATIENT

Blood + toxins

A typical adsorbent artificial cell

Toxins, drugs metabolites

membrane

ADSORBENTS

PLATELETS RBC, WBC

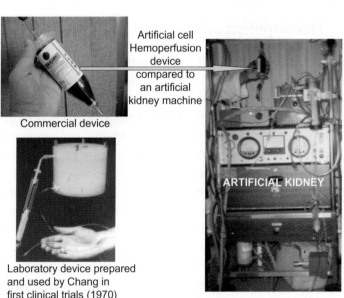

Artificial cell Hemoperfusion device compared to an artificial kidney machine

Commercial device

ARTIFICIAL KIDNEY

Laboratory device prepared and used by Chang in first clinical trials (1970)

Fig. 10.1. *Upper*: Basic principle of adsorbent artificial cells in hemoperfusion. *Lower left*: Laboratory and commercial hemoperfusion devices. *Lower right*: Much smaller and simpler hemoperfusion device compared to an artificial kidney machine.

dialysis units would refer patients to me and I would then make the final decision whether to carry out the procedure and to do the treatment on the patients myself.

Treatment of patients with severe accidental or suicidal poisoning

The phase I safety study demonstrated the safe use of the hemoperfusion device in patients (Chang and Malave, 1970; Chang *et al.*, 1971a, Chang, 1972a) and as a result clinical trials on efficacy followed. Clinical trials on its use in suicidal or accidental poisoning were more straight forward. The efficacy could be shown after the one or two treatments and the results could be assessed both quantitatively and clinically. Thus, clinical trials were successfully carried out in the first three patients (Chang *et al.*, 1973a, 1973b). Altogether, 11 adult patients (Chang 1975g, 1976c, 1980e) and one pediatric patient participated (Chang *et al.*, 1980a) all with conclusive clinical results. These will be discussed in more detail in a later section. The successful clinical trials led to the routine use of this approach around the world for the treatment of drug poisoning (Better *et al.*, 1979; Biberstein *et al.*, 1983; Cohan *et al.*, 1982; Verpooten *et al.*, 1984; Gelfand *et al.*, 1977; Gibson *et al.*, 1978; Lorch and Garella, 1979; Winchester, 1996; Diaz-Buxo *et al.*, 1978; Sideman and Chang, 1980b; Piskin and Chang, 1982; Chang and Zheng, 1983; Chang and Ho, 1985; Chang and Nicolaev, 1987; Chang and Odaka, 1991; Klinmann *et al.*, 1990; Kawasaki, 2000; Lin *et al.*, 2002; Lopez-Lago *et al.*, 2002; Peng *et al.*, 2004; Chinese Symposium on Hemoperfusion, 2005).

The method of preparing artificial cells containing an adsorbent has been published in reproducible format for all to use (Chang, 1972a, 1976g). As a result, a number of countries, for instance, Canada, China, USA, United Kingdom, Japan, Sweden, Italy, Spain and USSR were able to produce their own variations of industrial hemoperfusion devices with the usual variations in efficacy and biocompatibility. While the inferior ones are no longer in use, those that are safe and effective are being used in routine clinical treatment of accidental and suicidal poisoning. For some unknown reasons, in some areas of the world, these devices are extremely expensive and as a result, their use is not widespread. In other areas of the world, especially where hemodialysis machines are not easily accessible, inexpensive but safe and efficient devices are being produced and used extensively, saving thousands of patients with potentially fatal accidental or suicidal poisoning.

Hemoperfusion for removal of unwanted or toxic substances from blood under other conditions

Hemoperfusion has been an established routine clinical method for the treatment of patients with severe suicidal and accidental poisoning for many years. Its ability to remove unwanted or toxic substances from the blood also comes in useful in other clinical conditions. This includes its use in liver failure, kidney failure and use as an immunosorbent. In liver failure and kidney failure, hemoperfusion carries out only part of the functions of these organs, mainly in the removal of toxic or unwanted substances from the blood.

Treatment of patients with terminal kidney failure

Hemoperfusion in terminal kidney failure patients results in the improvements of patients' well-being and uremic symptoms (Chang *et al.*, 1971a, 1972g, 1974). It efficiently removes uremic wastes and toxins including the larger "middle" molecules (Chang, 1972e; Chang, and Migchelsen, 1973; Chang and Lister, 1980; 1981). "Middle molecule" is a term used by nephrologists to denote molecules in the molecular weight range of 300 to 15,000. However, hemoperfusion does not remove electrolytes, water or urea. Thus, it has been used in series with hemodialyzers as hemodialyzers were, at that time not effective in removing the larger "middle" molecules that were thought to be uremic toxins.

Clinicians around the world who use the better commercially prepared hemoperfusion devices in series with hemodialyzers reported clinical improvements in well-being of patients, nerve conduction velocity, pruritis, pericarditis, peripheral neuropathy as well as reduction in treatment time (Chang *et al.*, 1975, 1982b; Martin *et al.*, 1979; Inou *et al.*, 1979; Odaka *et al.*, 1980; Agishi *et al.*,

1980; Stefoni *et al.*, 1980). Since then, the quality of hemodialysis membranes has improved and the present high flux membranes are superior to the standard dialysis membrane in removing the larger "middle" molecules. Even then, the clearance is still much less than compared to the better hemoperfusion devices. Hemoperfusion devices are usually manufactured by manufacturers of hemodialysis machines and membranes. In those countries with strong hemodialysis companies, the hemoperfusion devices are extremely expensive. On the other hand, in those countries with no large dialysis industries, hemoperfusion devices are not expensive and therefore continued to be used in renal failure patients. In all the countries, hemoperfusion continues to be commonly used for uremic patients with aluminium or iron overload. This is based on the earlier clinical demonstration of the use of deferoxamine to bind these heavy metals and of hemoperfusion to remove the complex (Chang and Barre, 1983; Chang, 1986d; Chang *et al.*, 1984a, 1989a; Hakim *et al.*, 1985; Winchester 1996).

In an attempt to eliminate the need for the bulky and expensive hemodialysis machine, hemoperfusion has been used in series with a small ultrafiltrator with oral adsorbents to control potassium and phosphates (Chang *et al.*, 1975, 1977b, 1979a; Chang, 1976c). An urea removal system is being developed to complete the hemoperfusion-ultrafiltrator approach.

Treatment of patients with hepatic coma

The effectiveness of hemoperfusion devices in removing toxins from the circulating blood has also been shown in hepatic coma patients. Thus, the first use of the device in a grade IV hepatic coma patient resulted in the recovery of consciousness in the patient within an hour (Chang, 1972b). As will be described later, this was followed by extensive clinical trials in hepatic coma patients around the world, with effective removal of hepatic toxins and resulting in recovery of consiousness in a significant number of patients (Chang, 1972b, 1975e, 1975j, 1976c, 1982a; Gazzard *et al.*, 1974; Blume *et al.*, 1976; Chang *et al.*, 1977a; Bartels *et al.*, 1977, 1981; Silk and Williams, 1978; Gelfand *et al.*, 1978; Gimson *et al.*, 1978; Odaka *et al.*, 1978; Amano *et al.*, 1978; Maeda *et al.*, 1980; Cordoatri *et al.*, 1982; William, 1983;

O'Grady *et al.*, 1988). However, liver is a complex organ with many other functions besides detoxification. Thus, a detoxifier alone cannot fully support liver function. As mentioned in the previous chapter, hemoperfusion could play an important role in removing hepatic toxins to allow the use of artificial cells containing hepatocytes and stem cells for liver regeneration.

Protein-coated artificial cells in immunoadsorption

Albumin can bind tightly to the ultrathin collodion membrane of adsorbent artificial cells, and was initially used to increase the blood compatibility of the adsorbent artificial cells for hemoperfusion (Chang, 1969a). This albumin coating has also been applied to synthetic immunosorbents, resulting in blood compatible synthetic blood group immunosorbents (Chang, 1980d). The albumin-coated synthetic adsorbent has been applied clinically for removing blood group antibodies from plasma for bone marrow transplantation (Bensinger *et al.*, 1981). In addition, albumin-coated collodion activated charcoal (ACAC) was found to effectively remove antibodies to albumin in animal studies (Terman *et al.*, 1977). This principle has become a basis of one line of research in which other types of antigens or antibodies are applied to the collodion coating of the activated charcoal to form immunosorbents. Other immonosorbents based on the same principle have also been developed for the treatment of human systemic lupus erythematosus, modification of hyperacute renal xenograft rejection, removal of antiHLA antibodies in transplant candidates, and treatment of familial hypercholesterolemia with monoclonal antibodies to remove low-density lipoproteins (Terman, 1980; Terman *et al.*, 1979a, 1979b; Hakim *et al.*, 1990; Wingard *et al*, 1991; Yang *et al.*, 2004).

10.3.   Basic Principle of Artificial Cell in Hemoperfusion

Permeability and transport characteristics

An artificial cell membrane has an ultrathin membrane of less than 0.05 micron as compared to the 2.0 micron thickness of the dialysis membrane. Furthermore, the small size of artificial cells means that

Table 1. Analysis of Mass Transfer

	30 ml of Artificial Cells	1 Hemodialyzer Artificial Kidney
Total surface area	2 m^2	1 to 2 m^2
Membrane thickness	0.02 μ	2.00 μ
Mass transfer	100–200 $\times$	1

30 ml of artificial cells can have a total surface membrane area of 2 m^2 as compared to the 1 to 2 m^2 of a whole dialysis machine (Table 1, Fig. 10.1). This means that 30 ml of artificial cells can have a theoretical mass transfer that is 100 to 200 times that of a whole artificial kidney machine (Chang, 1966, 1972a) (Table 1).

We also analyzed the effects of varying the diameter and total volume of artificial cells over a wide range. Figure 10. 2 compares the mass transfer of artificial cells with that of a standard hemodialysis

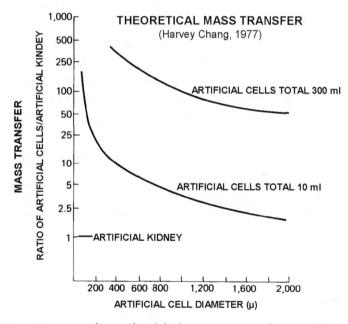

Fig. 10.2. Mass transfer artificial kidney represented as 1. Ratio of mass transfer of 10 ml and 300 ml of artificial cells of different diameters. Hemoperfusion uses artificial cells of about 100 micron diameter.

Artificial Cells

machine. Here, the mass transfer of a standard hemodialysis machine is taken as unity and comparison is made of 10 ml and 300 ml of artificial cells of different mean diameters. The mass transfer of even a 10 ml of artificial cells is many times that of the standard hemodialysis machine. With 300 ml of artificial cells, the mass transfer is even higher.

Experimental analysis

An experimental study of typical artificial cells has shown that the equivalent pore radius is 18 Å (Chang, 1965, 1972a). This means that the membrane would be permeable not only to the smaller waste metabolites but also to the "middle" molecules in the 1000–25,000 molecular weight range. A detailed analysis of the rate of movement of different molecules has also been carried out (Chang and Poznansky, 1968a) (Table 2). These results show that metabolites normally present in the body can equilibrate very rapidly across the artificial cell membranes. This being the case, 30 ml of artificial cells retained in a shunt perfused by circulating blood will have a mass transfer equivalent to that of a whole artificial kidney machine (Chang, 1966) (Figs. 10.1 and 10.2). However, in the case of the artificial kidney machine, once solutes cross the membrane they are "washed" away by 100–200 liters of dialysis fluid. For artificial cells with a total internal volume of 30 ml, within a very short time, the concentration gradient

Table 2. Experimental Permeability Data for Artificial Cells

Solutes	Half-time for Equilibration (Seconds)	Permeability Constant (P) (cm/sec)	Solute Permeability Coefficient (W) (moles/dyne-sec.)
Urea	4.3	2.01×10^{-4}	8.23×10^{-15}
Creatinine	17.5	0.61×10^{-4}	2.52×10^{-15}
Uric Acid	42.5	0.19×10^{-4}	0.77×10^{-15}
Creatine	16.6	0.75×10^{-4}	3.08×10^{-15}
Glucose	26.2	0.54×10^{-4}	2.17×10^{-15}
Sucrose	35.5	0.37×10^{-4}	1.62×10^{-15}
Actylsalicylate	39.0	0.32×10^{-4}	1.31×10^{-15}
Trtiated water	< 1.0	N/A	N/A

will disappear for any further solute diffusion. However, artificial cells are not meant to function as dialyzers. They are made to function as microscopic bioreactors, so that, solutes entering the artificial cells can be rapidly converted or removed, thus maintaining a concentration gradient. Thus, our earlier study showed that artificial cells containing urease in a hemoperfusion chamber can rapidly lower blood urea by converting it to ammonium (Chang 1966). Adsorbents, like activated charcoal, placed inside artificial cell can also remove solutes that equilibrate rapidly inside (Chang, 1966). Why do we choose to use activated charcoal?

10.4. Artificial Cells Containing Activated Charcoal in Hemoperfusion

Why use activated charcoal for artificial cell hemoperfusion?

Activated charcoal has the ability to adsorb and remove a large spectrum of molecules. Since ancient times, the Chinese and Greeks have used this for removing toxic substances. Activated charcoal has also been placed in a hemoperfusion device for direct perfusion by blood (Yatzidas, 1964). However, Dunea and Kolff (1965) showed in animal studies that this resulted in: 1) embolism due to the release of charcoal powder into the body, and 2) damage and removal of blood cells especially platelets.

By putting activated charcoal inside artificial cells one can prevent the release of harmful charcoal particles into the body and at the same time prevent its adverse effects on the blood cells (Chang, 1966) (Fig. 10.1). At the same time, unwanted molecules can move rapidly into the artificial cells and be removed by the activated charcoal.

In order to put this into practice, artificial cells were first prepared to contain activated charcoal powder (Chang, 1966). However, in a hemoperfusion device, the ultrathin membrane of the artificial cells cannot withstand the larger pressure gradient during perfusion. Furthermore, the flexibility of the ultrathin membrane increases flow resistance and also packing and obstruction at the exit port of the hemoperfusion device. A more successful method is to coat the ultrathin membrane directly onto activated charcoal granules (Chang, 1969a). Thus, there is no breakage of the ultrathin membrane, no increase in flow resistance and no obstruction to flow. This forms the basis of all the subsequent successful approaches. Later, industrial productions of medical grade spherical activated charcoal help to further improve this method.

Effects on embolism

In vitro studies show that there is no particle release since there is no significant differences in particle counts between the samples from the fluid entering the hemoperfusion devices and that from the fluid leaving the device (Chang and Malave, 1970). In animal studies, blood smear from blood leaving the device did not show any embolic particles. Histological studies in a total of more than 20 dogs did not show any evidence of embolism, even in artificial cell hemoperfusion devices stored for up to one month (Chang and Malave, 1970).

Effects of hemoperfusion on platelets

Figures 10.3 and 10.4 show that uncoated activated charcoal after contact with heparinized blood, when observed under a scanning electron microscope, showed fibrin entrapment of red blood cells, leukocytes and platelets. On the other hand, artificial cells in the form of albumin-cellulose nitrate-coated activated charcoal granules did not show fibrin entrapment of formed elements of blood (Chang, 1974b). This scanning electron microscopic study supported the finding that albumin-cellulose nitrate-coated activated charcoal, used in both animal studies (Chang, 1969; 1974b; Chang and Malave, 1970) and patients (Chang *et al.*, 1972a), prevents the depletion of platelets from the blood. There is no hemolysis or removal of red blood cells by the albumin-cellulose nitrate-coated activated charcoal. Unlike other polymers, cellulose nitrate can bind albumin tightly by adsorption even in the presence of plasma. There were no significant changes in post-hemoperfusion plasma hemoglobin levels or leukocyte levels (Chang and Malave, 1970).

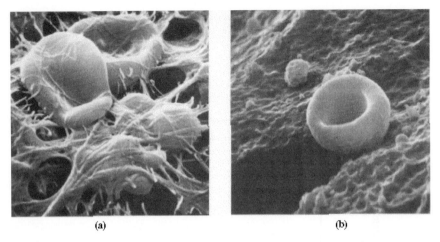

(a)   (b)

Fig. 10.3. *Left:* Uncoated activated charcoal after contact with heparinized blood observed under a scanning electron microscope, showed fibrin entrapment of red blood cells, leukocytes and platelets. *Right:* On the other hand, artificial cells in the form of albumin-cellulose nitrate-coated activated charcoal granules did not show fibrin entrapment of formed elements of blood.

Clearance of the ACAC artificial cell artificial kidney

On the basis of the mass transfer analysis (Table 1 and Fig. 10.2) one should not be too surprised at the very high clearance value of artificial cells containing activated charcoal. Activated charcoal actively removes and adsorb substances that rapidly cross the membrane. Furthermore, the albumin coating of the artificial cell membranes also takes part in the removal of protein-bound molecules in the circulation. As a result, the clearance of the different metabolites or drugs is many times higher than that for the standard hemodialysis machines.

For instance, as shown in Table 3, creatinine clearance is 230 ml/min as compared to 120 ml/min for standard hemodialysis machines. What is even more important is the clearance for "middle" molecules which is 140 ml/min as compared to 30 ml/min with the standard hemodialysis machines (available up to 1980). The clearance of drugs like phenobarbiturate, methaprylon, methaqualone and glutethimide of between 210 to 230 ml/min is even more striking when compared

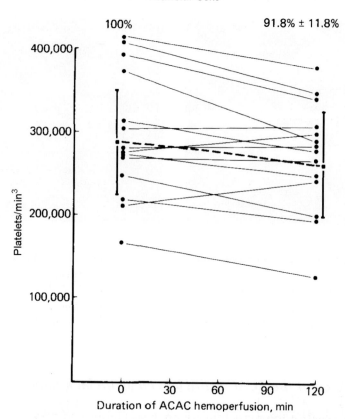

Fig. 10.4. Platelet levels in patients before and after 2 h of hemoperfusion using albumin-cellulose nitrated coated activated charcoal artificial cells. There is no significant decrease in platelet level. Earlier report elsewhere using nature activated charcoal without membrane coating resulted in marked decrease in platelet levels.

to the much lower clearance in the standard hemodialysis machines. The clearance of protein-bound molecules, as glutethimide, is many times higher using artificial cell hemoperfusion. On the other hand, it should be noted that the clearance depends a great deal on the ability of the charcoal to adsorb the material that crosses the artificial cell membranes. Thus, in the case of blood urea and electrolytes which are not adsorbed to any extent by the charcoal, very small clearance is obtained.

Table 3. Clearance: Hemoperfusion and Dialysis

Clearance ml/min (QB 300 ml/min)	Artificial Cell HP	Artificial Kidney HD 1980
Creatinine	230 ml/min	120 ml/min
Middle molecules (300–1500 mw)	140 ml/min	30 ml/min

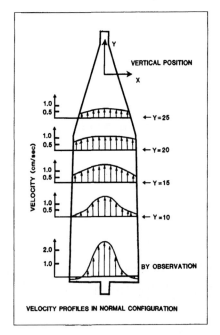 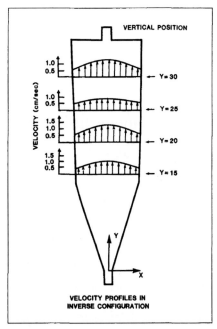

(From Victor Chang et al, 1987)

Fig. 10.5. Industrial scaled up ACAC hemoperfusion device. *Left:* Flow pattern when inflow is into the larger end of the device. *Right:* Inflow into the narrow end of the device resulted in improved flow hyrodynamic with less channeling and less stagnation of flow closer to the wall of the device.

Improvements in hydrodynamics

The laboratory ACAC hemoperfusion device as shown in Fig. 10.1 has configurations and inflow characteristics that allow for good hydrodynamics. However, this configuration is not convenient for industrial scale up. For industrial scale up, the configuration is also shown in Fig. 10.1. In this configuration, it is necessary to analyze the optimal hydrodynamics (Victor Chang *et al.*, 1987a). Figure 10.5 shows the hydrodynamic measurements of flow with the configuration of the industrial scaled up ACAC hemoperfusion device. When inflow enters the larger end of the device (left figure), there is channeling and stagnation of flow near the wall. This decreases the efficiency of extraction of unwanted molecules from the fluid. Furthermore, this can contribute to deposition of platelets and white cells. When inflow enters the narrow end of the device, there is improved flow hyrodynamic with less channeling and less stagnation of flow close to the wall of the device.

10.5.   Hemoperfusion in Acute Suicidal or Accidental Poisoning

Preclinical studies

Animal studies in this laboratory showed the effectiveness of the hemoperfusion system in the treatment of acute intoxication from glutethimide, pentobarbital and salicylate (Chang, 1972a). This led to our clinical trials in patients.

First clinical trials in acute poisoning

Other physicians at that time were not willing to carry out the procedure because of the potential for liability. As a result, I had to prepare the hemoperfusion devices in my laboratory at McGill, and to bring each of them to the hospital to treat the patients myself. The arrangement was that the nephrologists would refer patients to me. I would then make the final decision based on the following criteria. For the first 3 adult patients, I would only agree to carry out the procedures if the patients were seriously intoxicated, with a systemic drug level well above the lethal levels or who had serious complications requiring rapid recovery from the effects of the drugs. For example, in one case, before treatment the patient had cardiac arrests on two occasions with possible pulmonary embolism requiring heparinization. Her blood pressure was not obtainable by auscultation and her methyprylon level was twice the lethal level. In this particular case, the desirability of rapid elimination of the drug and its effects is obvious. In another case, a combined intoxication existed — the methyprylon level was three times, and the methaqualone level at least twice, the lethal level; furthermore, it is well known that glutethimide is cleared poorly by the standard artificial kidney available at that time. In the case of a pediatric patient with accidental theophylline overdose, irreversible brain damage would have been the result if treatment was not started immediately. In all these cases, the potential benefits to the patients would have to be high.

Result of clinical trials on 11 adult patients with suicidal or accidental drug poisoning

The first three patients are described in more detail below (Chang *et al.*, 1973a, 1973b), to followed by the other adult patients (Chang 1975g, 1976c, 1980e) and then the pediatric patient.

Case 1. A 50-year-old female, previously treated for depression, was admitted consequent upon her taking several drugs. She was comatose and her blood pressure was 90/60, pulse 80/min and regular, respiration 12/min and temperature, 98.4°F. Her corneal reflexes were absent. She was treated by intubation followed by gastric lavage containing activated charcoal. Heparin, 5000 units IV q4h, was started since pulmonary embolism was suspected. Four hours later two cardiac arrests occurred and were reversed with cardiac massage and direct current electroshock. A subsequent chest X-ray showed fracture of four ribs. Ventilatory assistance was required. Ten hours after admission her blood pressure could not be determined by auscultation. At this time her blood methyprylon level was 9.4 mg/dl and six hours later it was 9.6 mg/dl. Direct measurement of arterial blood pressure from the AV shunt gave a systolic pressure of 80 mm Hg.

The patient was referred to me and hemoperfusion was carried out. For the first 20 min the systolic blood pressure was maintained at 90–100 mm Hg with levarterenol. Then the levarterenol was discontinued

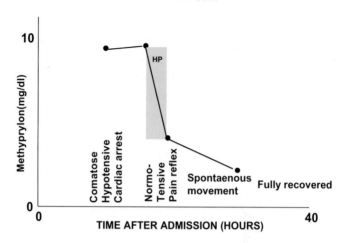

Fig. 10.6. Clinical and laboratory results of hemoperfusion in a patient with severe suicidal methryprylon overdose.

and the pressure remained at 130–150 mm Hg throughout the 3 h of hemoperfusion. The initial arterial blood methyprylon level was 9.6 mg/dl. Hemoperfusion for 2 h lowered this to 4.1 mg/dl, i.e. below the dangerous level of 6 mg/dl (Fig. 10.6). An initial clearance of 250 ml/min was obtained. As a result, the recovery was dramatic and straightforward. For example, voluntary respiration became more regular and increased to 40/min, with tidal volume increasing to 325 ml during hemoperfusion. Corneal reflexes and voluntary eye opening and blinking returned and response to painful stimuli appeared. Limb movements were detected 27 h after admission and 4 h later, the patient responded to verbal commands. She no longer required ventilatory assistance and started talking 4 h later.

Case 2. A 26-year-old female was admitted after ingesting a large amount of glutethimide. She was comatose and areflexic. Her blood pressure was 90/60 and she required ventilatory support. Six hours after admission, her systolic blood pressure was less than 70 mm Hg and isoproterenol was infused. Twenty-four hours after admission, her condition was unchanged. Her rectal temperature was 87°F, blood

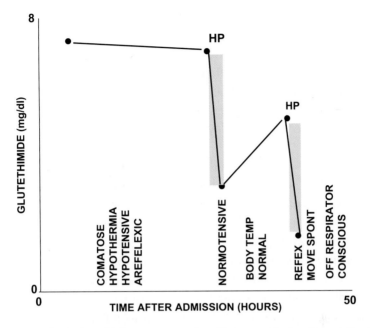

Fig. 10.7. Clinical and laboratory results of hemoperfusion in a patient with severe suicidal glutethimide overdose.

pressure 90/50, and she still required isoproterenol and ventilatory support. At that time, she had clinical and X-ray signs of bilateral bronchopneumonia. Her blood glutethimide level was 7 mg/dl.

At 27 h after admission, she was referred to me. She was treated by hemoperfusion with 300 g of the ACAC microcapsules for 2 h. In the course of the hemoperfusion, her blood pressure rose from 90/50 to 120/80 in 30 min without isoproterenol infusion. Her blood glutethimide level fell to 3 mg/dl, but 10 h after completion of the hemoperfusion, it rose to 5 mg/dl (Fig. 10.7). Except for a very slight pupillary reflex and gag reflex, she remained comatose. Her blood pressure was 110/55. Since glutethimide has a high lipid coefficient, a large part of the drug accumulated in the lipoid tissue. The rebound in its level after hemoperfusion likely reflects a redistribution of glutethimide from the lipoid tissue to the bloodstream. A second hemoperfusion with another 300 g of ACAC microcapsules was carried out. Within 1 h, a 2+ patellar reflex and a slight plantar reflex were elicited. After 1½ h of hemoperfusion, her tidal volume increased from 250 ml (pretreatment) to 475 ml. In addition, she exhibited some spontaneous limb movements. Her blood pressure increased to 130/75. The posthemoperfusion blood glutethimide level was 1.5 mg/dl (Fig. 10.6). Two hours after completion of the second hemoperfusion, she no longer required ventilatory support. She was up and about the following day.

Case 3. A 27-year-old male was admitted consequent upon his ingesting a mixture of medications. He was comatose and unresponsive to painful stimuli and required ventilatory assistance. He also showed signs of methaqualone intoxication: tonic convulsions, hypotension, cardiac failure, myoclonia, bleeding tendency, and hypothermia. Gastric lavage was carried out and forced diuresis started. Eight hours after admission, his blood methyprylon level was 18.3 mg/dl, 3 times the "dangerous" level of 6 mg/dl. A 60-min ACAC hemoperfusion was carried out at a blood flow rate of 300 ml/min. The methyprylon level fell to 12.9 mg/dl after the 60-min treatment and by 32 h after admission, it was 9.0 mg/dl having been lowered from three times the dangerous level to just below the dangerous level. Despite this, there was no marked immediate improvement in the patient's clinical condition. At that time, the laboratory result showed that he also had a methaqualone level of 6.5 mg/dl, a level which is more than twice the dangerous level of 2.5 mg/dl. A second ACAC hemoperfusion was given over a period of 2 h with a blood flow rate of 300 ml/min. The methyprylon and methaqualone level fell to 1.8 mg/dl and 2.7 mg/dl, respectively. By 48 h his pupils reacted sluggishly to light, and the gag reflex, deep pain response and bowel sounds were present. At 96 h his breathing was independent; all reflexes became normal. By 130 h after admission, he was oriented, rational and up and about.

Case 4. This patient was admitted comatose, areflexic, with fixed dilated pupils, and requiring respiratory assist and pressor drug to maintain her blood pressure. She had taken a large mixture of drugs, including glutethimide, thioridazine, salicylate, phenobarbital, and tetracycline, in addition to a large intake of alcohol. She was hemodialyzed three times in the first 48 h with no change in her condition. She was hemoperfused for 1–1/2 h with 300 g of ACAC, followed immediately by another 300 g of ACAC hemoperfusion column for 1 h. With this hemoperfusion, her corneal reflex returned and her blood pressure could be maintained without pressor drug. In the next 2 days, she was treated by two hemodialyses but without any changes in her clinical condition. A third hemoperfusion was carried out 10 h after the last hemodialysis. With this hemoperfusion, gag reflex and cough reflex returned and there was occasional spontaneous breathing. This was followed by a fourth hemoperfusion after which spontaneous breathing returned. The following day, she was responsive and following this, she recovered and was up and about. Her phenobarbital level was 20 mg% on admission. Clearance obtained was 228.6 mL/min. Her glutethimide level on admission was 22.62 mg/dl. Clearance obtained was 251 mL/min.

Case 5. A 79-year-old patient was admitted with a history of ingestion of phenobarbital with a blood level of 30 mg%. The patient was comatose, areflexic, and required ventilatory assist when hemoperfusion was carried out. With 2 h of hemoperfusion, pupil reflex, corneal reflex, and patellar reflex returned. Later, the patient started to trigger the respirator, had spontaneous movement and recovered shortly after this. Clearance obtained for hemoperfusion at a blood flow rate of 200 mL/min was 188 mL/min.

Case 6. This patient was admitted with grade IV coma due to phenobarbital intoxication. The patient received ACAC hemoperfusion at 180–200 mL/min blood flow rate. The clearance was 162 mL/min. Hemoperfusion resulted in recovery of the patient.

Case 7. This 21-year-old patient was admitted with a history of ingestion of glutethimide and phenobarbital. She was admitted with grade IV coma. There was X-ray evidence of right upper lobe atelectasis and right lower lobe consolidation. One 2 h hemoperfusion resulted in improvement of the patient and complete recovery and discharge subsequently.

Case 8. This 22-year-old man was admitted with a history of overdose of salicylate and sodium bromide. The salicylate level was 52.85 mg% when hemoperfusion was started. The clearance obtained at the flow rate of 300 mL/min was 165.9 mL/min. Hemoperfusion of 2 h lowered the systemic salicylate level to 37.76 mg%. At initiation of hemoperfusion, the patient was in grade II coma responding to painful stimuli. After 1 h of hemoperfusion, the patient recovered consciousness, complained of thirst and hunger, wanted to get up and was also carrying on a conversation. He recovered after this.

Case 9. This 51-year-old patient had taken an overdose of methyprylon. The patient was comatose, areflexic, hypotensive, and needed ventilatory assistance. He received 2 h of hemoperfusion with ACAC hemoperfusion at a blood flow rate of 160 mL/min. Two hours after hemoperfusion was started there was bilateral patellar reflex 2+, ankle reflex 2+, light reflex, and some voluntary movement. Shortly afterwards ventilatory assistance was not required and he recovered shortly after this.

Case 10. This 30-year-old patient was admitted with a history of having taken phencyclidine. Phencyclidine was qualitatively demonstrated in the blood and urine of the patient. The patient was comatose, areflexic, hypotensive, and a 2 h ACAC hemoperfusion was carried out. After the hemoperfusion, the patient still remained comatose. Three days later the patient was off the respiratory and started to have voluntary movements and then recovered after this. It is not possible to measure the phencyclidine level quantitatively, in order to calculate clearance or changes in level because the method available at that time could only give a qualitative assessment of the presence of phencyclidine in the blood and urine.

Case 11. This 55-year-old patient was admitted after he had taken a bottle of Seconal (secobarbital). Shortly after admission, he developed respiratory arrest. He was resuscitated, put on a respirator, and his blood pressure was maintained with pressor drug. He developed ventricular tachycardia which responded to lidocaine. His secobarbital level was 6 mg%. Two hours of hemoperfusion was carried out. At

Table 4. First Series of Clinical Trials in 11 Adult Patients with Suicidal Drug Overdose. (Chang *et al.* 1973a, 1973b; Chang, 1975g; Chang, 1980e)

#	Coma	Drugs	Clearance ml/min	Hemoperfusions Number	Outcome
1	IV	Methyprylon	230	2	Recovery
2	IV	Glutethimide	150	1	Recovery
3	IV	Methyprylon	230	2	Recovery
		& Methaqualone	230		
4	IV	Glutethimide	230	4	Recovery
		& Phenobarbital	228		
5	IV	Phenobarbital	180	1	Recovery
6	IV	Phenobarbital	162	1	Recovery
7	IV	Glutethimide	—	1	Recovery
8	II	Salicylate	150	1	Recovery
9	IV	Methyprylon	160	1	Recovery
10	IV	phencyclidine	—	1	Recovery
11	IV	Secobarbital	—	1	Recovery

the beginning, the patient was comatose with no reflexes. After 1 h of hemoperfusion, there was 2+ achilles tendon reflex. And 2 h after hemoperfusion, both patellar achilles reflexes were present. Two days later he recovered completely and was transferred out of the intensive care unit. This hemoperfusion was carried out with the Hemosorba system and the clearance obtained was 200 mL/min.

Pediatric patient with accidental theophylline overdose

A 3-year, 10-month female. child with a two-year history of asthma developed tachynea, respiratory difficulty and fever. She was accidentally given 750 mg of theophylline, an amount 10 times in excess of that prescribed. She became restless, developed tachycardia, and vomited coffee-ground material positive for occult blood. Her serum theophylline level was 67 μg/ml 1 h and 45 min after being given the drug. She was transferred to the intensive care unit. The hospital immediately referred the patient to me for possible treatment with ACAC hemoperfusion. It was decided to carry out the procedure immediately because: 1) The theophylline level varied from 67 to

74 μg/ml and 40 μg/ml being considered a potentially lethal dose;
(2) It is known that if the level is not lowered quickly, the child would
suffer irreversible brain damage — as in an earlier case when the child
was referred to us too late; and (3) Available dialysis systems at that
time were not effective in removing theophylline.

A 3 h hemoperfusion procedure was carried out (Chang *et al.*,
1980a). The maximal blood flow possible in this 15.4 kg child was
60–75 ml/min in the first hour. After the first hour, the flow rate was
increased to 90 ml/min, and this rate was maintained for the remainder
of the treatment. Before hemoperfusion, the child had a tachycardia
of 187/min and a blood pressure of 120/70 mm Hg. As hemoperfusion
lowered the systemic theophylline level, the heart rate decreased to
145/min at one hour, 135/min at two hours and was 127/min at the
end of the procedure (Fig. 10.8). Her blood pressure was unchanged
throughout the hemoperfusion.

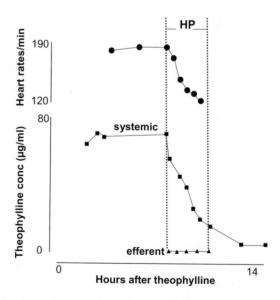

Fig. 10.8. Clinical and laboratory results of hemoperfusion in a 3-year-old
pediatric patient with severe accidental theophylline overdose. Her level
was well above the dangerous level of 40 μg/ml that could have resulted
in irreversible brain damage. Hemoperfusion rapidly lowers this level.

Figure 10.8 shows that the blood level of theophylline before HP was 74 μg/ml. Hemoperfusion lowered the level within 1.5 h to 40 μg/ml and within 3 h to 15 μg/ml. Hemoperfusion was discontinued. The theophylline level in the efferent from the hemoperfusion device was zero showing a 100% extract of the drug on one circulation through the device. The removal of the drug from the tissue compartment must also have been high since before hemoperfuion the blood level remained high and stable but after hemoperfusion was completed, the serum level continued to remain at the low level. The child recovered uneventfully.

Routine clinical uses in patients around the world

Activated charcoal is a well-known adsorbent for a large number of drugs and toxins. Artificial cells containing activated charcoal, as demonstrated here, take advantage of this adsorbing property and at the same time prevents the two major problems of charcoal embolism and reduction in platelets. The results of these initial clinical trials have led to the use of this approach around the world for the treatment of patients with accidental or suicidal poisoning (Better *et al.*, 1979; Biberstein *et al.*, 1983; Cohan *et al.*, 1982; Verpooten *et al.*, 1984; Gelfand *et al.*, 1977; Gibson *et al.*, 1978; Lorch & Garella, 1979; Winchester, 1996; Diaz-Buxo *et al.*, 1978; Sideman & Chang, 1980b; Piskin & Chang, 1982; Chang & Zheng, 1983; Chang & Ho, 1985; Chang & Nicolaev, 1987; Chang & Odaka, 1991; Klinmann *et al.*, 1990; Chang & Odaka, 1991; Kawasaki, 2000; Lin *et al.*, 2002; Lopez-Lago *et al.*, 2002; Peng *et al.*, 2004; Chinese symposium in hemoperfusion, 2005).

The methods of preparation of artificial cells containing activated charcoal have been published in reproducible for all to use (Chang, 1972a). As a result, a number of countries have started to produce their own industrial hemoperfusion devices. As would be expected, there are variations and extension of Chang's original method, resulting in variations in efficacy and biocompatibility. While the inferior ones are no longer in use, those that are safe and effective are being used in routine clinical treatment of accidental and suicidal poisoning. In

Table 5. Clinical Criteria for Hemoperfusion (modified from Winchester, 1996)

1. Progressive deterioration despite intensive care.
2. Severe intoxication with mid-brain dysfunction.
3. Development of complications of coma.
4. Impairment of normal drug excretory function.
5. Intoxication with agents producing metabolic and/or delayed effects.
6. Intoxication with an extractable drug which can be removed at a greater rate than endogenous elimination

some areas of the world, for some unknown reasons, these devices are extremely expensive and as a result are not in widespread use. In other areas of the world, especially where hemodialysis machines are not easily accessible, inexpensive but safe and efficient devices are being produced and used extensively. Nephrologists have been able to establish the following criteria for using this in acute poisoning (Table 6), based on the extensive experience in the use of adsorbent artificial cells in hemoperfusion for acute poisoning.

Because of its efficient adsorbing properties for numerous drugs (Table 5), the system may be even more useful in cases of acute intoxication involving a number or drugs, especially when the exact drugs are not known at the time that treatment is required. Being portable and having a constant extracorporeal volume, it is particularly useful in areas where hemodialysis machines are not easily accessible or in pediatric practices.

10.6.   Hemoperfusion in Terminal Renal Failure Patients

Hemodialysis machines are effective in the treatment of chronic renal failure patients. However, at that time there were not enough machines; furthermore, it was extremely expensive. Only a small number of patients could afford treatment in a few countries. In most other countries, the high cost and unavailability of machines posed a barrier to their use in terminal renal failure patients. Since hemoperfusion is efficient in removing toxin or unwanted waste from the blood, we carried out studies into its possible use in uremic patients.

Table 6. Drugs and Chemicals removed with Hemoperfusion (modified from Winchester 1996)

amobarbital	carbromal	ampicillin	digoxin	amanitin
butabarbital	chloral hydrate	carmustine	diltiazem	chlordane
hexabarbital	chlorpromazine	chloramphenicol	flecainide	demeton sulfoxide
pentobarbital	diphenhydramine	chloroquine	metoprolol	dimethoate
phenobarbital	ethchlorvynol	clindamycin	n-acetylprocainamide	diquat
quinalbital	glutethimide	dapsone	procainamide	methylparathion
secobarbital	meprobamate	doxorubicin	quinidine	nitrostigmine
thiopental	methaqualone	gentamicin		parathion
yinalbital	methsuximide	isoniazid		phalloidin
	methyprylon			polychlorinated biphenyls
			acetaminophen	paraquat
aminophylline	promazine	carbon tetrachloride	acetylsalicylic acid	
cimetidine	promethazine	ethylene oxide	colchicine	aluminum-with chelating agent
phenols		trichloroethane	d-propoxyphyene	iron-with chelating agent
theophylline		xylene	methylsalicylate	
			phenylbutazone	
			salicylic acid	

This author has carried out 55 hemoperfusion procedures in 14 patients (Table 7). Initially, only one hemoperfusion procedure was carried out on each of four terminal renal failure patients. Having shown the safety of this procedure, two hemoperfusions were carried out on the 5th patient and three procedures were then carried out on each of the next three patients. After this, a 72-year-old female, who could not be managed by peritoneal dialysis or hemodialysis was placed on an 8 month regime of hemoperfusions, supplemented with hemodialysis as required for the removal of urea, water, and electrolytes. Long-term hemoperfusion in this patient was safe and effective and her condition was stabilized sufficiently for her to be maintained on standard dialysis for the next 10 years. This led to the inclusion of other patients for long-term hemoperfusion.

Until commercial devices became available, the number of patients who could be treated was limited. This is because I had to prepare the device and then carry out the procedures in the patients. To allow for objective clinical assessment, I was responsible for the preparation of the hemoperfusion device and the carrying out of the procedures on patients. A group of nephrologists would refer the patients to me and they would independently assess the clinical and laboratory results.

Table 7. Chronic Renal Failure

Patients	Type of Procedure in Addition to Hemodialysis	Number of Procedures	Duration of Clinical Trial
1	Hemoperfusion	1	Single
2	Hemoperfusion	1	Single
3	Hemoperfusion	1	Single
4	Hemoperfusion	1	Single
5	Hemoperfusion	2	2 weeks
6	Hemoperfusion	3	3 weeks
7	Hemoperfusion	3	3 weeks
8	Hemoperfusion	3	3 weeks
9	Hemoperfusion	25	8 months
10	Hemoperfusion	13	4 months
11	Hemoperfusion	26	6 months
12	Conjoint hemoperfusion-hemodialysis	13	6 months
13	Conjoint hemoperfusion-ultrafiltration	6	3 months
14	Conjoint hemoperfusion-ultrafiltration	27	6 months

The patients were otherwise treated as any other hemodialysis patient in the McGill teaching hospital dialysis units, where medications, frequencies of transfusions, and number of supplemental hemodialyses were determined independently by the nephrologists. The standard chronic hemodialysis program at that time was for each patient to receive an average of two 6 h-hemodialyses each week, usually on the EX 01 hemodialyzer.

10.7. Hemoperfusion Alone Supplemented by Hemodialysis

General clinical results

No adverse effects from hemoperfusions were observed in the patients. There was little or no change in blood pressure throughout the hemoperfusion procedure, even in patients who normally experienced severe hypotensive episodes when on the EX 01 hemodialyzers. Immediately after the 2 h hemoperfusion, the patients felt much less fatigued than when they underwent the 6 h hemodialysis with the EX 01 artificial kidney. This is especially in a patient who normally stayed home in the evening to rest after the 6 h EX 01 hemodialysis treatment. On the other hand, 1 h after the 2 h ACAC hemoperfusion, she would go out for dinner, social visits, or the theater. At the initiation of the clinical trial, each patient receiving a hemoperfusion was required to stay in the hospital overnight for observation before being allowed to return home the next day. When the procedure was found to be safe, the patients treated with the hemoperfusion procedure were allowed to go home 1 h after hemoperfusion. The patients' general well-being was at least as good on the weekly 2 h ACAC hemoperfusion and 6 h hemodialysis (EX 01) as on the twice weekly 6 h EX 01 hemodialysis. The following are examples of patients on longer term treatment with hemoperfusion.

First patient on long-term hemoperfusion

This is a 72-year-old female with chronic renal failure and congestive heart failure. She was admitted with complaints of nausea, vomiting, low back pain, diarrhea and hiccups. She was bedridden in the hospital, with a BUN of 186 mg/dl,creatinine of 24 mg/dl,and a 24 h creatinine excretion of 80 mg. She was placed on peritoneal dialysis but this resulted in massive intra-abdominal bleeding and severe shock. Hemodialysis treatment resulted in hypotension. She was referred to this author since there was no other way to treat her condition. At that time, because of lack of facilities, patients of her age were not generally accepted into long-term hemodialysis program. Thus, I first had the assurance of the dialysis unit that if she improved on the hemoperfusion program, the dialysis unit would accept her on the standard dialysis program. For the next 50 days, she underwent hemoperfusion procedures, each lasting for 2 h. During this 50 days, she received only one hemodialysis for the removal of water and electrolytes. After that, she continued for a total of eight months on hemoperfusion combined with hemodialysis. The schedule of treatment and response for the first 50 days of her 8-month treatment on hemoperfusion are summarized in Fig. 10.9.

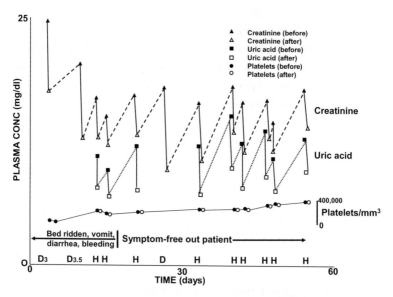

Fig. 10.9. First terminal renal failure patient on long-term hemoperfusion. Graph shows the results of the first 50 days of the 8-month treatment. (D — dialysis; H — hemoperfusion.)

With this regime, her uremic symptoms of nausea, vomiting, diarrhea, and hiccups cleared. Surprisingly, after two weeks on this regime, she felt well enough to be allowed by the nephrologists to leave the hospital and return home. During that time, she no longer had uremic symptoms and was able to do shopping and housework. At the beginning, she was asked to stay in the hospital for 24 h after each procedure. Later, she only came to the hospital 1 h before her hemoperfusion and then returned home 1 h after completion of the hemoperfusion. There were no side effects throughout all her hemoperfusions. There was little or no change in her blood pressure, and no other clinical side effects. This is contrary to her standard hemodialysis treatment with the EX 01 when she had hypotensive episodes with hemodialysis especially during fluid removal.

After the 50 days period, the dialysis unit admitted her to on a half time basis with one 6-h hemodialysis per week, and a 2 h hemoperfusion per week. During the 8-month period, she received a total of 24 hemoperfusion procedures and all were uneventful with no side effects. The only problem was fluid retention, as hemoperfusion does not remove fluid. After a total of 8 months on the hemoperfusion program, she was placed on a standard two weekly 6-h hemodialysis program for the next 10 years until she passed away at over 80 years of age. Independent inquiry by the nephrologists showed that her feeling of well-being was the same when she was on an average weekly regime of 2 h hemoperfusion plus 6-h hemodialysis comparable to her twice a week 6-h hemodialysis. Thus, 2 h on hemoperfusion seems to be as effective as 6 h of hemodialysis.

A 51-year-old female patient had a long history of hypertension and chronic renal failure caused by chronic pyelonephritis. Her urine output remained at about 500 ml/day with creatinine excretion of 0.4 gm/24 h. She was started on occasional peritoneal dialysis and then on 6 h twice weekly EX 01 hemodialysis while waiting for renal transplantation. She developed severe pruritis which disturbed her sleep at night and which was a nuisance during the day. Because of this, she was started on hemoperfusion to treat her pruritis. She continued on the 6-h twice weekly hemodialysis program with the addition of a 2 h weekly hemoperfusion. This treatment resulted in a marked decrease of her pruritis. In addition, her acceptance of the hemoperfusion procedure was such that she was placed on the long-term hemoperfusion regime of one 2 h hemoperfusion plus one 6-h EX 01 hemodialysis weekly. In addition to her diminished pruritis she found that with hemoperfusion she could come in 1 h before hemoperfusion, have a 2 h ACAC hemoperfusion and after the treatment, felt well enough to go directly to social evenings, dancing, or playing cards with her friends. On the other hand, with the EX 01 hemodialysis, she felt tired and had to rest at home in the evening. Furthermore, she experienced cramps in her legs during or soon after each 6-h EX 01 hemodialysis. She remained symptom-free and with no pruritis on the hemoperfusion and hemodialysis program for 4 months. After this, she returned to the standard hemodialysis program because one hemodialysis each week was not enough to control her body fluid.

A 53-year-old male with chronic renal failure was started on a 6 h twice weekly EX 01 hemodialysis. His daily urine output was 1 liter and creatinine excretion was 0.5 gm, respectively. He was started on a weekly 2 h hemoperfusion and 6-h EX 01 hemodialysis program. Being a very active business man, he accepted this program because it meant that on the day of treatment with hemoperfusion, he could work in the morning, have his lunch and come in for treatment, then return in the same afternoon, sometimes to work further in the evening. At the end of three weeks, he received a renal transplant.

A 21-year-old male was admitted to the hospital with symptoms of cramps, headaches, nausea, vomiting and diarrhea. His creatinine clearance was 1.6 ml/min and urinary output, 700 ml/day. This patient, with spina bifida at birth, was paraplegic with a neurogenic bladder, recurring urinary tract infection, and acute pyelonephritis. At that time, there was no opening for him in the chronic hemodialysis program and he was not considered suitable for renal transplantation. The patient was treated with 2 h of hemoperfusion weekly, supplemented by very occasional hemodialysis as needed for control of fluid and electrolytes. This treatment resulted in marked improvement of the patient's symptoms — except for occasional diarrhea. He was well enough to return to work. Three months later, an opening became available in the chronic hemodialysis unit and the patient was treated with an average of once a week 2 h hemoperfusion plus once a week 6 h hemodialysis. After a total of 6 months on hemoperfusion, the patient continued to be symptom-free. There was improvement in nerve conduction velocity. Each 2 h hemoperfusion lowered the serum middle molecule (300–1500 mol. wt.) level by 55%. He functioned well and was able to work and swim. The only problem was that one hemodialysis per week was not enough to prevent fluid retention. Thus, after 6 months on this program he was placed on the standard twice a week hemodialysis program.

Hemoperfusion and removal of uremic metabolites

Since then other artificial cell-based hemoperfusion devices have been produced industrially. Figure 10.10 shows that the better hemoperfusion devices are much more effective than the dialysers available at that time for the removal of uremic metabolites as shown by the clearance for creatinine, uric acid and "middle" molecules. Also, with the better hemoperfusion system like ACAC and Adsorba 300C, there was no significant effect on the platelet levels.

10.8.   Conjoint Hemoperfusion-Hemodialysis

The above results show that hemoperfusion is capable of removing toxin or unwanted waste metabolite and maintain patients symptom-free. On the other hand, the above clinical results also show that hemoperfusion, even with occasional hemodialysis, cannot deal with electrolytes or fluid retention.

It has been strongly proposed that "middle" molecular weight substances, including β-2-microglobulin (11,800 m.w.), may be responsible for the uremic symptoms. However, the hemodialysis membrane available at that time had a very low clearance for "middle" molecules and required up to 8 h for effective removal. On the other hand, the better hemoperfusion devices are much more efficient in removing "middle" molecules (Tables 2 and 7) (Chang, 1972e; Chang and Migchelsen, 1973; Chang and Lister, 1980; 1981).

Artificial Cells

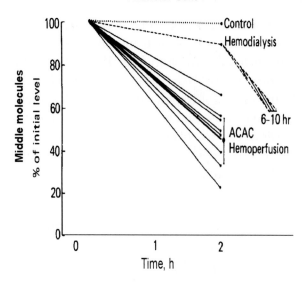

Fig. 10.10. Clinical studies in uremic patients show that middle molecules can be lowered rapidly after 2 h of ACAC hemoperfusion. On the other hand, hemodialysis machines available at that time was not effective.

Table 8. Characteristic of Artificial Cell-based Hemoperfusion Devices

	ACAC	Hemacol	Adsorba 300C
Hemoperfusion system	Albumin-collodion coated (300 g) (0.05–0.5 micron)	Acrylic hydrogel coated (300 g) (3–5 micron)	Cellulose acetate coated (300 g) (3–5 micron)
Max. Clearance (ml/min)			
Creatinine	235 (Q_B 300)	181 (Q_B 300)	185 (Q_B 200)
Uric acid	230 (Q_B 300)	116 (Q_B 300)	186 (Q_B 200)
middle molecules			
300–1500	144 (Q_B 300)	as in dialyzer	no data
1000–2000 (peak 7c)	no data	no data	120 (Q_B 300)
2000–5000 PTH (free)	93 (Q_B 300)	no data	no data

For the above reasons, clinical investigators around the world started to carry out clinical trials using hemoperfusion in series with hemodialysis for patients with dialysis resistant uremic symptoms and also to reduce the time needed for treatment. (Chang *et al.*, 1975;

Table 9. Effects of Hemoperfusion in Uremic Patients

Authors	Clinical Improvements in the following Symptoms
Chang *et al.*, 1971–1974	Pruritis, well-being, nausea, vomiting
Chang *et al.*, 1975	Peripheral neuropathy
Odaka *et al.*, 1978	Pericarditis, symptom free
Chang *et al.*, 1979	Hematocrit
Martin *et al.*, 1979	Pericarditis
Inou *et al.*, 1979	Peripheral neuropathy
Odaka *et al.*, 1980	Pericarditis
Agishi *et al.*, 1980	Peripheral neuropathy
Stefoni *et al.*, 1980	Peripheral neuropathy, pruritis, pericarditis, well-being

1982b; Martin *et al.*, 1979; Inou *et al.*, 1979; Odaka *et al.*, 1980; Agishi *et al.*, 1980; Stefoni *et al.*, 1980). The use of the better commercially prepared hemoperfusion devices in series with hemodialyzers resulted in clinical improvements in patients' well being, nerve conduction velocity, pruritis, pericarditis, peripheral neuropathy and what is also important, reduction in treatment time (Table 9) (Chang *et al.*, 1975; 1982b; Martin *et al.*, 1979; Inou *et al.*, 1979; Odaka *et al.*, 1980; Agishi *et al.*, 1980; Stefoni *et al.*, 1980). These clinical trial results led many centers to use the combined hemoperfusion-hemodialysis, especially in those patients with resistant uremic symptoms under the standard hemodialysis program.

Since then, the quality of hemodialysis membranes has improved so that the present high flux membranes are better than the standard dialysis membrane in removing the larger "middle" molecules. Even then, the clearance is still much less compared to the better hemoperfusion devices. Hemoperfusion devices are usually manufactured by manufacturers of hemodialysis machines and membrane. In those countries with strong hemodialysis companies, the hemoperfusion devices are extremely expensive. On the other hand, in these countries with no large dialysis industries, hemoperfusion devices are not expensive and therefore continued to be used in renal failure patients. In all the countries, hemoperfusion continues to be commonly used for uremic patients with aluminium or iron overload. This is based on the earlier clinical demonstration of the use of deferoxamine to bind these heavy metals and of hemoperfusion to remove the complex (Chang and Barre, 1983; Chang, 1986d; Chang *et al.*, 1984a, 1989a; Hakim *et al.*, 1985; Winchester, 1996).

10.9. Miniaturized Artificial Kidney Based on Hemoperfusion-Ultrafiltration

A third approach is to construct a truly miniaturized artificial kidney based on hemoperfusion in series with small ultrafiltrator (Chang *et al.*, 1975, 1979a, 1977b, Chang, 1976c) (Fig. 10.9). In clinical studies, a small Amicon ultratiltrator was used (Fig. 10.11). Hydrostatic pressure from the blood pump alone gave effective ultrafiltration. Dialysis fluid is not required, the ultrafiltrate flowing directly into a measuring cylinder.

This ultrafiltrator in series with ACAC hemoperfusion was studied in two patients, one for three and the other for six months. The patient, treated for 6 months, had a creatinine clearance of 0.2 mL/min

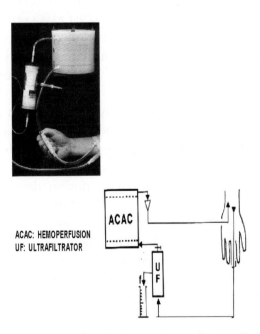

ACAC: HEMOPERFUSION
UF: ULTRAFILTRATOR

Fig. 10.11. ACAC hemoperfusion in series with a small ultrafiltrator.

and a urine volume of 50 ml/day. He was followed for a 6-month control period of thrice weekly hemodialysis treatments. This was followed by a 6-month test period when a 2.5 h hemoperfusion-ultrafiltration replaced one of the 6 h three weekly hemodialysis treatments (Fig. 10.12).

The patient felt well throughout. Hematocrit increased slightly from 25.15% ± 3.61 to 26.6% ± 1.73. Fluid retentions diminished as body weight fell from 82.6 ±1.2 to the optimal dry weight of 78.0 ± 0.8 kg and hypertension improved. More fluid could be removed without side effects in 2.5 h of hemoperfusion-ultrafiltration compared to 6 h of hemodialysis. "Middle" molecule clearance in this patient was 113 mL/min at a blood flow rate (QB) of 200 ml/min and 160 mL/min at

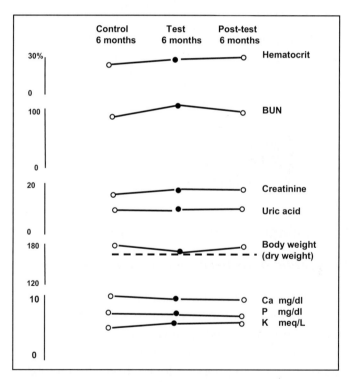

Fig. 10.12. Laboratory data of patient. 6-month control period followed by 6-month test period. During the 6-month test period, one of the weekly 6 h hemodialysis treatment was replaced by 2.5 h of treatment using the miniaturized hemoperfusion-ultrafiltration device.

Table 10. Patient on Hemoperfusion-Ultrafiltraion

Clearance (ml/min) (Blood Flow: 300 ml/min)	Hemoperfusion	Hemoperufsion-Ultrafiltration
Creatinine	230	235
Uric acid	235	235
300–1500 MW	120	134
PTH free	73	81
Fluid removal/2 h	0	2500–2700 ml
NaCl removal/2 h	0	17.0–18.4 g

creatinine clearance: 0.2ml/min; urine output: 50 ml/24 h

a QB of 300. This was significantly higher than for hemodialysis. Nerve conduction velocity did not change. Hemoperfusion-ultrafiltration was more effective than hemodialysis in removing "middle" molecules, creatinine, uric acid, sodium chloride and water (Table 10). On the other hand, it did not remove sufficient potassium and urea since after 6 months on this regime, there was an increase in mean blood urea nitrogen from 86 ± 14 mg/dl to 105 ± 18 mg/dl and potassium from 5.2 ± 0.57 mEq/l to 5.8 ± 0.08 mEq/l. Oral adsorbents can remove potassium and phosphate. An urea removal system is being developed to complete this miniaturized artificial kidney system. For this, artificial cells containing multi-enzyme systems and artificial cells containing urea removing microorganisms are being investigated.

10.10.   Hemoperfusion in Liver Failure

First observation of recovery of consciousness in hepatic coma

As shown in acute poisoning patients in the above section, adsorbent artificial cell hemoperfusion is effective in removing suicidal and accidental overdoses of medications in patients. It is also effective in removing other toxic substances from the circulating blood. Its effectiveness in removing hepatic toxins from the circulating blood was first shown in a grade IV hepatic coma patient, with recovery of consciousness within an hour after the initiation of hemoperfusion (Chang, 1972b).

A 50-year-old female was admitted with a history of alcohol abuse, onset of jaundice, fatigue, nausea, vomiting, and dark urine. There was no history of contact with hepatitis or of intravenous or intramuscular medication. On admission, she had spider nevi and ascites. The diagnosis was acute alcoholic hepatitis. Her condition deteriorated after admission and she became comatose and unresponsive. After remaining comatose for two days, her condition was considered as terminal and with the insistence of her relatives she was referred by her physician to me for possible hemoperfusion since nothing else could be done. One hour after hemoperfusion, she started to regain consciousness and began to recognize her relative and answer questions in sentences. Hemoperfusion was carried out for a total of 80 min. She remained conscious for about an hour after the end of the hemoperfusion, but lapsed into coma again. Three days later she was still comatose, and a second hemoperfusion was initiated. At the start of perfusion she was comatose; an hour later she looked at people when spoken to and there were increased voluntary movements and response to pain. She did not, however, recover full consciousness as in the first hemoperfusion. Shortly after this, a liver biopsy specimen showed cirrhosis, acute hepatitis, with small foci of regeneration. Thereafter, consciousness fluctuated between stupor and coma. A third hemoperfusion was carried out. Before this hemoperfusion, E.E.G. background activity was irregular and diffuse, with continuous theta and delta activity and occasional high-amplitude single slow-wave and rare complex biphasic or triphasic configurations. At this time, the patient was semi-comatose and did not respond to questions. An hour and a half after hemoperfusion the patient's consciousness improved and she started to answer questions; she also complained of thirst and heaviness in the leg. E.E.G. recording after hemoperfusion showed a minimum improvement in the background activity.

Biochemical evaluation of the treatment of hepatic coma was difficult, because the precise pathological mechanism was not known. On the other band, the three hemoperfusions each produced a clinical response in consciousness. In one case, when the E.E.G. was recorded there was a slight improvement in background activity. Since there was no change in the blood-ammonia level after hemoperfusion, we

Table 11. Hepatic Coma-related Chemicals Removed by Hemoperfusion

Aromatic branched chain amino acids
Fatty acids — oleic, hexanoic,
    octanoic N-valeric
Inhibitor of Na-K ATPase
Inhibitor of hepatic regeneration
Mercaptan
Middle molecules
Phenols
Protein bound molecules
Others

could not attribute the improvement in consciousness to the removal of blood-ammonia. However, whereas standard hemodialyzers at that time were not efficient in removing large molecules from blood, hemoperfusion was very efficient in removing molecules up to 5000 molecular weight (Chang, 1972e; Chang and Migchelsen, 1973; Chang and Lister, 1980; 1981) as well as protein bound molecules. Hemoperfusion is also effective in removing other chemicals related to hepatic coma (Table 11).

Results around the world on effect of hemoperfusion on hepatic coma

The above results led to further studies around the world in a large number of patients (Gazzard *et al.*, 1974; Chang, 1976c; 1982a; Blume *et al.*, 1976; Bartels *et al.*, 1977; 1981; Silk and Williams, 1978; Gelfand *et al.*, 1978; Gimson *et al.*, 1978; Odaka *et al.*, 1978; Amano *et al.*, 1978; Maeda *et al.* 1980; Cordoatri *et al.*, 1982; William, 1983; O'Grady *et al.*, 1988). These results support the finding that a significant number of hepatic coma patient recovered consciousness. However, this did not translate into improved survival rates. Table 12 is a summary of those reports that included the number of patients recovering from consciousness in addition to survival rates.

However, these clinical results are not conclusive since the pathogenesis of hepatic coma and survival rate is complicated. In addition to hepatic toxins, other factors like brain edema and

Table 12. Effect of Hemoperfusion on Fulminant Hepatic Coma and Survival Rate

Reference	No. of Patients	Consciousness Recovered	Survival
Chang (1972b)	1	recovered with each of 3 hemoperfusions	Until HP discontinued
Gazzard *et al.* (1974)	31	48%	39%
Chang (1976c)	6	66%	16%
Gelfand *et al.* (1978)	10	90%	40%
Gimson *et al.* (1978)	31 (Grade III)	68%	65%
	45 (Grade IV)	22%	20%
Cordopatri *et al.* (1982)	2	100%	100%

standard of care also play important roles (William, 1983; O'Grady *et al.*, 1988). Furthermore, survival rates are related to age, etiology, grade of coma, and other complications, especially brain edema. In addition, hemoperfusion can only remove toxins from the body but the liver is a very complex organ with many other metabolic and synthetic functions. We should not expect hemoperfusion alone to be a complete artificial liver. Hemoperfusion will more likely be an important component of a more complex system (Chang, 1983a, 1983e, 1986c, 1992b).

Control studies in galactosamine-induced hepatic failure rats

We, therefore, carried out detailed control studies in galactosamine-induced hepatic failure rats to allow for a more valid statistical analysis (Chang *et al.*, 1978; Chang, 1971a, 1978b; Chirito *et al.*, 1978, 1979; Mohsini *et al.*, 1980; Tabata and Chang, 1980). With this model we can avoid variations in age, etiology and grades of coma. However, this model may not be the counterpart of fulminant hepatic failure in human. In addition, the survival rate in this model of grade III coma is more like that of grade IV coma in human. This research included detailed studies in regard to the time of initiation of treatment; the need for essential factors; comparison of hemoperfusion with exchange transfusion, liver perfusion, and hormones; and other approaches,

Male Wistar rats weighing 275–309 gm and 47–67 days old were used. Forty-eight hours after galactosamine injection (1.1 gm/kg), the rats in grade III coma were used in this study. Grade III coma rats were those which "sleep most of the time, but arousable." Of the 356 rats, 122 reached grade III coma 48 h after injection with galactosamine. To avoid variations between batches, all the animals in grade III coma in each batch were randomly divided into equal numbers of test and control animals. Similar studies were also carried out in rats in grade II coma. The results of all the studies (Chang *et al.*, 1978; Mohsini *et al.*, 1980; Tabata and Chang, 1980) are very briefly summarized in Tables 13 and 14.

In grade II coma rats, either one hemoperfusion or one liver perfusion resulted in significant increase in survival time and survival rates as compared to the control group (Table 1). However, cross circulation did not have any significant effects.

In grade III coma rats, carrying out one hemoperfusion, liver perfusion or blood exchange transfusion alone did not result in any significant increase in survival time or survival rates. On the other hand, the combined use of one hemoperfusion and one blood exchange transfusion resulted in significant increase in survival time but not survival rate. Two hemoperfusions also resulted in significant increase in survival time but not survival rates (Table 2). Another approach was carried out with two hemoperfusions after one blood exchange, with the second hemoperfusion carried out 4 h after the first. When this

Table 13. Survival Time and Rates and Grades of Coma in Rats Treated by Hemoperfusion (ACAC) Alone or In Combination

Coma	Treatment	Survival Time	Survival Rate
II	ACAC hemoperfusion	↑	↑
II	Liver perfusion	↑	↑
II	Cross-circulation	NS	NS
III	ACAC hemoperfusion	↑	NS
III	Liver perfusion	NS	NS
III	Blood exchange	NS	NS
III	Blood exchange + ACAC	↑	NS
III	ACAC + glucagon + insulin	NS	NS
III	Blood exchange + 2ACAC	↑	↑

Table 14. Galactosamine-induced Hepatic Failure Rats Treated by Hemo-perfusion (ACAC) Alone or In Combination

Treatment (No. of Animals)	Survival Time (hrs)	Significance	Survival Rate	Significance (%)
1. Control (8)	53.90 ± 0.75	$p < 0.05$	0	N.S.
ACAC (8)	58.96 ± 1.97		0	
2. Control (9)	52.85 ± 1.03	N.S.	0	N.S.
ACAC & Insulin, Glucagon (9)	52.66 ± 1.96		0	
3. Control (9)	56.20 ± 2.44	$p < 0.05$	11.1	N.S.
2 ACAC (9)	63.42 ± 2.16		22.2	
4. Control (8)	53.90 ± 0.75	N.S.	0	N.S.
X-Blood (8)	55.38 ± 2.54		12.5	
5. Control (15)	59.38 ± 1.60	$p < 0.001$	0	N.S.
ACAC & X-Blood (15)	72.80 ± 2.85		20.0	
6. Control (11)	54.86 ± 2.43	$p < 0.01$	0	$p < 0.05$
2 ACAC & X-Blood (11)	65.48 ± 2.08		36.4	

regime was used, both the survival time and the survival rate of grade III coma rats were significantly increased ($p < 0.01$ and $p < 0.05$, respectively) (Table 14).

Our research in fulminant hepatic rat models shows that hemoperfusion started in the earlier grades of coma can result in significant increase in both survival time and survival rates (Chang *et al.*, 1978; Chirito *et al.*, 1979). This has been supported by the result of another group that treated patients in grade III coma rather than in grade IV coma as shown in Table 12 (Gimson *et al.*, 1978). The use of galactosamine rats for hemoperfusion has also been carried out by another group (Niu *et al.*, 1978). Their results corroborated our findings.

Our laboratory research also shows that hemoperfusion alone increases the survival time but not the survival rates in grade III hepatic coma rats. This may correspond to the transient recovery of consciousness in grade IV hepatic coma patients. Fresh blood exchange transfusion alone did not improve the survival time or survival rate of FHF grade III coma rats. Two ACAC hemoperfusion combined with fresh blood exchange transfusion significantly improved the survival time of FHF grade III coma rats, but not the survival rates. An additional hemoperfusion after the first hemoperfusion combined with blood exchange transfusion improved both the survival time and survival rates in grade III coma rats. This seems to suggest that hemoperfusion could be a part of a more complete liver support system. Thus, studies are ongoing in using hemoperfusion to remove hepatic toxins to allow the liver to regenerate. One line of research is to combine this with the use of hepatocytes perfusion system (Rozga *et al.*, 1993; Sussman *et al.*, 1994). Another approach is to combine this with artificial cells containing hepatocytes or stem cells as discussed in the previous chapter.

10.11.   Immunoadsorption

Protein coated artificial cells in immunoadsorption

Albumin can bind tightly to the ultrathin collodion membrane of adsorbent artificial cells (Chang, 1969a). This was initially used to increase the blood compatibility of the adsorbent artificial cells for hemoperfusion (Chang, 1969a). We also applied this albumin coating to synthetic immunosorbents resulting in blood compatible synthetic blood group immunosorbents (Chang, 1980d). This albumin-coated synthetic adsorbent has been applied clinically for removing blood group antibodies from plasma for bone marrow transplantation (Bensinger *et al.*, 1981). In addition, albumin-coated collodion activated charcoal (ACAC) was found to effectively remove antibodies to albumin in animal studies (Terman *et al.*, 1977). This has become the basis of one line of research in which other types of antigens or antibodies are applied to the collodion coating of the activated charcoal to form immunosorbents. Other immonosorbents based on this principle have also been developed for the treatment of human systemic lupus erythrematosus, removal of antiHLA antibodies in transplant candidates, treatment of familial hypercholesterolemia with monoclonal antibodies to low-density lipoproteins and other conditions (Terman, 1980; Terman *et al.*, 1979a; 1979b; Hakim *et al.*, 1990; Wingard *et al.*, 1991; Yang *et al.*, 2004)

10.12. Detailed Procedures for the Laboratory Preparation of ACAC Hemoperfusion Device

The extracorporeal shunt chamber

The laboratory version of the extracorporeal shunt chamber (Fig. 10.1) is prepared from high density polypropylene with internal dimensions of 10-cm diameter and 8-cm height. A 10-cm diameter steel mesh screen with 100-micron openings placed on either side of the shunt chamber retains the microcapsules but allows the free flow of blood. Blood from patients enters the chamber at a right angle to the screen and is thus distributed evenly through the 10-cm diameter screen mesh before coming into direct contact with the microcapsules. The extracorporeal shunt chambers and the screens are re-usable, but careful washing and sterilization are required. Thus, after each clinical hemoperfusion, the microcapsules are discarded from the shunt chambers, and the chambers and screens are washed clean. Both the new and reused chambers and screens are completely submerged in a 1 vol% sodium hypochlorite for 14 h. After this, the shunt chambers and screens are rinsed for 30 min with distilled water. They are allowed to dry, then coated with silicone spray (Manostat Co.) and stored in the sterilization room. On the day of the preparation of the microcapsule artificial kidney, the shunt chambers and screens are placed in a 1 vol% sodium hypochlorite solution for 4 h and then rinsed for 30 min with distilled water, and finally rinsed with sterile pyrogen-free water and then used immediately.

Preparation of activated charcoal

Medical-grade spherical charcoal is available for the preparation of hemoperfusion devices. If this is not available, the laboratory experiment can also be carried out using crude activated charcoal granules (Fisher Scientific Co., Coconut activated charcoal, 6 to

14 mesh, #5-685). However, this would involve more preparations as follows. Sieves of 10 mesh (2 mm openings) and 20 mesh (0.85 mm openings) are used to divide into three diameter ranges: 1) those retained by the 10-mesh sieve, 2) those retained by the 20-mesh sieve, and 3) those passing through the 20-mesh sieve to be discarded. Each 400 g portion of sieved charcoal is placed in a 3-liter beaker and covered with a 20-mesh sieve. Tap water is run through the sieve into the container at a high speed for 2 h. This is sufficient to stir up the charcoal particles to wash away soluble material or fine powder. The washed activated charcoal granules are then placed in a 1-liter beaker and autoclaved for 1 h at 121°C and 15 psi. In all the autoclaving, only autoclaves with internally generated steam (e.g. Castle-Fisher Scientific Co.) are used, and pyrogen-free sterile water is used to generate the steam. After autoclaving, a second washing is carried out for 2 h as before to remove any soluble material or fine powder which might result from the autoclaving. Each portion of the charcoal is then wrapped in clean cloth and secured with an autoclave tape. It is then autoclaved for 30 min at 121°C and 15 psi, then placed in a well-ventilated 50°C oven for at least 48 h to dry completely. Each wrapped portion is then kept in the preparation room until ready for use. The microencapsulation of the charcoal by coating with cellulose nitrate and albumin is carried out as described in the detailed and updated version below.

Preparation of artificial cells containing activated charcoal

A polymer solution is made up by dissolving 20 ml of collodion (USP 4 g cellulose nitrate in 100 ml of 1 alcohol:3 ether) in an organic solvent consisting of 30 ml of alcohol and 400 ml of ether. 450 ml of polymer solution is poured into a beaker containing 100 g of mesh-10 and 300 g of mesh-20 activated charcoal granules previously prepared and sterilized as described above. The suspension is stirred manually with a metal stirring rod to ensure that all the granules are immersed in the solution. The slightly wet polymer-coated activated charcoal is spread out in a 2 × 1 foot tray. This is then placed in a ventilated oven for 5 h at 50°C. After this, the polymer-coated activated charcoal granules are placed in 2 liters of pyrogen-free distilled water. The suspension is carefully transferred into a 20-mesh sieve (20-cm in diameter and 5-cm in depth) and washed carefully with pyrogen-free water until all fine particles are removed. Care should be taken not to be too vigorous at this step to avoid breakage. The silicone-coated high density polypropylene chamber prepared as described above is filled with pyrogen-free distilled water, and the polymer-coated granules are added to fill the chamber. The chamber is closed, drained, stoppered, wrapped, sealed and placed in an autoclave for 30 min at 121°C and 15 psi. At the end of this, the shunt is allowed to cool in its sealed, sterilized wrapping, if possible in a refrigerator. Using sterile techniques, the chamber is perfused with 4 liters of cold pyrogen-free sterile saline (Baxter Laboratories, for human use) at about 200 ml/min by gravity. Albumin coating is carried out when the temperatures of the efferent washing from the chamber is less than 10°C. A 1g/dl albumin solution in sterilized saline is prepared from human albumin (Cutter Co.). 200 ml of this albumin solution is used to displace the same volume of saline from the chamber containing the polymer-coated activated charcoal. The chamber is then sealed by sterile techniques and kept at 4°C for 15 h for the albumin to coat the polymer membrane. No autoclaving is carried out after albumin coating. After preparation, the microcapsule shunts are kept in sterile wrapping in the refrigerator at 4°C. In the present study, they can be used after being kept in the refrigerator for up to two months. Just before use, the albumin solution is displaced with 4 liters of saline (Baxter Laboratories, for human infusion) without allowing the albumin-polymer-coated activated charcoal to come into contact with air at any time. Rabbit pyrogen testing is carried out using the displaced saline. In the five years of following the above procedure of washing and sterilization, there have been no positive rabbit pyrogen tests nor have there been any pyrogenic reactions in patients.

Procedure for hemoperfusion

Just before use in patients, the same general procedure of saline washing of standard hemodialyzers is carried out. Thus, the shunt chamber is perfused with about 3 liters of saline, the last liter containing 2000 U of heparin. The same amount of heparin that the patient normally receives at the start of standard hemodialysis treatment is given. The remaining heparin is best given as a slow infusion of 2500 to 3000 U/h into the afferent blood line in the shunt chamber. To start the hemoperfusion, the shunt chamber saline (about 200 ml) is displaced by the patient's blood before the outlet tubing is connected to the patient. However, in the treatment of patients with severe drug intoxication or hepatic coma, where there is severe hypotension, it is advisable to prime the ACAC system with 7 g/100ml albumin in saline, and to start the hemoperfusion without displacing the albumin solution. The clotting time of blood is maintained at about 50 min. Two hours after the initiation of hemoperfusion, the procedure is terminated when another 700 U of heparin is injected into the shunt chamber and the blood in the chamber is returned to the patient by air displacement, followed by washing with 100 ml of saline containing 100 U of heparin. In the present series of studies, 300 g of ACAC microcapsules are used for 2 h in each hemoperfusion.

CHAPTER 11

Perspectives on the Future of Artificial Cells as Suggested by Past Research

11.1. Introduction

In the last 50 years, there have been numerous extensions and developments of the idea of artificial cells around the world. However, as shown by the examples in the last 10 chapters, we have barely touched the surface of the full potential of this idea. Many other extensions and variations in the membrane material, the configurations and the contents are possible. The current international interest in the transfer of technology for industrial development will help accelerate these developments. To do this, there is a need for patents that are based on novel ideas. A characteristic of such developments is that a new development or extension of "artificial cells" tends to be hidden under numerous new names. Some of these are nanoparticle, nanotubule, lipid vesicle, liposome, polymer tethered lipid, polymersome, microcapsule, bioencapsulation, nanocapule, nanosensor, macroenapsulation, polyhemoglobin, conjugated hemoglobin, etc. The result is a fragmentation of the field of artificial cells into different subdivisions, subdisciplines as well as scientific societies that do not interact with one another. Now is the time for researchers from the different areas in this interdisciplinary field to come together to move the field forward. This monograph is an attempt to bring some of these areas together. As the whole field is now so broad and extensive, that it is not possible to cover everything in one monograph. This final chapter can only touch on some of the other extensions and future developments of the idea of artificial cells that have not been discussed in the previous chapters. There will be more books published on the subject under this book series.

11.2. Membrane Material

At one time, biological cell membrane was thought to be a simple bilayer lipid membrane. However, extensive research by many investigators is beginning to show that it is much more complicated. Indeed, there are indications that protein components play an important structural and functional role (Fig. 11.1). Biological cell membrane is complex and there is still much basic research to be done. Artificial cell membrane is much more simple minded. However, since it can be formed using different types of synthetic or biological material, there are great potential for variations and extensions. The following are some examples (Fig. 11.1): 1) polymer and crosslinked protein membranes; 2) lipid membranes; 3) lipid-polymer complex membranes; 4) albumin, antitgen, antibody or protein complexed membranes; 5) lipid-polymer complexed membrane with cyclic carrier; 6) lipid-polymer complexed membrane with Na-K ATPase and other transport carriers; 7) lipid membrane with transport carriers; and 8) others.

11.3. Polymeric Membrane

Porosity of polymeric membrane of artificial cells

The first artificial cells prepared have ultrathin membranes of cellulose nitrate, polyamide, silastic, polystyrene or crosslinked protein (Chang, 1957 to 1972). A silastic membrane is only permeable to gases like oxygen and carbon dioxide and lipophilic molecules. Those with cellulose nitrate or polyamide membranes are permeable to hydrophilic molecules. For instance, experimental analysis shows that a polyamide membrane has an equivalent pore radius of 18–19 Å (Chang, 1965,1972a) (Fig. 11.2). This means that polyamide membranes and cellulose nitrate membranes would exclude protein molecules but allow polypeptides and other smaller molecules to pass through rapidly.

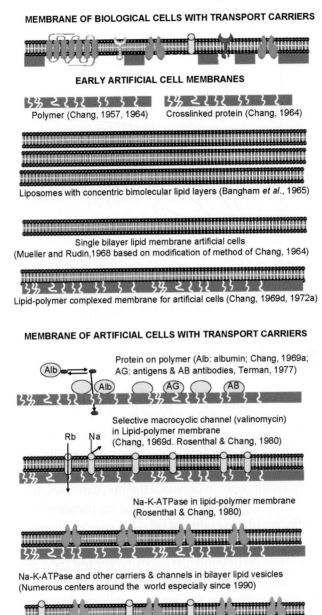

MEMBRANE OF BIOLOGICAL CELLS WITH TRANSPORT CARRIERS

EARLY ARTIFICIAL CELL MEMBRANES

Polymer (Chang, 1957, 1964)   Crosslinked protein (Chang, 1964)

Liposomes with concentric bimolecular lipid layers (Bangham *et al.*, 1965)

Single bilayer lipid membrane artificial cells
(Mueller and Rudin,1968 based on modification of method of Chang, 1964)

Lipid-polymer complexed membrane for artificial cells (Chang, 1969d, 1972a)

MEMBRANE OF ARTIFICIAL CELLS WITH TRANSPORT CARRIERS

Protein on polymer (Alb: albumin; Chang, 1969a;
AG: antigens & AB antibodies, Terman, 1977)

Selective macrocyclic channel (valinomycin)
in Lipid-polymer membrane
(Chang, 1969d. Rosenthal & Chang, 1980)

Na-K-ATPase in lipid-polymer membrane
(Rosenthal & Chang, 1980)

Na-K-ATPase and other carriers & channels in bilayer lipid vesicles
(Numerous centers around the world especially since 1990)

Fig. 11.1. Schematic representation of biological membrane (*top*). Examples of artificial cell membranes: polymer and cross-linked protein membrane; lipid membranes; lipid-polymer membrane; membrane with albumin, antigen, antibody or other proteins; lipid-polymer membrane with cyclic carrier; lipid-polymer membrane with Na-K-ATPase transport carriers; lipid membrane with transport carriers.

Artificial Cells

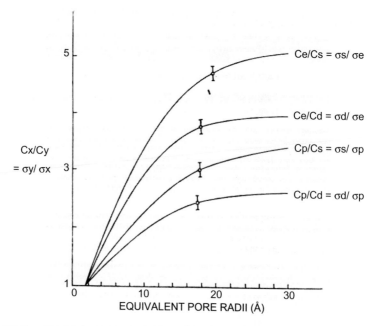

Fig. 11.2. Equivalent pore radius of polyamide membrane artificial cells obtained experimentally, as shown in above graph, is 18–19 Å.

Polymeric membranes with unlimited variations in porosity and molecular weight cut off are available (Fig. 11.3). At the one, polymeric membranes like silastic or polyethylene membrane are not porous and only allow gases and lipophilic molecules to pass through. On the other, macroporous polymeric membranes allow very large protein molecules to pass through. In between the two extremes, different types of polymeric artificial cell membranes can be prepared with different porosity and molecular weight cut off. This would allow for variations in porosity and permeability characteristics to suit different conditions. In addition, we can go beyond just porosity. For instance, one can even fulfill the requirements for those applications that require the removal of large lipophilic molecules but not the smaller hydrophilic molecules or moisture. For instance, adsorbent artificial cells are normally formed by the direct coating of porous ultrathin membranes onto active charocoal microspheres to remove hydrophilic molecules (Chang, 1969a). However, coating with a polyethylene membrane results in

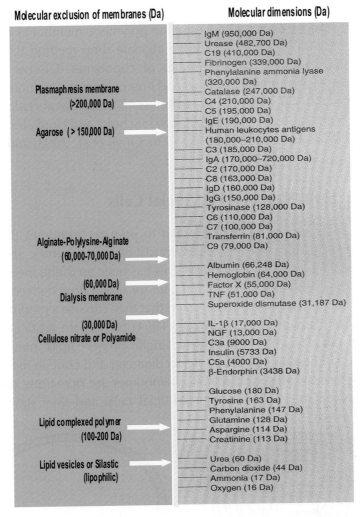

Fig. 11.3. Possible variations in porosity of the polymeric membranes of artificial cells.

adsorbent artificial cells that can exclude small hydrophilic molecules (e.g. creatinine) and rapidly remove larger lipophilic molecules (e.g. vitamin B) (Sipehia, Bannard and Chang, 1986).

The methods of preparing polymeric membrane artificial cell include: solvent evaporation with emulsion or drop (Chang, 1957, 1964); interfacial polymerization with emulsion or drop (Chang,

1964, 1965, 1972a); solvent evaporation with double emulsion (Chang, 1965, 1972a, 1976a); ultrathin polymeric membrane coating (Chang, 1969a); alginate-polylysine-alginate drop method (Lim and Sun, 1980); self-directed assemblies of copolymer (Discher *et al.*, 1999); and many other extensions and improvements (Calafiore *et al.*, 1999; De Vos *et al.*, 2002; Duvivier-Kali *et al.*, 2001; Sakai *et al.*, 2001; Cruise *et al.*, 1999; Hunkeler *et al.*, 2001; Schuldt and Hunkeler, 2000; Uludag *et al.*, 2000; Dionne *et al.*, 1996; Orive *et al.*, 2003; and many others).

11.4. Lipid Membrane Artificial Cells

Bangham *et al.* (1965) were the first to prepare liposomes that are liquid crystal microspheres, each consisting of concentric shells of bimolecular lipid layers (Fig. 11.3). This has been an useful membrane model for membrane research and later for use in drug delivery.

Mueller and Rudin (1968) reported that they can use a modification of the standard procedure (Chang, 1964) to prepare artificial cells having only 60–100 Å thickness single bilayer lipid membrane (Fig. 11.4). These are of about 90 μ in diameter and contain red blood cell hemolysate. The bilayer lipid membranes are not strong enough to support these 90 μ-diameter artificial cells and they stay intact

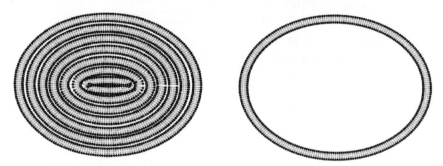

Fig. 11.4. *Left*: Liposomes that are liquid crystal microspheres with concentric shells of bimolecular lipid layers (Bangham *et al.*, 1965). *Right*: bilayer lipid membrane artificial cells containing hemoglobin of Mueller and Rudin (1968) who wrote that they used a modification of the method of Chang (1964) to prepare this.

for only a short time. As described below, one solution is to tether the lipid membrane onto the ultrathin polymeric membrane of the artificial cell to form lipid-polymer membrane artificial cells (Chang, 1969d, 1972a). Other workers (Deamer and Bangham, 1976) used a modification of the "ether evaporation method" (Chang 1957, 1964) to form smaller single bilayer (unilamellar) lipid membrane artificial cells. They call these lipid membrane artificial cells, unilamellar liposomes or lipid vesicles. As will be discussed later, this forms the basis of the types of lipid vesicle being currently used in drug delivery systems (Torcillin, 2005).

11.5. Artificial Cells with Lipid-Polymer Membrane and Incorporation of Macrocyclic Carrier, NA-K-ATPase and Other Carriers

The exact molecular organization of biological cell membranes is not known. However, it is known that biological membranes are not composed of lipids alone and there is increasing evidence that the protein components play an important structural and functional role. Korn (1968) even suggested that in the biosynthesis of biological cell membranes the protein unit of the membrane is formed first, followed by the addition of the lipid components.

There are three types of lipid-polymer membrane artificial cells first reported (Fig. 11.5): 1) lipid-polymer membrane with ultrathin lipid component tethered to the polymeric membrane of artificial cells (Chang, 1969d, 1972a); 2) addition of macrocyclic carrier (valinomycin) to the lipid-polymer artificial cell membrane (Rosenthal and Chang, 1971, 1980); 3) addition of transport carrier, Na-K-ATPase, to the polymer artificial cell membrane (Rosenthal and Chang, 1980). A number of ongoing studies by different groups will also be discussed below.

Artificial cells with lipid-polymer membrane

As described in detail in Appendix II, we complex an equimolar lecithin-cholesterol mixture to the artificial cells membrane (polymeric

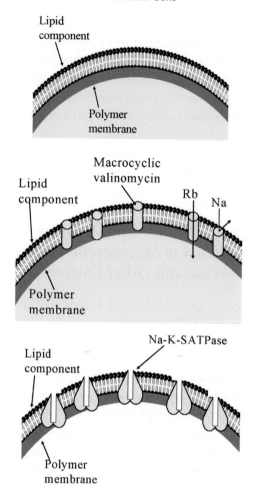

Fig. 11.5. *Upper*: lipid-polymer membrane artificial cells (Chang, 1969d, 1972a); *Middle*: addition of macrocyclic carrier (valinomycin) to the lipid-polymer membrane of artificial cells (Rosenthal and Chang, 1971, 1980), (this basic principle is now being used extensively in nanobiosensors); *Lower*: addition of transport carrier, Na-K-ATPase (Rosenthal and Chang 1980).

or crosslinked protein) (Fig. 11.5) (Chang, 1969d, 1972a). For artificial cells without lipid on the membrane, sodium influx is very high across the porous polymeric membrane, 6×10^{-6} mM/h-mm^2 (Fig. 11.6). Complexing the polymeric membrane with an

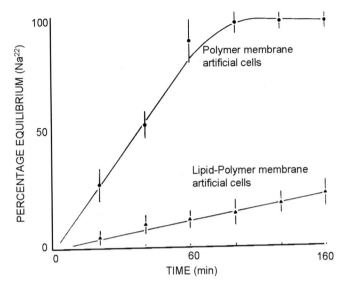

Fig. 11.6. Equilibration of Na$^+$ across large polymeric membrane artificial cells as compared to that across large lipid-polymer membrane artificial cells (Chang, 1969d, 1972a).

equimolar lecithin-cholesterol mixture reduces the sodium influx to 4×10^{-6} mM/h-mm^2 (Fig. 11.6). Surprisingly, this sodium flux across the lipid-polymer membrane artificial cells is of the same order of magnitude as that observed by Stein for most natural membranes (Stein, 1967). On the other hand, a study by another group (Pagano and Thompson 1968) showed that across bilayer lipid membranes, the sodium influx is of orders of magnitude smaller than that of biological cell membranes. Thus, lipid-polymer membranes function more like biological membranes as compared to lipid bilayers alone.

Incorporation of channels into lipid-polymer membrane of artificial cells

However, lipid-polymer membrane artificial cells by themselves do not show any selectivity between sodium and rubidium (Fig. 11.7a, 11.7b). We incorporated valinomycin in the lipid-polymer membrane by exposing them to valinomycin at a concentration of 5×10^{-6} M.

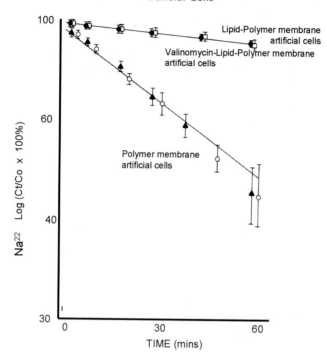

Fig. 11.7a. The effect of lipid and valinomycin on the time-course of decrease in Na$^+$ concentrations in the aqueous suspending medium. Co — original concentration; Ct — concentration at time (t). Mean and S.D. based on 4–12 experiments.

This results in selectivity so that the rubidium influx is enhanced but the sodium influx remains unchanged (Fig. 11.7a, 11.7b) (Rosenthal and Chang, 1971, 1980). Thus, the permeability coefficients of lipid-polymer membrane artificial cells to rubidium is 1.45 ± SD 0.56 × 10^{-6} cm/sec. Valinomycin at a concentration of 5 × 10^{-6} M significantly increases this to 4.91 ± 1.6 × 10^{-6} cm/sec. However, valinomycin has no effect on the sodium influx and it remains unchanged at 3.8 ± 0.3 × 10^{-6} cm/sec.

This simple study only shows the potential of the incorporation of channel forming material in the membrane of artificial cells. However, this is just the beginning and much can be learned from the ongoing research of molecular biologists. For instance, ion conducting channels in cell membranes are complex protein structures, but

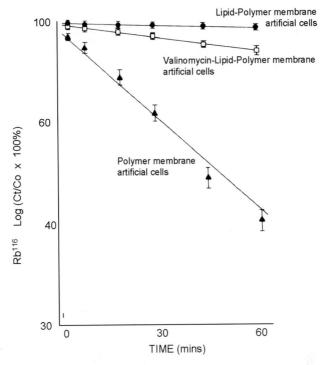

Fig. 11.7b. The effect of lipid and valinomycin on the time-course of decrease in Rb^+ concentrations in the aqueous suspending medium. Co — original concentration; Ct — concentration at time (t). Mean and S.D. based on 4–12 experiments.

Bayley (1999) showed that only a bundle of α-helices or nanotubular β-strands is needed to carry out the function. They can retain their function even when most of the protein portion is removed from the complex protein structure. This means that researchers can design synthetic channels or redesign natural channels by changing the amino acid sequence and helix orientation. One of the simplest form is a neutral peptide, Ac-(Leu-Ser-Ser-Leu-Leu-Ser-Leu)$_3$CONH$_2$, that can function as a voltage-gated channel. Cyclic peptides can form hollow cylindrical nanotubules with possible variations in the lumen of the channel to allow for transport of different types of solutes based on size (Hartgerink J. D. *et al.*, 1998). In the case of lipid membrane artificial cells, lipid vesicles, exposure to α-hemolysin or its genetically engineered mutants can increase the permeability to large single stranded nucleic acid (Bayley,1997). This is an example of how progress in molecular biology can be applied to artificial cells. The large amount of work being carried out by molecular biologists can be tapped for formation on different types of channels in artificial cell membranes.

Incorporation of Na-K-ATPase in the membrane of artificial cells

The feasibility of incorporating Na-K-ATPase in the membrane of artificial cells was studied (Rosenthal and Chang, 1980). The details of the method can be found in Appendix II. Briefly, large polyamide membrane artificial cells are first formed. Then Na-K-ATPase is extracted from red blood cell membranes and covalently crosslinked to the artificial cell membranes. The ouabain-sensitive ATPase activity per sq. cm of the surface area of the artificial cells is $90–250 \times 10^{-5} \mu$moles Pi/h/cm^2 (Table 11.1). This is 1–2 orders of magnitude greater than the ouabain-sensitive ATPase activity of human red blood cell membranes of $6 \times 10^{-5} \mu$moles Pi/h/cm^2 (obtained by Dunham and Glynn, 1961).

Thus, it is possible to incorporate in artificial cell membranes large amounts of Na-K-ATPase normally found associated with biological membranes. Furthermore, the incorporated enzyme retains its enzyme activity. However, this is only a feasibility study and much remains to be done regarding the orientation, transport characteristics, and association with lipids and other membrane components. The need for ATP recycling in artificial cell is now possible (Campbell and Chang, 1975) and this will be discussed in a later section in this chapter. Again,

Table 11.1 Incorporation of Na-K-ATPase into Artificial Cell Membranes

Cell type	Ouabain-sensitive ATPase activity per cm^2 of surface area
Red blood cells	$6 \times 10^{-5} \mu$moles Pi/h/cm^2
Artificial cells with Na-K-ATPase	$90–250 \times 10^{-5} \mu$moles Pi/h/cm^2

this area will benefit greatly from the ongoing research and progress in the molecular biology of the structure and function of different transport mechanisms. A few examples are highlighted here.

Melkikh and Seleznev (2005) used nonequilibrium statistical models of the active transport of ions in biomembranes. They used this to study models of active transport of ions in biomembranes of different cells, including neurons, muscular cells, bacteria, plants and mitochondria. They found the values of the membrane potential and ion concentrations to be in qualitative agreement with the experimental data. They discussed the requirements for models of active transport of ions in biomembranes (Melkikh and Seleznev 2006). The factors include resting potential, intracellular concentrations of ions, ATP-ADP chemical potentials, reversibility of the ionic pump, correlation between theoretical and experimental data, pump efficiency and effects of blocking the active transport. Barwe *et al.* (2007) looked at the Janus model of the NA-K-ATPase beta-subunit. Na-K-ATPase is an oligomer consisting of alpha and beta-subunits. The Na-K-ATPase beta-subunit (Na-K-beta) takes part in both the regulation of ion transport activity, and in cell-cell adhesion. These results provide a structural basis for understanding how Na-K-beta links ion transport and cell-cell adhesion. Pitard *et al.* (1996) studied ATP synthesis using the F_0F_1 ATP synthase from thermophilic Bacillus PS3 reconstituted into liposomes with bacteriorhodopsin. There is also increasing interest in using the principle of the lipid-polymer membrane artificial cells (Chang, 1969a, 1972; Rosenthal and Chang, 1971, 1980) as a model of biological membrane for studying the structure and function of membrane protein and receptors. Thus, Wagner and Tamm (2000) developed a polymer supported lipid bilayer system to reconstitute integral membrane proteins in a laterally mobile form. When reconstituted in bilayers tethered on quartz, the proteins are not mobile. However, on the lipid-polymer system, they found that the two proteins studied can become laterally mobile. They proposed that the lipid-polymer model allows for the interaction of these proteins with the underlying polymer.

Nanobiosensors

Nanobiosensor is an area that is of increasing interest and different approaches are being investigated (Scheller *et al.*, 2001). One of the many approaches is to use biosensors where the lipid bilayer is "tethered" on ultrathin polymeric support to form a lipid-polymer complex. In this form, different "channels" can be inserted into the membrane to allow for selective movement of specific solute for detection, somewhat similar to the principle reported earlier (Chang, 1969d, 1972a; Rosenthal and Chang, 1971, 1980). Another approach that is also possible is to encapsulate enzymes inside artificial cells of microscopic or nanodimension. This way, the product of enzymatic reaction can be followed by florescence or other methods. The ability to prepare artificial cells with intracellular compartmentation (Chang, 1965, 1972a; Chang *et al.*, 1966) would allow multistep enzyme reactions to occur and be detected separately. Depending on the type of reactions being followed, one can use either polymeric membrane artificial cells, lipid membrane artificial cells or lipid-polymer membrane artificial cells.

11.6.   Surface Properties of Artificial Cell Membranes

As discussed under the section on blood substitutes, the surface property of artificial cell membranes is important in terms of their biocompatibility and retention in the circulation, as it is in terms of blood compatibility in hemoperfusion systems. Artificial cells for delivery systems also need to have different types of surface properties. Cellulose nitrate membrane artificial cells (Chang, 1957, 1964) contain negative charges due to the carboxyl groups in the polymer. Polyamide nylon membrane artificial cells (Chang *et al.*, 1963, 1966; Chang, 1964) contain both carboxyl and amino groups. With the proper selection of polymer materials, membranes with the desired fixed charges can be obtained. For example, artificial cell membranes with strong negative charge groups have been prepared by using different amounts of sulfonated diamine such as 4,4'-diamine-2,2'–diphenyldisulfonic acid when preparing polyamide

VARIATIONS IN SURFACE PROPERTIES OF ARTIFICIAL CELLS

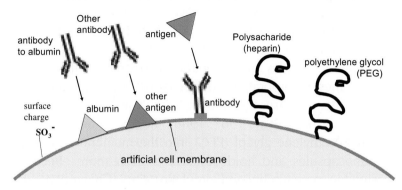

Fig. 11.8. Surface properties of artificial cell membranes can be varied by 1) incorporation of negative or positive charge; 2) incorporation of albumin to increase blood compatibility; 3) incorporation of antigens to bind antibodies or antibodies to bind antigen; 4) incorporation of polysaccharide like heparin or polyethylene glycol (PEG) to increase compatibility or retention time in circulation.

membrane artificial cells (Chang, 1964, 1965, 1972a). The principle of incorporation of surface charges has also been used the preparation of lipid membrane artificial cells, lipid vesicles, drug delivery systems (Gregoriadis, 1976; Torchilin, 2005).

Albumin can bind tightly to the ultrathin collodion membrane of adsorbent artificial cells. This is initially used to increase the blood compatibility of the adsorbent artificial cells for hemoperfusion (Chang, 1969a). This albumin coating has also been applied to synthetic immunosorbents, resulting in blood compatible synthetic blood group immunosorbents (Chang, 1980d). In addition, Terman *et al.* (1977) showed in animal studies that albumin-coated collodion activated charcoal (ACAC) can remove antibodies to albumin. This has become a basis of one line of his research in which other types of antigens or antibodies are applied to the collodion coating of the artificial cells to form immunosorbents. Other immunosorbents based on this principle have also been developed for the treatment of human systemic lupus erythematosus; removal of antiHLA antibodies in transplant candidates; treatment of familial hypercholesterolemia with monoclonal antibodies to low-density lipoproteins (Terman, 1980;

Terman *et al.*, 1979a, 1979b; Hakim *et al.*, 1990; Wingard *et al.*, 1991; Yang *et al.*, 2004). Antibodies have also been incorporated in the surface of lipid liposomes to allow for drug targeting to cells bearing the corresponding antigen (Torchilin, 2005).

Mucopolysaccharides are important components of the surface of cell membranes. Incorporating a polyscharide, heparin to the artificial cell membrane, increases the biocompatibility and circulation time of artificial cells (Chang *et al.*, 1967, 1968). Incorporation of a synthetic polymer, polyethylene glycol (PEG) into the membrane of artificial cells, nanocapsules and lipid vesicles, is even more effective for increasing biocompatibility and circulation time (Chang *et al.*, 2003; LaVan *et al.*, 2002; Torchilin 2005).

11.7. Drug Delivery

Artificial cell-based drug delivery systems differ from the usual artificial cells discussed up to now. Artificial cells retain the bioreactive material inside them to act on external molecules diffusing into the artificial cells or to release their secretions. For instance, enzyme is retained in an artificial cell to act on the substrate diffusing in while the product diffuses out. Another example is artificial cells containing islet that release insulin. Unlike artificial cells, a drug delivery system separates the drug from the external environment initially, but in order for the system to function, the contents need to be released as and when or where it is needed.

General

This monograph is not about drug delivery, an extremely large and broad area and which are covered in many excellent reviews and books; only the artificial cells related area will be very briefly summarized below. Submicron lipid membrane artificial cells, i.e. lipid vesicles have been developed for sometime now, and are best for use in the delivery of drugs and medications, including chemotherapeutic agents. With the recent interest in biotechnology and gene therapy, there is increasing need for delivery system that can carry sufficient amounts of macromolecules. Thus, biodegradable

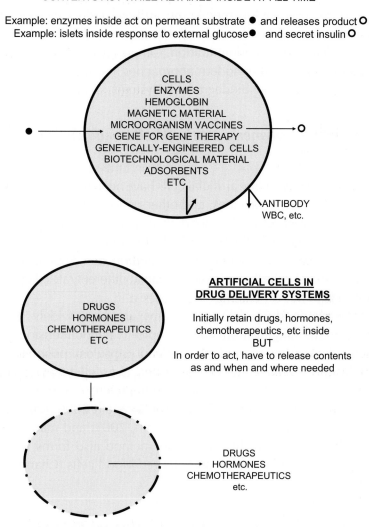

STANDARD ARTIFICIAL CELLS:
CONTENTS ACT WHILE RETAINED INSIDE AT ALL TIME

Example: enzymes inside act on permeant substrate ● and releases product ○
Example: islets inside response to external glucose● and secret insulin ○

CELLS
ENZYMES
HEMOGLOBIN
MAGNETIC MATERIAL
MICROORGANISM VACCINES
GENE FOR GENE THERAPY
GENETICALLY-ENGINEERED CELLS
BIOTECHNOLOGICAL MATERIAL
ADSORBENTS
ETC

ANTIBODY
WBC, etc.

ARTIFICIAL CELLS IN
DRUG DELIVERY SYSTEMS

DRUGS
HORMONES
CHEMOTHERAPEUTICS
ETC

Initially retain drugs, hormones,
chemotherapeutics, etc inside
BUT
In order to act, have to release contents
as and when and where needed

DRUGS
HORMONES
CHEMOTHERAPEUTICS
etc.

Fig. 11.9. Standard artificial cells compared to extension of artificial cells for use in drug delivery.

polymeric artificial cells especially those of nano dimenions are of increasing importance. These nanodimension biodegradable polymeric artificial cells are disguised under different names, including nanocapsules, nanoparticles, nanotubules, polymersomes, etc.

Polymeric semipermeable microcapsules

Luzzi (1970) used nylon membrane artificial cells, microcapsules, prepared as reported earlier (Chang 1964) to microencapsulate drugs for slow release in oral adminitration. Others have also extended this approach. However, the modern approaches in drug delivery systems are based on injectable biodegradable systems.

Biodegradable polymeric artificial cells, nanoparticles, nanocapsules

Biodegradable membrane artificial cells have been prepared to contain enzymes, hormones, vaccines, and other biologicals (Chang, 1976a). The detailed method is described in Appendix II. The polylactide polymer can degrade in the body into lactic acid and finally into water and carbon dioxide. Variations in preparation (see Appendix II) can result in artificial cells that releases insulin at different rates (Fig. 11.10) (Chang, 1976a).

Biodegradable drug delivery systems are now widely used in different forms ranging from microscopic to nanodimensions. They are also known as nanoparticles, nanocapsules, polymersomes, nanotubules, etc. Langer' group has written an excellent review on this topic (LaVan *et al.*, 2002). Examples of other references include books and reviews (Malsch, 2005; Gupta and Kompell, 2006; Ranade and Hollinger, 2003; Pardridge, 2002; Duncan, 2003; and many others). As described earlier in this book, this method also forms the basis for preparing nanodimension artificial red blood cells (Chang, 2005, 2006).

Liposomes evolved into lipid vesicles that are lipid membrane artificial cells

Bangham first reported the preparation of liposomes each consisting of microspheres of hundreds of concentric lipid bilayers — multi-lamellar (Fig. 11.4) (Bangham *et al.*, 1965). They used these as a membrane model for basic membrane research. Back in the 1960s, I encouraged a

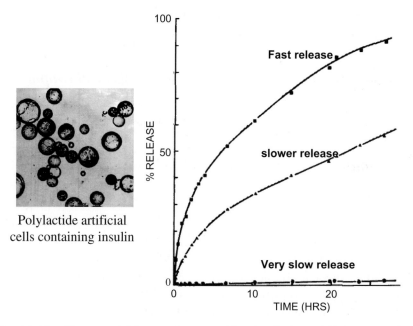

Polylactide artificial
cells containing insulin

Fig. 11.10. Biodegradable membrane artificial cells have been prepared to contain enzymes, hormones, vaccines, and other biologicals (Chang, 1976a). This figure summarizes the result of polylactide artificial cells prepared using the double emulsion method (Appendix II). Variations in the molecular weight of polylactide and thickness of the membrane can result in artificial cells that release insulin at different rates. Faster release occurs during encapsulation of insulin solution at high concentration, while very slow release occurs during encapsulation of insulin crystals. (Chang, 1976a).

newly graduated McGill Ph.D., Gregoriadis, to look into different ways of forming artificial cells for the delivery of biologics. His subsequent research in England resulted in the first report on the use of liposomes as drug delivery systems that has opened a whole new approach (see 1976 book edited by Gregoriadis). The large amount of lipid in the original multi-lamellar liposome limits the amount of water soluble drugs that can be enclosed. Thus, the basic principle and method of preparing artificial cells using ether as the dispersing phase (Chang 1957, 1964) was extended by researchers into what they called an "ether evaporation method" to form single bilayer (unilamellar) lipid membrane liposomes (Deamer DW and Bangham AD, 1976). These lipid membrane artificial cells, i.e. lipid vesicles, have since been extensively studied for use as drug delivery systems (Torchilin, 2005).

Polymer(PEG)-lipid membrane artificial cells or PEG-lipid vesicles

The original liposome drug carrier consists of a lipid with a water soluble drug entrapped in the aqueous phase, and a water-insoluble drug in the lipid phase. It is good to see that researchers have applied a number of the results of our earlier basic research on surface charge, lipid-polymer membrane, surface modification, and magnetic targeting into the original lipid vesicles for drug delivery. These and other important developments being carried out by many researchers in the field of drug delivery have resulted in important progress in the field. Thus, surface charges are incorporated in liposome for possible targeting of drug and more recently the use of positively charged lipid to complex with DNA. Then a biodegradable polymer, polyethylene glycol (PEG), is incorporated in the liposome surface to result in longer circulation time. Further developments led to the incorporation of antibodies in the lipid membrane to allow for targeting to cells with the corresponding antigens. Magnetic particles loading into liposome allows for magnetic targeting. Thus, lipid vesicles are getting to be more like the lipid-polymer membrane artificial cells (Chang, 1969d, 1972a) and no longer pure lipid vesicles. One major advantage of lipid vesicles is its ability to fuse with cellular membranes or membranes of intracellular organelles. This allows for much versatility in their ability to deliver drugs to different sites of the cells (Fig. 11.11).

As a result of all these developments, and in particular the preparation of polymer(PEG)-lipid membrane artificial cells, there has been much recent progress. A number of drugs in PEG-lipid vesicles has already been approved for clinical use or use in clinical trials. The following are brief examples of some of the drugs and more details are available elsewhere (Torchilin, 2005): "lipid vesicles" containing daunorbicin for Kaposi's sarcoma; doxurbicin as combination therapy for recurrent breast cancer; doxorubicin for refractory Kaposi's sarcoma and ovarian cancer and breast cancer; amphotericin B for fungal

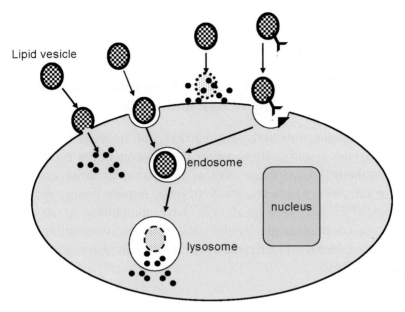

Fig. 11.11. Routes of drug delivery of lipid vesicles; these can fuse with the cell membrane to release the drug into the cell. The lipid resides can undergo endocytosis to form endosomes that can enter lysosomes where the drug is released. Thus can also release the drug outside the cell. Lipid vesicles with a surface antibody can bind to the antigen receptor and enter the cell as endosomes.

infections; cytarabine for lymphomatous meningitis; vincristine for non-Hodgkins lymphoma; lurtotecan for ovarian cancer; nystatin for fungal infection; and many others.

Polymersomes: polymeric membrane artificial cells

Discher's group (Discher *et al.*, 1999, Photos *et al.*, 2003) attempted to increase the strength of the PEG-lipid-membrane artificial cells through self-assembling of copolymers to form a membrane of PEG polymer. This significantly increases the circulation time and strength as compared to PEG-lipid membrane artificial cells (Discher *et al.*, 1999; Photos *et al.*, 2003). Thus, liposome has evolved into lipid membrane artificial cells, then polymer(PEG)-lipid membrane artificial cells and finally back to the original polymeric membrane artificial cells (Chang 1964) disguised under the name of polymersomes.

11.8. Artificial Cells Containing Multienzyme Systems with Recycling of ATP and NADH

Most metabolic functions are carried out in cells by complex multienzyme systems. Thus, artificial cells containing multienzyme systems would have more uses as compared to those containing single enzymes. However, multienzymes require energy from ATP or NAD(P)H and biological cells have the ability to regenerate these for continuous use by the cells. Can we make artificial cells containing useful multienzyme systems that can also regenerate ATP and NAD(P)H?

Artificial cells containing multienzyme system for recycling of ATP

One of the most important cofactor that is recycled in biological cells is ATP. We therefore started by studying artificial cells containing hexokinase and pyruvate kinase for the recycling of ATP (Fig. 11.12) (Campbell and Chang, 1975). Our study showed that this system can effectively recycle ATP as analyzed by the continuous conversion of glucose into glucose-6-phosphate and phosphoenol pyruvate into pyruvate.

Artificial cells containing multienzyme system for recycling of NAD(P)H

Artificial cells containing yeast alcohol dehydrogenase and malate dehydrogenase (Campbell and Chang, 1975, 1976) (Fig.11.12). NAD$^+$ and NADH can equilibrate rapidly across the membrane to be recycled in the artificial cells. Recycling of NAD$^+$ and NADH follows as shown by the production of malate (Fig. 11.13). When the recyling is blocked, production of malate ceases.

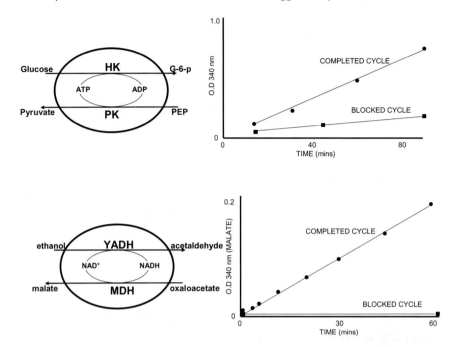

Fig. 11.12. *Upper*: Polymer membrane artificial cells containing hexokinase and pyruvate kinase recyle ATP in the presence of glucose (Campbell and Chang, 1975). Absence of glucose blocks the recyle. *Lower*: Polymer membrane artificial cells containing yeast alcohol dehydrogenase and malate dehydrogenase can recycle NADH as shown by the continuous production of malate (Campbell and Chang, 1975,1976). In the absence of recycle, malate production ceases.

Artificial cells containing urease, glutamate dehydrogenase and glucose-6-phosphate dehydrogenase

Since the needed NADH can be recycled, the next study was to see whether we could use a multienzyme system for the removal of urea. First, we used a model system of polymer membrane artificial cells loaded with three enzymes: urease, glutamate dehydrogenase and glucose-6-phosphate dehydrogenase (Fig. 11.13) (Cousineau and Chang, 1977; Chang and Malouf, 1979; Chang *et al.*, 1979b). Urease converts urea into ammonia. Then glutamate dehydrogenase converts

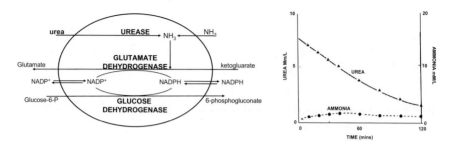

Fig. 11.13. Polymer membrane artificial cells containing urease, glutamate dehydrogenase, glucose-6-phosphate dehydrogenase for the multistep enzyme reaction of conversion of urea into amino acid glutamic acid and for the recyling of NADPH. Urea is lowered without any increase in ammonia level, as ammonia is converted into glutamate (Cousineau and Chang, 1977; Chang and Malouf, 1979; Chang *et al.*, 1979b).

ammonia into glutamic acid. Glucose-6-P dehydrogenase serves to recycle of NADPH.

The rate of conversion of urea into glutamate and the recycling of NADPH could be followed by the rate of formation of 6-phosphogluconate, decrease in urea level, or the formation of glutamic acid. Ammonium acetate could be used in this reaction instead of urea. Figure 11.13 shows that the area level is lowered without any increase in ammonia level, as ammonia is efficiently converted into glutamate. Glutamate is formed at the same rate irrespective of whether urea or ammonium acetate is used as the substrate. This suggests that the rate-limiting step in this sequential reaction is in the conversion of ammonia into glutamic acid.

Dextran-NADH retained and recyled inside artificial cells (Fig. 11.4)

The above studies show that artificial cells containing multienzyme systems can carry out sequential enzymatic reactions to convert urea into glutamate and at the same time recycle the required NADH. The above study is a model experiment to test this feasibility. However, in potential medical or engineering uses, one would prefer to have the cofactor retained and recycled inside the artificial cells. This way, it will not be necessary to have NADH supplied from outside. The first step is to see how we can retain the NADH inside the artificial cells without decreasing the permeability to other substrates and products. We use Mosbach's method (Larsson and Mosbach, 1971) to covalently link NADH to a macromolecule to form a soluble NADH-dextran. We found that the resulting NADH-dextran can be retained inside the artificial cells where it is recycled (Fig. 11.14) (Campbell and Chang, 1978; Grunwald and Chang, 1978,1979,1981).

The artificial cells are retained in a flow reactor perfused by the substrates. The recycling activity is followed by the production of malate (Table 11.2). The recycling activity of the shunt is high as 65% of the oxaloacetic acid is converted into malic acid in one passage through the shunt. The dextran-NAD$^+$ present within the

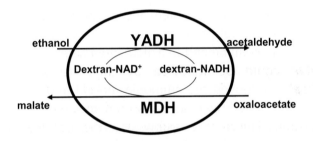

Fig. 11.14 and Table 11.2. Artificial cells containing alcohol dehydrogenase, malate dehydrogenase and oluble NADH-dextran. NADH-dextran is retained inside the artificial cells where it is recycled (Campbell and Chang, 1978; Grunwald and Chang, 1978,1979,1981). The soluble NADH-dextran is formed using Mosbach's method (Larsson and Mosbach, 1971). Table 11.2 shows the rates of regeneration of NAD$^+$.

Table 11.2 Recycling of Dextran-NAD$^+$ in Multienzyme Polymer Membrane Artificial Cells

Storage (days)	NAD$^+$ Regenrated (cycles/h)	Amount in Artificial Cells (μmole/h/ml)
0	120.0	150.0
1	117.5	147.0
4	84.0	105.0
7	37.3	46.5

microcapsules is regenerated two times each minute during the reaction. The shunt showed good stability, with the reaction rate remaining constant for the first hour and 83% of the original activity being retained even after 3 h of continuous reaction. The same shunt could be stored and reused several times. It is important to use purified hemoglobin instead of crude hemoglobin for coencapsulation. Crude hemoglobin contains impurities that interfere with the recycling.

Multienzyme with NAD-dextran for conversion of waste, urea, into useful essential amino acids

We first showed that artificial cells containing multienzyme systems and NAD^+-dextran can convert urea or ammonia into glutamate (Gu and Chang, 1988b, 1988c). The next step was to see if we could use a more complex multienzyme system and NAD^+-dextran to convert the waste, urea, into useful products, essential L-branched-chain amino acids, L-leucine, L-valine and L- isoleucine. This is based on artificial cells containing urease to convert urea into ammonia and leucine dehyrogenase to convert the ammonia formed into essential amino acids. Glucose dehydrogenase is also included for the recycling of NAD^+-dextran using glucose as source of energy for cofactor recycling (Gu and Chang, 1988a). Other enzymes for recycling of NAD^+-dextran include yeast alcohol dehydrogenase that uses alcohol as source of energy (Gu and Chang, 1990a); malate dehydrogenase that uses malate as a source of energy (Gu and Chang, 1990b) and L-lactic dehydrogenase that uses L-lactic acid as a source of energy (Gu and Chang, 1991). The results obtained using artificial cells containing leucine dehyrogenase, urease, glucose dehydrogenase and NAD^+-dextran is shown below (Fig. 11.15) to illustrate this approach.

Recycling of free NADH retained within the lipid-polymer membrane artificial cells for conversion of urea into amino acid

We have studied another way to retain NADH inside artificial cells without crosslinking it to dextran or other macromolecules. The

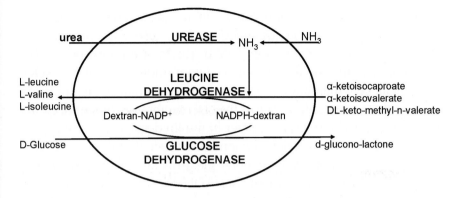

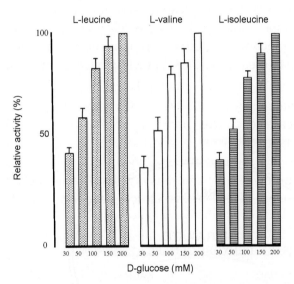

Fig. 11.15. *Upper*: Artificial cells containing urease, leucine dehydrogenase, glucose dehydrogenase, dextran-NADP$^+$ to convert the waste, urea, into useful essential L-branched-chain amino acids, L-leucine, L-valine and L-isoleucine. Glucose dehydrogenas, uses glucose as a source of energy to recycle NAD$^+$-dextran (Gu and Chang, 1988a). *Lower*: The effect of D-glucose concentrations on the production rates of L-leucine, L-valine and L-isoleucine. The production rates for 200mM D-glucose are used as the 100% control.

membrane of biological cells can retain free NADH without the need to be crosslinked to macromolecules. Thus, use our lipid-polymer membrane artificial cell (Chang, 1969d, 1972a) that is only permeable to ammonia and urea but not to larger molecules like NADH. Biological cell membranes have transport mechanism to transport glucose that cannot diffuse across the cell membranes. Since we do not yet have a glucose transport carrier in the lipid-polymer member, we have to replace glucose dehydrogenase with alcohol dehydrogenase. This way, external ethanol can enter the artificial cells and serve as a source of energy for the recycling of NADH (Fig. 11.16) (Yu and Chang, 1981a, 1981b,1982a, 1982b; Ilan & Chang, 1986).

The standard lipid-polymer complexed artificial cell membrane is not permeable to urea (Table 11.3). By decreasing the proportion of cholesterol in the lecithin-cholesterol component, the membrane becomes permeable to urea (Table11.3) but with no leakage of NAD^+ and ketoglutarate.

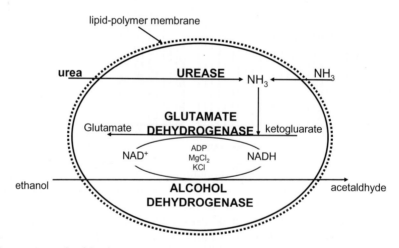

Fig. 11.16 and Table 11.3. Recycling of free NADH retained within the lipid-polymer membrane artificial cells for the conversion of urea and ammonia into glutamate. The lipid-polymer membrane artificial cells contain three enzymes, urease, glutamate dehydrogenase and alcohol dehydrogenase. Table 11.3 shows the effects of permeability to urea and rate of conversion of urea into glutamic acid (Yu and Chang, 1981a, 1981b, 1982a, 1982b, Ilan and Chang, 1986).

Table 11.3 Multienzyme Lipid-Polymer Membrane Artificial Cells Convert Urea into Glutamate with NADH Recycling

Permeability to Urea	Glutamic Acid (μmol/L/3h)
Low	0
Medium	2.42
High	6.93

Summary

The above basic research shows that it is possible to prepare artificial cells containing multistep enzyme systems with recycling of ATP and NAD(P)H. Further developments may lead to a number of possible applications. One example is to complete the urea removal component of the miniaturized hemoperfusion-ultrafiltration system for treating kidney failure. We have already applied this basic information in the preparation of nano artificial red blood cells that contain multienzyme system with cofactor recycling to convert methemoglobin back to hemoglobin (Chang, 2005i, 2006a). Potentially, any combinations of enzyme systems can be enclosed within each artificial cell. The lipid-polyamide membrane artificial cell can also be used in basic research as a cell model.

11.9. Artificial Cells Containing Microsomes, Cytosol Ribosomes and Polymerases

Introduction

Chapter 5 shows that it is possible to prepare complete artificial red blood cells containing hemoglobin and all the complex enzyme systems of its biological counterpart. In addition, safety and efficacy in animal studies have also been shown. The next step is to prepare artificial cells that have the properties of more complex biological cells.

Artificial cells containing liver microsomes and cytosol

We isolated microsomes and cytosol from rat liver and encapsulated these into cellulose nitrate membrane artificial cells (Yuan and Chang, 1986). The method is based on the standard method (Chang, 1964, 1972a), with the updated method (Chang, 2005a) available in Appendix II. NADPH-cytochrome C reductase and lactate dehydrogenate are used as the marker enzymes for, respectively, microsmes and cytosol. Table 11.4 shows that after encapsulation into artificial cells, NADPH-cytochrome C reductase, marker enzyme for microsmes, retained 24.5% of its original activity. Lactate dehyrogenase, enzyme marker for cytosol, retained 11% of its original activity after microencapulation into artificial cells. It is important to carry out the preparation at 4°C instead of 23°C.

The result shows that artificial cells can be used to encapsulate subcellular organelles. Liver cells contain lysomes which when released can destroy the subcellular organelles. By preparing the homogenate of the liver cells and rapidly removing the lysosome fraction, we can reconstitute the other organelles and cytosol into an artificial cell free from the possible proteolytic effects of the lysosmomes.

Table 11.4 Artificial Cells Containing Liver Microsomes and Cytosol NADPH- Cytochrome C Reductase as Marker for Microsomes Lactate Dehydrogenase as Marker for Cytosol

	Original Activity (units)	Activity Recovered (units)	Activity Recovered (%)
Microsome enzyme marker	8550	2150	24.5%
Cytosol enzyme marker	2150	260	9.0%

Towards a "living" artificial cell containing polymerases, ribosomes and transcription/translation system

An even more exciting and ambitious effort is the recent serious attempts to develop a "living" artificial cell. A "living" artificial cell is defined by Deamer (2005) as an artificial cell that can capture energy, maintain ion gradients and contain macromolecules. Macromolecules must be able to grow by polymerization and evolve in a way that speeds up growth. They must be able to store information and possess the ability to mutate and direct the growth of catalytic polymers. They must also be able to reproduce themselves and divide. Studies on artificial cells described earlier in this monograph suggest that many of these criteria can be met. However, preparing artificial cells that can reproduce themselves and divide will be extremely difficult. Some of the exciting studies on "living" artificial cells are highlighted in the next paragraph.

Szostak, at Harvard, have been carrying out a number of studies (Hanczyc *et al.*, 2003; Chen *et al.*, 2004; Josephson *et al.*, 2005). For example, they have prepared a simplified model artificial cell by putting simple RNA enzymes inside the cells, and showing that the cells can conduct their characteristic activities. They are also carrying out the modification of the cellular translational system to allow for the ribosomal translation of molecules for medical uses. Monnard and Deamer (2001) prepared models for primitive cellular life by encapsulating T7 RNA polymerases and templates into lipid membrane artificial cells. This enzyme was able to synthesize an RNA transcript from the DNA template that required four nucleoside triphosphates. Since the lipid membrane was not permeable to nucleoside triphosphates, they had to use DMPC liposomes at a temperature of $23\,^{\circ}\mathrm{C}$. Despite this, permeation was still very slow, thus limiting the rate of RNA synthesis. Other researchers encapsulated a complex polymerase system into lipid membrane artificial cells and showed that the PCR reaction could be carried out (Oberholzer *et al.*, 1995). Since the substrate could not diffuse across the lipid membrane, it had to be loaded together with the whole system. The result was that there was only enough substrate to produce a very small amount of the product. They have also encapsulated ribosomes into lipid membrane artificial cells and obtained some translation product. Noireaux and Libchaber (2004) from Rockefeller University were able to engineer a DNA plasmid to express proteins and encapsulate this into lipid membrane artificial cells. However, they could only produce proteins for a few hours because the source material inside the artificial cells could not be replaced. They solved this problem by incorporating a channel-forming protein, alpha hemolysin, into the lipid artificial cells membrane. This way, the artificial membrane could become permeable to larger molecules like nucleotides. As a result, they could function for up to four days. The use of selectively permeable crosslinked protein membrane or polymeric membrane artificial cells as described in Section 11.3 above (Chang, 1964, 1972a, 2005i) can result in an even more efficient supply of external substrates.

In another study, Griffiths and Tawfik (2000) used compartment-alization to load the transcription/translation system into a water-in-oil emulsion. This way, each gene could occupy a separate water emulsion to carry out its function. A more efficient approach may be possible by using the polymeric compartmental system of artificial cell where each artificial cell contained "subcellular compartments" with ultrathin polymeric membranes (Chang 1965, 1972a; Chang *et al.*, 1966, 2005i) (Fig. 2.4). This way, biologically active systems could be enclosed separately or in combination into each of these artificial subcellular compartments. Since the permeability of each of the compartments could be individually and separately adjusted, this might allow for more efficient stepwise functions.

The most difficult steps will be attempts to prepare artificial cells that can reproduce themselves and divide. To do this, the artificial cells have to contain genes and enzymes that can be replicated and shared among daughter cells. One possible way is to use a self-replicating ribozyme that can duplicate itself and act as both genetic material and the catalyst for replication. Johnston *et al.* (2001) used a ribozyme to catalyze the replication of another RNA template molecule. Their next step is to verify that the RNA can continue with the process of replication. Rasmussen *et al.* (2003) of Los Alamos National Laboratory used computer simulation to design a micelle with information coding and metabolic machinery on its exterior. They used DNA-mimics conjugated to a light-sensitive molecule. In theory, light would trigger the conversion of substrates into peptide nucleic acid (PNA) based on the PNA template. These are incorporated into the micelle that would grow and divide. The next step is to test this theoretical simulation experimentally. Ishikawa *et al.* (2004) of the University of Osaka, Japan used a lipid membrane artificial cell containing a bacterial plasmid and SP6 RNA polymerase, which acted as a driving force for the production of a T7 RNA polymerase needed to catalyze the transcription of a gene. Thus, this membrane-bound system shows both gene expression and a two-stage genetic cascade of events. Information regarding the minimal number of genes needed for a "living" artificial cell is important. Kobayashi *et al.* (2003) identified the minimum set of genes in *Bacillus subtilis*, narrowing the number to 271 and only 4% of these have unknown functions. It would help if this number can be narrowed down even further to allow for the preparation of a primitive "living" artificial cell.

All these exciting and ambitious efforts show that many of the criteria for a "living" artificial cell can possibly be met. The most difficult criterion is the ability of the "living" artificial cell to reproduce itself and divide. Instead of using lipid membrane artificial cells alone, a combination with crosslinked-protein membrane or polymeric membrane artificial cells could be used — either individually or in combination.

11.10. New Generations of Computer System and Nanoscale Robotics Based on Artificial Cells

Researchers in this field suggest that next generation self-repairing computer and robotics technology requires the use of intelligent technical systems based on artificial cells. Toward this end, the European Commission is supporting an integrated program of "PROGRAMMABLE ARTIFICIAL CELL EVOLUTION" (PACE) (http://wills.gex.gmd.de/bmcmyp/Data/PACE/Public/). This is aimed at a new generation of embedded information technology using programmable, self-assembling artificial cells. To achieve this, the PACE project focusses on the intelligent technical (IT) potential of artificial cells that are truly "artificial" and not replicates of biological cells. This project consists of a consortium of 13 partners and two cooperating groups from eight European countries and the USA. The group includes expertise in complex systems, embedded systems, robotics, evolution, statistics, chemical kinetics, physical simulation, microfluidics, organic and bioorganic chemistry, computer interfaces and control systems. With this interdisciplinary approach, they plan to prepare microscopic artificial cells as chemical information processing systems. They suggest that for the next-generation robotics and nanoscale technologies to be realized, they would need distributed processing systems with self-organizing properties in the form of artificial cells. These will be designed and assembled automatically from non-living materials.

They have developed a computer-programmable closed-loop controller, based on microfluidic technology. This can be used to observe and control the local physical environment of cell-sized chemical subsystems. They have also shown that an artificial peptide nucleic acid (PNA) can synthesize or direct the information in the synthesis of another copy of itself — a step towards self-replication. They have also developed simulation models to describe artificial cells on an intermediate physical level to follow the control, stability and breakdown of the subsystems. They are also studying the use of artificial cells as a novel platform for self-assembling microscopic robots, both in simulation and experiments. On an ethical level, they emphasize that "What we are talking about is not chemically making a copy of an existing artificial cell, but chemically synthesizing something which we could call an artificial cell."

11.11. The Future of Artificial Cells

Nature's creations are complex, intricate and ingenious. For ages, humans have tried to reproduce exact replicates of some of these without success. Success only comes to those who are humble enough to use the principles learned from nature to produce much simpler systems instead of trying to replicate nature.

A good example is in the many unsuccessful and often fatal attempts made by humans to fly like birds. The first initial success is when the Chinese, instead of trying to fly like birds, use the principle to build very simple kites that later can even carry humans. These evolve into gliders. Since then, each major progress in flight is related to major progresses in other areas. Thus, the invention of the motor engine in Europe allows the gliders to evolve into the first propeller driven planes first in the US and then Europe. Centuries ago, the Chinese invented fireworks and rocket-propelled weapons. This principle has been applied to jet engine that allows the propeller planes to evolve into the first jet-planes in Germany. These have now reached enormous size and speed. Further developments allows human to travel to space first in Russia and then the USA. This is an example of how progress is made on both an interdisciplinary and international basis. Much delay can be the result if there is not enough interaction between the different disciplines and countries.

For many years, scientists have carried out tremendous amounts of research on the structural and functional basis of biological cells. However, attempts to reproduce exact replicas of biological cells have not been successful. On the other hand, the purpose of the very first humble "artificial cells" (Chang, 1957) was not to reproduce biological cells, but to use available basic knowledge to prepare very simple systems for possible uses in medicine and other areas. Each major progress in other areas has led to stepwise progress in artificial cells. First, there was the coming of age of polymer chemistry and biomaterials. Then there was the recognition of the importance and developments in biotechnology. There is the present ongoing progress in molecular biology and genomics that will contribute to a quantum leap in the area of artificial cells, including the recent efforts on "living" artificial cells. One can expect that there will be important future progress in other areas that will contribute to unlimited progress in the area of artificial cells. One example is in the use of artificial cells for next generation computer and nanoscale robotics.

The following prediction in my 1972 monograph on "Artificial Cells" is already out of date: *"Artificial Cell is not a specific physical entity. It is an idea involving the preparation of artificial structures of cellular dimensions for possible replacement or supplement of deficient cell functions. It is clear that different approaches can be used to demonstrate this idea"*. In the last 50 years (Chang, 1957), artificial cells have progressed way beyond this 1972 prediction. Artificial cells can now be of macro, micro, nano and molecular dimensions.

There are also unlimited possibilities in variations for the artificial cell membranes and contents. Searching for "artificial cells" on the Internet (for example, www.google.com) gives more than 50,000 hits in all areas of artificial cells. Even then, we have only just touched the surface of the enormous potential of artificial cells. One hopes that the many arbitrary subdivisions of "artificial cells" under the guise of different names can be brought together! When this takes place, the result of the pooling of talents, specialized know-how in this very interdisciplinary and international area will lead to progress beyond anyone's imagination.

1957 Report on "Method for Preparing Artificial Hemoglobin Corpuscles"

1957 Research Report for Physiology 43b for partial fulfillment of honors B.Sc. in Physiology McGill University

Thomas M. S. Chang

(Figures scanned from original report with new fonts added for clarity — text reformatted but with no change from original, except errors in spelling, grammar and typing.)

The red corpuscle, once it has passed the reticulocyte stage, is almost without metabolic activity and is little more than a hemoglobin containing shell[1]. This pointed towards the possibility that perhaps an artificially produced hemoglobin corpuscle may be used as a physiologically functioning red blood cell. With this possibility in mind, first, a survey of the literature concerning erythrocytes, hemoglobin and possible artificial membranes was made; then possible methods of getting hemoglobin into corpuscles of microscopic sizes were tried.

ERYTHROCYTES

1. Structure

The normal human erythrocyte is a biconcave disc, with an average diameter in dried films of 7.2 μ. The thickness of cell is l μ at the centre and 2.4 μ near the edge. Although the cell membrane is not elastic, the cell can become distorted in order to pass through narrow capillaries, also, the concave faces of the corpuscle can be distended to a convex shape when the cell content increases as in the case of hypotonic environment. The membrane is of lipoprotein, and when examined with electromicroscope appears to be finely granular without folds. The mean corpuscular volume of the cell has an average of 86 cubic μ.

2. Function

Hemoglobin of the erythrocyte carries out the functions of transportation of O_2 to the tissues and that of transportation of CO_2 from the tissues to the lung to be excreted. This respiratory function of the hemoglobin of the erythrocyte is the most important function. Hemoglobin and electrolytes inside cell also act as a buffering system to maintain the normal pH of 7.3–7.4. 60% of buffering action of whole blood is due to red cell, the other 40% to plasma. Carbonic anhydrase is highly concentrated in erythrocytes. This is important since the reaction $CO_2 + H_2O = H_2CO_3$ is speeded up 2500 times, resulting in the transportation of CO_2 to the lung to be excreted.

3. Enzymes and inclusion bodies

Besides carbonic anhydrase, metabolic enzymes, dehydrogenases and pyridine nucleotides; there are various other enzymes whose functions are at the presence obscure. Also there are inclusion bodies in circulating erythrocytes: Punctate basophilia, Cabot rings, Howell-Jolly bodies, siderocyte, Heinz bodies. Quite a number of these are associated with diseases, especially the Howell-Jolly bodies which are associated with severe anemia of both the iron deficiency and the pernicious type, and with leukemia, and are particularly common after splenectomy. This is mentioned, because in an artificially produced hemoglobin corpuscle it is possible to include in the cell content all sorts of combinations of enzyme systems and inclusion bodies. This way, the function and activity of the various enzymes and inclusion bodies in the erythrocyte may be studied.

HEMOGLOBIN

1. Molecular structure

Hemoglobin is the red coloring matter of the red blood cell. It is a conjugated protein, consisting of a protein part called globin and an iron-pyrrol compound known as hematin or heme[5]. Work by Svodberg, Adair and others indicates that hemoglobin has a molecular weight of 66,800 and that the molecule of hemoglobin contains four molecules of hematin. The hemoglobin molecule is 64 Å in length, 48 Å in width and 36 Å in thickness.

2. Hemoglobin content of the blood

Hemoglobin is held within the stroma of the erythrocyte. The normal hemoglobin content of the blood is usually measured colorimetrically depending upon the oxygen combining capacity or the blood. The present assessment of normality is between 15 and 16 gm per 100 ml for males and with a lower value for females. Knowing the hemoglobin content of the blood the mean corpuscular hemoglobin (M.C.H.) and the mean corpuscular hemoglobin concentration (M.C.H.C.) can be calculated from the following formulae:

M.C.H. = (Hb in gms/1000 ml blood)/Red cells in millions per mm^3.

This is expressed in μu and the normal value for male being 29 μu (with a range or 27 to 32 μu) M.C.H.C. = (Hb in gms per 100ml blood × 100)/(VOL in ml packed cells per 100ml blood) This normal value is about 33% with a range of 32–38% and shows no variation with sex.

3. Methemoglobin

The ferrous iron in hemoglobin is readily oxidized by oxidizing agents to the ferric state, which is methemoglobin. Methemoglobin is brown in solution and has an absorption band with a maximum at 634 mu. In normal human blood it is approximately 2% of the total Hb. Now, since methemoglobin does not have the power to combine with O_2, it is important in the preparation of hemoglobin corpuscles to preserve the hemoglobin in the ferrous state. There are various enzyme systems in the red blood cell for the purpose of reduction of methemoglobin to hemoglobin. These are: triosephosphate and lactate dehydrogenating system[3], dehydrogenases and pyridine nucleotides. Other ways of reducing methemoglobin to hemoglobin are by the action of sodium sulphide, amyl nitrite in phosphate-saline solution, $Na_2S_2O_4$, ascorbic acid, or alkaline solution prepared from $TiCl_3$. All these will readily convert methemoglobin to hemoglobin which does combine with O_2 and CO_2.

4. Derivatives of hemoglobin

In the handling of hemoglobin during the preparation of the artificial hemoglobin corpuscles, it is important to be able to tell whether the hemoglobin inside the membrane is still hemoglobin, or if not, it is helpful to be able to know what substance it is changed into. The derivatives of hemoglobin may be OHb, methemoglobin, CO-hemoglobin, nitric oxide hemoglobin, hemochromogen, hematin, alkali-hematin, and hematoporphyrin. Each of these possesses a specific absorption spectrum which serves as an aid in their detection[2].

MEMBRANE

1. Reason for using collodion membrane

After much searching around, collodion is chosen for the membrane, since it has been studied quite extensively by a number of physiologists.[6,7,9,10,11,12] It is also being widely used by bacteriologists as membranes for ultrafiltration[8]. It is used despite the fact that being a plastic material, it cannot be used as the membrane of cells to be perfused into the circulation since once injected, it will remain as such in the circulation. However, since the problem of finding a suitable membrane would already be a project in itself and since the amount of time available for this project is limited, collodion was used as the membrane. Of course, a lipoprotein membrane would be most suitable.

2. Studies made on collodion membrane

Much can be made used of the information obtained by the physiologists and bacteriologists concerning the permeability of collodion to various molecules and methods of preparing collodion membrane with any desired pore size. In 1907, Bechhold discovered that porous membranes of collodion could be precipitated by water from non-aqeous solution. His first paper (Bechhold H. Z. Physik Chem 60,257,1907; 64,328,1908) is still among the best on this subject, here he sets forth many principles and applications. Elford[8] classified colloidal membranes into two main types:

1. Microgel: It is an acetic collodion membrane which has relatively course structure with pores of microscopic order.
2. Ultragel: Which is ether alcohol collodion membranes. The structure or this type or membrane is not resolved by microscopic apparatus. Ultaviolet light microphotography show that it is built up of particulate matter.

He also found that dilution of the ether alcohol Collodion with ether gives a thinner and less permeable membrane, whereas thickness and permeability increases with addition of amyl alcohol. Bauer[6] found that the addition or 1 ml of glacial acetic acid to 200 ml of diluted collodion solution would reduce average pore size to 200 mμ. The addition of 2 ml reduce the pore size to l00 mμ, and the addition of 3 ml reduces pore size further to 15 mμ. Field made collodion membranes which held back potassium chloride and sucrose. Kistler[9] found that collodion membrane can be made as dense as desired by filtering through them solution of cellulose or of collodion.

3. Final constituents of collodion solution used in this experiment

The collodion solution purchased is the commercially prepared Collodion U.S.P. XI. It is a solution of 4 gms of pyroxylin (chiefly tetranitropcellulose) in 100 ml of a mixture of 1 vol alcohol and 3 vol ether. Its specific gravity is 0.765–0.775. When exposed in thin layers it evaporates and leaves a tough somewhat opaque and permeable film. During the experiment it was found that the hemoglobin in solution was precipitated by the addition of this collodion solution. To decide the substance causing this precipitation, hemoglobin solution was added to each of pure ether, alcohol and it was found that whereas Hb is precipitated by alcohol, it is not precipitated when added to ether. The collodion solution was then evaporated to dryness, and then redissolved in ether, when hemoglobin is added to this ether collodion solution, hemoglobin remains in solution without precipitating. Thus it is concluded that it is the alcohol that precipitated the hemoglobin, and from here on, only ether collodion solution was used.

METHOD

The basic problem is how to make a hemoglobin corpuscle of about 7 μ diameter with a thin and permeable collodion membrane. Various methods were tried.

Method I

The arrangement for this method is as shown in the diagram (Fig.1).

A jet of air is sent into the funnel leading to a very narrow nozzle. Hemoglobin crystals, which are finely ground to microscopic sizes, are put in at X, so that the powder is sent through a stream of collodion. It was hoped that, each crystal in passing through the stream of Collodion would be coated with a coating of collodion which on drying would form a membrane.

Three problems arise immediately in attempting this method. First, using this method, the amount of Hb crystals required is quite enormous, and attempts to crystallize a large amount of hemoglobin from the solution was not too successful.

Secondly, a stream of collodion exposed to the air would be evaporated very rapidly, and in order to maintain such a stream approximately 1000 ml/3 mins is required, and the amount available is only sufficient for very small scale experimentation. However, the most severe problem is that for the hemoglobin crystals to be sent through

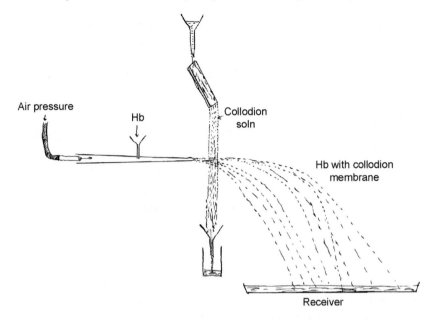

Fig. 1. Arrangement of Apparatus.

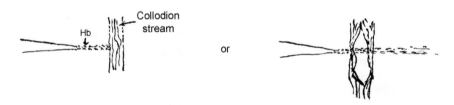

Fig. 2.

the stream of collodion, it would be necessary for the jet of air to be blown through the collodion at the same time. It was found that if the force of the air jet was not strong enough, then it would not be able to pass through the stream of collodion solution. However, when the force is increased, nothing can penetrate the collodion stream until a hole is made by the jet of air in the collodion stream, and the crystals would then be sent through by the jet of air without coming into contact with the stream of collodion. A sort of an all-or-none effect.

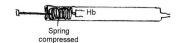 

Spring
compressed

Spring
released

Hb crystals

Fig. 3.

The solution to this problem is to send the crystals through the stream of collodion by means other than air pressure. One possible way is to eject the crystals through the use of a very powerful spring, and one was made as shown.

This way, the crystals are not carried by a jet of air, and it is hoped that therefore each crystal might be able to pass through the stream of collodion individually due to their force of momentum. However, further experiments shows that this method will turn out to be rather elaborate, since in order for the fine crystals to attain the necessary speed to penetrate the collodion stream, they will have to be ejected in a vacuum to avoid the resistance of air to such small particles. However, instead of increasing the speed of the crystals, it would be just as good to decrease the viscosity of the collodion stream. Thus another method was attempted.

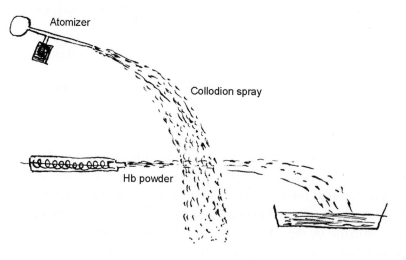

Fig. 4.

Method II

Instead of using a stream of collodion, a spray of collodion is used. Here the viscosity of the collodion is extremely small.

But here, there is a serious problem, since it takes a considerable amount of time for the collodion membrane to be perfectly set. When hemoglobin covered by a membrane of collodion which has not set touches water, the hemoglobin would just simply dissolve in water. Another problem is that, the droplets are so close together that once they are coated with collodion they would stick together.

Experimentation with collodion

Knowing that one of the important problem is the 'setting' of the collodion as a membrane, a study was made of the way in which a collodion membrane is set. Two useful observations were made. First it was observed that a thin collodion layer sets slowly (10 to 15mins) after it comes in contact with water. Secondly, a film of collodion is attached to the surface of water when it comes in contact with it. With these facts in mind, it was felt that it might be more feasible to approach the problem from a different angle.

Method III

Collodion was placed on top of a liquid which is immiscible with water and which has a specific gravity greater than collodion but less than saline. (In this case paraffin oil was used.) A small drop of hemoglobin solution was dropped about 1/2 feet from the surface of the collodion solution. This height is to give the hemoglobin droplet the neccessary force to penetrate the collodion solution. As the hemoglobin droplet passes through the collodion solution a film of collodion is attached to its surface. Due to the force of gravity, this droplet falls into the paraffin oil through the collodion layer. The purpose of the paraffin oil is to allow the collodion membrane time to set on the droplet of the hemoglobin solution since if this droplet was to enter water right away the membrane would rupture since it is still immaturely set. About 30 minutes was allowed for the droplet to remain in the paraffin oil. At

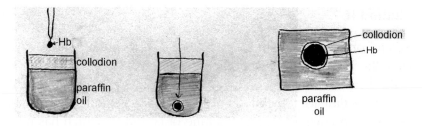

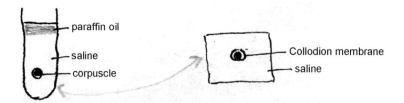

Fig. 5.

the end of this time the membrane was perfectly set, and the collodion layer together with as much as possible of the paraffin oil was removed. The remaining paraffin oil, together the hemoglobin corpuscle was then emulsified in saline, the paraffin oil having been emulsified and float to the surface leaving behind the hemoglobin corpuscle in saline.

With this method, it was successful in preparing hemoglobin corpuscles with size of about 0.2 mm in diameter. The trouble with this method being that the surface tension between collodion and paraffin oil was such that only hemoglobin droplets of appreciable size would be able to penetrate the collodion-paraffin oil interface.

Method IV — successful method

With the above principle, methods for obtaining hemoglobin corpuscles of microscopic sizes were then attempted.

Introduction: A possibility that comes into mind was the method of emulsion. Books on colloidal chemistry were read. For a while the method of emulsion appeared to be rather impractical in this case, since in the emulsions used by the chemists, it is only possible to emulsify one liquid in another and the emulsion would only remain in the liquid in which it is emulsified. While on the verge of discarding the method for emulsion, the principle found in using collodion layer and paraffin layer comes into mind. Many attempts were then made to combine this principle with the principle of emulsion.

After many fruitless and unsuccessful attempts, a method was found. This method is based entirely on the principle of the previous method.

Procedure

1. Place one part of concentrated hemoglobin solution in 10 parts of ether. Shaking this gives a dispersion of hemoglobin in ether, but this dispersion is unstable, since as soon as shaking is stopped ether and hemoglobin separate into two layers almost immediately
2. Collodion solution (ether-collodion) is put in drop by drop while shaking continuously. Collodion on entering the solution and coming into contact with the aqueous hemoglobin solution dispersed in ether, would form a layer of collodion around the hemoglobin droplets.(As shown by the observations obtained in method III). When enough collodion is adhered to the surface of the hemoglobin droplets, the dispersion becomes stable, i.e. the dispersion remain as such after shaking is discontinued, since now around each droplet is a thin layer of collodion acting as a sort of emulsifier. When about 1 part of collodion to 10 parts of solution has been added it was found that a collodion membrane with sufficient strength would be formed.
3. The problem now is to remove the hemoglobin droplets with the collodion membrane from the ether. Here, two problems were encountered. Collodion is soluble in ether, thus there is collodion in the ether containing the hemoglobin droplets. Now, collodion sets when comes in contact with water, thus water cannot be used to separate the droplets of hemoglobin from the ether since on the addition of water, the whole solution would "coagulate". Another problem is that since collodion is soluble in ether, the collodion membrane of the hemoglobin droplets would be incompletely set

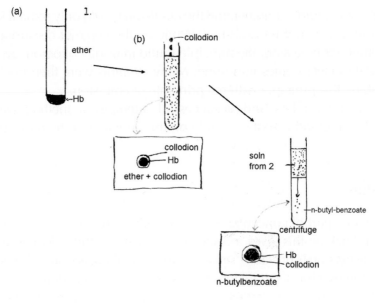

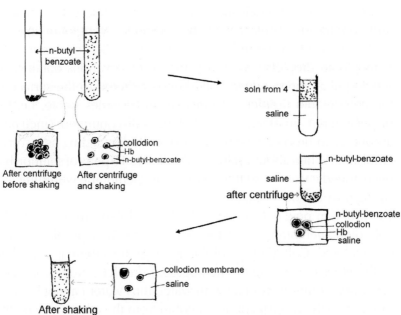

Fig. 6.

i.e. the part that is in contact with the hemoglobin solution is set, but the part in contact with ether is not set. To solve this problem, the ether containing the hemoglobin droplets with their membrane, is placed on top of n-butyl-benzoate which has been recommended as a good emulsifier. This is put in for centrifuging, so that the microscopic hemoglobin droplets could pass through the interface into the n-butyl-benzoate.

(After centrifuge, the corpuscles are in the n-butyl-benzoate. Since the collodion outside being in contact with ether previous to centrifuging is not completely set and thus is still adhesive. As a result, they tend to stick together. Here is where n-butyl-benzoate comes into use. Being a good emulsifying liquid, on shaking this, the corpuscles would be dispersed in the n-butyl-benzoate. Quite a bit of shaking is required to disperse the corpuscles which have adhered together. This dispersion allows the corpuscles to be separated from one another and thus when the collodion membrane sets, they would not be stuck together. After leaving it in this condition for 15 mins the membrane would be set completely.)

5. The dispersion of corpuscles in n-butyl-benzoate is placed over saline and centrifuge. The hemoglobin corpuscles would then be centrifuged into the saline. Now, some n-butyl-benzoate would be stuck to the corpuscles resulting in the corpuscles having the appearance of being in an aggregation in the saline. Here, another advantage is made use of n-butyl-benzoate, since on shaking the corpuscles inside the saline, the n-butyl-benzoate which may have adhered to the surface of the corpuscle would be emulsified in saline leaving behind the corpuscles with their collodion membrane.

Proof that collodion membrane is formed

To ensure that the membrane of the corpuscle is collodion and not n-butyl-benzoatel the following experiment was done:

1. Disperse hemoglobin in n-butyl-benzoate
2. Centrifuge over saline as before.

3. Hemoglobin inside n-butyl-benzoate would be centrifuged into the saline, but when this is shaken as before, the n-butyl-benzoate is emulsified and hemoglobin is left behind and dissolves in saline. From this experiment it is concluded that the membrane in the previous method was not n-butyl-benzoate. Since, collodion is the only other alternative, it may be concluded that a membrane of collodion has been formed around each hemoglobin droplet.

RESULTS

i. Size of corpuscles

Microscopic studies show that the size of the corpuscles approximates that of the red blood cells. However, there are some which are either too big or too small. It is suggested that a solution with an uniform size of corpuscles could be obtained as follows: Using the ultrafiltration membrane of the bacteriologists, first the solution could be filtered through filter which allows all particles smaller than $8\,\mu$ to pass through — this gets rid of the large corpuscles. Following this the solution is filtered through filter which allows anything smaller than $5\,\mu$ to pass through. Thus what is left behind in the filter would be corpuscles with diameters of between $5\,\mu$ and $9\,\mu$.

ii. Content of corpuscles

Microscopic studies show that the corpuscles contain a colored material. Spectrometric studies show that these pigments are oxyhemoglobin. (This is observed after the alcohol-ether solvent of the collodion solution is replaced by ether.)

iii. Oxygen content

It was found that using the principle of oximetry, a dissociation curve was obtained. However, it must be mentioned that this particular experiment is extremely unreliable, since it is found that during the experiment there has been a considerable amount of hemolysis of the corpuscles due to too much shaking.

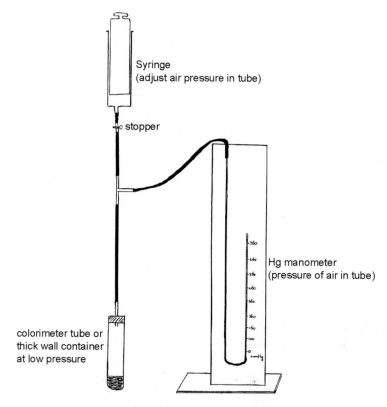

Fig. 7. Arrangement of Apparatus for Oxygen Dissociation Curve.

Method: The solution was placed in container(C) the atmospheric pressure in the container was adjusted by the syringe, and, at low pressures, by vacuum pump, to the desired pressure which is read from the manometer(M). The container was shook intermittently to allow for equilibrium to set up. After 20 minutes, the solution was read in the colorimeter with a wave length of 650 mu.

The dissociation curve obtained is then plotted as shown.

Actual amount of oxygen contained: Using a syringe with its needle inside a stop cork, a vacuum can be made in the syringe and the gas that evaporate from the solution can be measured. In one experiment the amount of gas obtained for that of a fully oxygenated solution of 0.6 ml is 0.1 ml, and that for a fully reduced one is about 0.02 ml

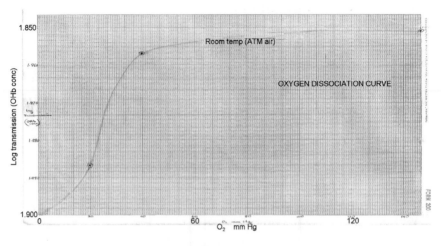

Fig. 8.

(very approximate.) It is realized that using such a small amount of solution, the results obtained is extremely unreliable. However, the solution of n-butyl-benzoate ordered has not as yet arrived and not sufficient amount was left to prepare anything more than 0.6 ml. Since it is quite inaccurate and impossible to do a dissociation curve with the possible range of gas of less than 0.08 ml, this was forced to be omitted.

(N.B. Many other possible experiments on this preparation could and should be done, but with the sudden realization that more than 120 hours have been unknowingly spent on this project at the expense of other work, and with the approaching of the final, it was necessary to put and end to this project — not without reluctance!!)

DISCUSSIONS

How far does this preparation fulfill the function of the red blood cells? Those prepared in this experiment are so crude that nothing can be said of them. However, further perfection may lead to the fulfillment of most of the main functions of the red blood cells, as discussed in the following sections.

Transportation of O_2 and CO_2

The kinetics of the formation of oxyhemoglobin from hemoglobin in the red blood cells is different from that in solution. With a pO_2 of 75 mm Hg and at 10–20°C, and with the same concentration, the half time for oxyhemoglobin saturation of the corpuscles is 0.05 seconds (i.e. for the erythrocytes), whereas that for hemoglobin solution the half time is 0.004 second. This difference in speed could be due to one or more of the following factors[5]

1. Presence of gradients of dissolved gas in the fluid surrounding the corpuscles thus leading to retardation owing to the limiting effect of diffusion.
2. Delay due to time of diffusion through membrane.
3. Dissolved gas that enter first combine with hemoglobin on the periphery of the erythrocyte and succeeding dissolving gas having to penetrate a finite distance into the corpuscle before combining with hemoglobin.

Thus in preparing artificial hemoglobin corpuscles, these factors have to be considered. Now, in the circulation the cells are well stirred and thus there is no appreciable concentration gradient of dissolved gas in the fluid surrounding the corpuscles, thus factor one can be disregarded. As far as factor 2 is concerned, that of delay due to time of diffusion through membrane, the artificial membrane used must be such that it allows easy passage of dissolved O_2 and CO_2. Also it should not be thicker than necessary, since according to Fick's equation:

$$\text{rate of diffusion} = D(C_1 - C_2)/dx$$

Where x is the thickness of the cell membrane. The studies of Weech and Michaelis[10], shows an interesting factor of permeability of collodion. He found that urea passes through the collodion membrane many times more rapidly than glycerine and about 250 times faster than for glucose. Also, he found that collodion is freely permeable to water molecules. From these observations it would appear that there shouldn't be too much trouble for O_2 or CO_2 to pass through the membrane. However, it would be interesting to find out the permeability of the collodion membrane to dissolved O_2 and CO_2, and to prepare a membrane (by method described under the section "Membrane") which affords the best permeability to O_2 and CO_2 without allowing hemoglobin molecules to leak through.

As far as factor three, that of dissolved gas which enter first combine with hemoglobin on the periphery of corpuscle, and succeeding dissolved gas that enter having to penetrate a finite distance into the corpuscles before combination with hemoglobin occurs. The solution is to have the greatest surface area for a given volume of hemoglobin solution, and to have a shape which is such that it gives the smallest depth. Of course, the biconcave disc shape of the erythrocytes is the best solution to this problem, since with this shape all the requirements mentioned are fulfilled. Would it be possible to make a biconcave hemoglobin corpuscle artificially? It has been observed under the microscope that some of the artificial corpuscles have a sort of biconcave shape-others have surface as shown:

This may have resulted from the fact that the collodion when set, tends to shrink slightly — Since the setting of membrane does not take place simultaneously in the surface of any single corpuscle. The part that still has not set is still very elastic, thus when one part of the membrane is set the volume of the corpuscle is decreased slightly, and a pressure is exerted on the enclosed solution. This pressure in turn is transmitted from the solution onto the part of the membrane which has not yet set, thus causing them to bulge out slightly. Thus it is a matter of pure chance for the biconcave, triconcave or tetraconcave and 'polyconcave' shape of the corpuscles. These shapes may in a way compensate for the lack of the actual biconcave disk shape of the erythrocytes.

Enzyme systems in the erythrocyte

As tar as the various enzyme systems present in the erythrocyte are concerned, it wouldn't be too much of a problem since it is possible to include any enzyme system in the hemoglobin solution when preparing the corpuscles. One of the most important enzyme in the erythrocyte is carbonic anhydrase in the way of transportation of CO_2, this can easily be included in the hemoglobin solution.

Buffering action of erythrocyte

Hemoglobin and potassium ions in the erythrocyte play an important part in the buffering of the pH of the body. Now, in the prepared corpuscles only hemoglobin is present for this action, and potassium ion is absent. (This may be simply a speculation, however, from the wide variety of potassium concentration in different species of animal and from the fact that hemoglobin may be present as a potassium salt — it is suspected that perhaps part of the potassium ions are kept inside the cells despite the higher concentration because of their loose ionic association with hemoglobin; the other part being kept in by the metabolic activity of the membrane. Even before this project was started, this question has always been present. Thus it would be interesting to see whether there is any concentration gradient of potassium ions in the case of an artificial membrane. If there is a gradient in concentration in the case of an inactive collodion membrane, then it may point towards the possibility of the potassium ions being present in an ionic association with the hemoglobin.)

From what has been said before, it would appear that further perfection may make it possible for the artificially produced corpuscles to meet the requirement of the erythrocytes as far as the physiological function is concerned. However, to perfect this for perfusing is quite another question. Predominant of the numerous problems would be the problem of finding a suitable membrane — a lipoprotein membrane. Although it is not a hopeless case, a tremendous amount of work has to be put in before anything could be perfected.

SUMMARY

I. Four methods for preparing artificial hemoglobin corpuscles were attempted.

II. One of the four methods was practical and appeared to be usable.

III. From this method, corpuscles approximating the size of erythrocytes were produced. The content of these corpuscles inside the collodion membrane was found to be oxyhemoglobin under spectrometric studies.

IV. An oxygen dissociation curve was obtained and the oxygen content of these preparations was found. But it is felt that the results obtained in IV is not reliable, since only one set of result was done on these, since there was not enough time for repeating these measurements.

BIBLIOGRAPHY

1. Fulton J.F.: 'Textbook of Physiology'
2. Hawk and Bergeum: 'Practical Physiological Chemistry'
3. Cambridge University Press: 'Hemoglobin'
4. Fructon: 'General Biochemistry'
5. Barcroft: 'Respiratory Function of Blood'
6. Bauer J.H: a Journey of General Physiology 18:145:1934
7. Davson H.: 'Permeability of Natural membranes'
8. Elford W.J.: Journey Pathology and Bacteriology 34:505:1931
9. Kistlera: Journey General Physiology 12:187:1928
10. Weech A.A. and Michaelis L: Journey General Physiology 12155:1928
11. Michaelis, Ellsworth R. and Weech A.A.:Journey General Physiology 1926-27 X 671
12. Michaelis and Perlzweig W.A.: Journey Gen. Physiology 1926-7X 575
13. Weiser: 'Colloids'
14. Alexander and Johnson: 'Colloidal Scinece'
15. Kruyt H.R. and van Kloter: 'Colloids'
16. McBain: 'Colloidal science'

Methods not Described in Detail in the 11 Chapters

The available methods of preparation of artificial cells are based on the original principle of emulsion for small dimension cells and the drop method for larger dimension cells. Appendix I is a reprint of this first report (Chang, 1957). There are now numerous extensions, improvements and developments of this principle and books can be written on these. Thus, Appendix II can only describe a few simpler laboratory examples. Once these are mastered, other methods can be easily reproduced from the literature.

A. EMULSION METHODS FOR ARTIFICIAL CELLS

A.1. Cellulose nitrate membrane artificial cells of micro dimensions

Cellulose nitrate membrane artificial cells are prepared using an updated procedure based on earlier publications (Chang 1957, 1964, 1965, 1972a, 2005a).

A.1.1. *Materials*

1. Hemoglobin solution containing enzymes: 15 g of hemoglobin (Bovine Hemoglobin Type 1, 2x crystallized, dialyzed and lyophilized) (Sigma; St. Louis, MO) is dissolved in 100 ml of distilled water and filtered through Whatman No. 42 (Whatman; Kent, UK) (*see* **Note**).

2. Water saturated ether: Shake analytical grade ether with distilled water in a separating funnel, and then leave the mixture standing for the two phases to separate so that the water can be discarded.
3. Fisher magnetic stirrer.
4. Cellulose nitrate solution: Spread 100 ml of USP Collodion (USP) in an evaporating dish in a well ventilated hood overnight. This allows the complete evaporation of its organic solvents leaving behind a dry thin sheet. Cut the thin sheet of polymer into small pieces and dissolve them in a 100 ml mixture containing 82.5 ml analytical grad ether and 17.5 ml analytical grade absolute alcohol (*see* **Note**).
5. Tween 20 solution. The 50% (v/v) concentration solution is prepared by mixing equal volumes of Tween 20 (Atlas Powder; Montreal, Canada) and distilled water, and then adjusting the pH to 7. For 1% (v/v) concentration solution, mix 1% of Tween 20 to the buffer solution used as the suspending media for the final microencapsulated enzyme system.

A.1.2. *Procedure for microscopic dimension artificial cells*

1. Enzymes and other materials to be microencapsulated are dissolved or suspended in 2.5 ml of the hemoglobin solution. The final pH is adjusted to 8.5 with Tris HCl, pH 8.5, and hemoglobin concentration adjusted to 100 g/l.
2. 2.5 ml of this solution are added to a 150 ml glass beaker, and 25 ml of water saturated ether is added.
3. The mixture is immediately stirred with a Fisher magnetic stirrer at 1200 rpm (setting of 5) for 5 sec.
4. While stirring is continued, 25 ml of a cellulose nitrate solution is added. Stirring is continued for another 60 sec.
5. The beaker is covered and allowed to stand unstirred at 4°C for 45 min.
6. The supernatant is decanted and 30 ml of n-butyl benzoate added. The mixture is stirred for 30 sec at the same magnetic stirrer setting.
7. The beaker is allowed to stand uncovered and unstirred at 4°C for 30 min. Then the butyl benzoate is removed completely after centrifugation at 350 g for 5 min.

8. 25 ml of the 50% (v/v) Tween solution are added. Stirring is started at a setting of 10 for 30 sec.
9. 25 ml of water is added and stirring is continued at a setting of 5 for 30 sec, then 200 ml of water is added.
10. The supernatant is removed and the artificial cells are washed 3 more times with 200 ml of the 1% Tween 20. The artificial cells are then suspended in a suitable buffer, e.g. phosphate buffer (pH 7.5). In properly prepared artificial cells, there should not be leakage of hemoglobin after the preparation (*see A.1.3. Comments*).

A.1.3. *Comments*

1. Hemoglobin at a concentration of 100 g/l is necessary for the successful preparation of cellulose nitrate membrane artificial cells. Furthermore, this high concentration of protein stabilizes the enzymes during the preparation and also during reaction and storage (Chang, 1971b). When the material (e.g. NADH) to be encapsulated is sensitive to the enzymes present in hemoglobins, highly purified hemoglobins are used. Purification using affinity chromatography on an NAD + sepharose column will be required.
2. The long-term stability of microencapsulated enzyme activity can be greatly increased by cross-linking with glutaraldehyde (Chang, 1971b). This is done at the expense of reduced initial enzyme activity.
3. When using cellulose nitrate artificial cells containing enzymes for oral administration, the permeability of the membrane may need to be decreased so as to prevent the entry of smaller tryptic enzymes. Permeability can be decreased by decreasing the proportion of alcohol used in dissolving the evaporated cellulose nitrate polymer.

A.2.   Polyamide membrane artificial cells of micro dimensions

Polyamide membrane artificial cells are prepared using an updated procedure based on earlier publications (Chang, 1964, 1965, 1972a, 2005a).

A.2.1. Materials

1. Span 85 organic solution: 0.5% (v/v) Span 85 (Atlas Powder; Montreal, Canada) in chloroform: cyclohexane (1:4).
2. Terephthaloyl organic solution: Add 100 mg of terephthaloyl chloride (ICN Pharmaceuticals Inc., Costa Mesa, CA, U.S.A.) to a 30 ml organic solution (chloroform:cyclohexane, 1:4) kept in an ice bath. Cover and stir with a magnetic stirrer for 4 h, and then filter with Whatman no. 7 paper (Whatman). Prepare just before use (*see A.2.3.Comments*).
3. Diamine-polyethyleneimine solution: Dissolve 0.378 g NaHCO₃ and 0.464 g 1.6-hexadiamine (J. T. Baker Chemical) in 5 ml distilled water that contains the material to be encapsulated. Adjust pH to 9. Add 2 ml 50% polyethyleneimine (ICN K+K Inc.) to the diamine solution, readjust pH to 9, and make up the final volume to 10 ml with distilled water. Prepare just before use (*see A.2.3. Comments*).
4. Hemoglobin solution 10 g/100 ml: Prepare as described above (A.2.1) for cellulose nitrate artificial cells, but the material to be encapsulated is dissolved in 5 ml of hemoglobin solution instead of distilled water. The final pH is adjusted to 9.

A.2.2. Procedure for preparing microscopic polyamide membrane artificial cells

Polyamide membrane artificial cells of 100 μM mean diameter are prepared using an updated method based on the earlier methods (Chang, 1964, 1965, 1972a; Chang, *et al.*, 1966).

1. Enzyme is added to 2.5 ml of the hemoglobin solution with pH and concentrations adjusted as in A.1.1.
2. 2.5 ml of the diamine-polyethyleneimine solution are added to the above solution and mixed for 10 sec in a 150 ml beaker placed in an ice bath.
3. 25 ml of Span 85 organic solution, prepared as described in Section A.2.1, are added and stirred in the Fisher magnetic stirrer at a speed setting of 2.5 for 60 sec.
4. 25 ml of terephthaloyl organic solution are added and the reaction is allowed to proceed for 3 min with the same stirring speed.

5. The supernatant is discarded and another 25 ml of the terephthaloyl organic solution are added.
6. The reaction is carried out with stirring for another 3 min. The supernatant is discarded.
7. Then, 50 ml of the Span 85 organic solution are added and stirred for 30 sec. The supernatant is discarded.

After this, the same procedure for the use of Tween 20 as described for cellulose nitrate artificial cells (A.1.1) is used for the transfer of the artificial cells into the buffer solution.

A.2.3. *Comments*

1. Failure in preparing good artificial cells is frequently due to the use of diamine or diacids that have been stored after they have been opened. A new unopened bottle will usually solve the problems. Unlike the cellulose nitrate artificial cells, in interfacial polymerization the hemoglobin solution can be replaced by a 10% polyetheleneimine solution adjusted to pH 9. However, the artificial cells prepared without hemoglobin may not be as sturdy. Cross-linking the microencapsulated enzymes with glutaraldehyde after the preparation of the enzyme artificial cells could also be carried out to increase the long-term stability of the enclosed enzymes (Chang, 1971b), although this will decrease the initial enzyme activity.
2. In multienzyme reactions requiring co-factor recycling, the cofactor can be crosslinked to dextran-70 and then encapsulated together with the enzymes. For example, NAD^+-N^6-[N-(6-aminohexyl)-acetamide] is coupled to dextran T-70, polyethyleneimine or dextran to form a water soluble NAD+ derivative and then encapsulated together with the multienzyme systems in the artificial cells. This way both the cellulose nitrate artificial cells and polyamide artificial cells can be used, and a high permeation to substrates and products is made possible. However, linking cofactor to soluble macromolecules will result in significant increases in steric hindrance and diffusion restrictions of the cofactor.

A.3. Lipid-polymer membrane artificial cells of micro dimensions that retain ATP and NAD(P)H

As described under B.2, large lipid-polyamide membrane artificial cells have been prepared (Chang 1969d, 1972a). Lipid-polyamide artificial cells of $100 \mu m$ mean diameter containing multienzyme systems, cofactors and alpha-ketoglutarate can also be prepared (YT Yu and Chang 1981) as a modification of A.2. above for microscopic polyamide artificial cells.

A.3.1. *Materials*

1. See A.2.1. above.
2. Glutamic dehydrogenase, bovine liver, type III, 40 U per mg (Sigma).
3. Alcohol dehydrogenase, yeast, 330 U per mg (Sigma).
4. Urease, 51 U per mg (Millipore).
5. Lipid-organic liquid: 1.4 g lecithin and 0.86 g cholesterol are added to 100 ml tetradecane and stirred for 4 h at room temperature. If a more permeable lipid membrane is required to allow urea to diffuse across, then the lipid compositions should be 0.43 g cholesterol and 0.7 g lecithin.

A.3.2. *Procedure*

The first part is similar to the procedure described in A.2.2. for polyamide artificial cells.

1. To 2 ml of the hemoglobin solution is added 12.5 mg glutamic dehydrogenase, 6.25 mg alcohol dehydrogenase, 0.5 mg urease, 1.18 mg ADP, and either NAD^+ (0.52 mg, 105 mg, 2.11 mg or 21.13 mg) or NADH (21.13 mg) dissolved in 0.25 ml of water. Finally, 56.5 mg alpha-ketoglutarate, 2.5 mg $MgCl_2$, and 0.93 mg KCl are added to the 0.25 ml solution.
2. 2.5 ml of the hemoglobin-enzyme solution so prepared is added to 2.5 ml of the diamine-polyethyleneime solution. The remaining steps are the same as described above (A.2.2) except that the Tween 20 steps are omitted here. Instead, after washing with the Span 85

organic solution, the following steps are carried out to apply the lipids to the polyamide membranes.

3. The artificial cells are rinsed twice with 10 ml of the lipid-organic liquid.
4. Then, another 10 ml of the lipid-organic liquid are added and the suspension is slowly rotated for 1 h at 4°C on a multi-purpose rotator.
5. After this, the supernatant is decanted and the lipid-polyamide membrane artificial cells are recovered and left in this form at 4°C without being suspended in an aqueous solution until it is added to the substrate solution just before the reaction.

The procedure takes practice and the artificial cells prepared must be tested for the absence of leakage of enzymes or cofactors before being used in experimental studies.

A.3.3. *Comments*

Lipid-polyamide membrane artificial cells containing multienzyme systems, cofactors and substrates can retain cofactors in the free form. Thus, analogous to the intracellular environments of red blood cells, free NADH or NADPH in solution inside the artificial cells is effectively recycled by the multistep enzyme systems which are also in solution. However, only lipophilic or very small hydrophilic molecules like urea can cross the membrane.

A.4. Double emulsion methods

This is based on the use of polymers that are soluble in organic solvents resulting in a solution that is not soluble in the aqueous phase. An aqueous solution containing a solution or suspension of the material to be encapsulated is added to a larger volume of this polymer solution. The aqueous phase is emulsified in the polymer solution. After this the polymer solution containing the aqueous emulsion is placed into a larger volume of aqueous phase containing an oil in water emulsifier to form a double emulsion. Each microdroplet of polymer solution contains a smaller emulsion of the aqueous material.

As the organic solvent is evaporated (polystyrene, polylactidie) or as polymerization takes place (e.g. silastic), the polymer solidified resulting in microspheres each containing an emulsion of the aqueous material. Instead of material dissolved or suspended in the original aqueous solution, crystals or power of the material can also be added directly to the polymer solution. This way, the resulting final microspheres will each contain crystals or powder. The original methods reported for polystyrene (Chang, 1965, 1972a), silastic (Chang, 1966), and biodegradable polymer (e.g. polylactide, Chang, 1976a) have been greatly extended and improved upon especially by those in the field of drug delivery. This is now a very extensive area, but is not directly related to the discussion of artificial cells in this monograph. Thus, no attempt is made here to describe the details of the methods.

B. DROP METHODS FOR LARGER ARTIFICIAL CELLS

B.1. Polymer membrane artificial cells

B.1.1. *Materials*

1. Hemoglobin solution: 10 g hemoglobin substrate (Worthington Co.) in 100 ml aqueous solution. Filter with Whatman #42 paper.
2. Diamine solution I: Solution containing 1,6-hexanediamine (0.38 M) (Eastman Kodak Co.), $NazC03$ (0.62 M), $NaHC03$ (0.17 M). Filter with Whatman #42 paper.
3. Diamine solution II: Solution containing 1,6-hexanediamine (0.1 M), NaOH (0.2 M). Filter with Whatman #42 paper.
4. Mixed organic solvent: cyclohexane-chloroform (4:1).
5. Sebacoyl chloride solution I: formed by adding 0.4 ml of sebacoyl chloride (Eastman Kodak Co.) to 100 ml of mixed organic solvent immediately before use. (Glass syringe used.)
6. Sebacoyl chloride solution II: formed by adding 0.4 ml sebacoyl chloride to 30 ml of the mixed organic solvent immediately before use. (Glass syringe used.)

7. Suspending aqueous solution containing NaCl (147 mM), CaClz (2.2 mM), and glucose (1 M).

B.1.2.   *Procedure*

1. Immediately before use, equal volumes of the hemoglobin solution and diamine solution are mixed and placed in a 30 ml glass syringe. The syringe is fitted with an 18-gauge stainless steel needle bent at right angles and placed in a Harvard infusion pump.
2. 80 ml of the sebacoyl chloride solution is added to a 140 mm diameter glass petri dish.
3. The tip of the 18-gauge needle is placed 5 mm above the surface of the sebacoyl chloride solution.
4. The infusion pump is operated at a flow rate of 2.6 ml/min for 1.6 min to produce 250 aqueous droplets. As the droplets fell into the sebacoyl chloride solution, a nylon membrane is formed around each droplet.
5. The petri dish containing the newly formed artificial cells is gently agitated by hand for 5 min. After this, 20 ml of the sebacoyl chloride solution containing 1 ml Span 85 (Atlas Powder Co.) is added.
6. The polymerization is allowed to continue with intermittent gentle agitation of the petri dish by hand for a further 10 min.
7. To stop the reaction, the sebacoyl chloride solution is decanted and the artificial cells are washed three times with 100 ml of cyclohexane.
8. The cyclohexane is discarded after 15 min and replaced with the same volume of fresh cyclohexane. This is repeated twice. After the last wash with cyclohexane, all the cyclohexane is removed by aspiration followed by evaporation with an air current.
9. The artificial cells are then suspended in 100 ml of the suspending aqueous solution and washed three times with the same solution.
10. The artificial cells are then used immediately for the flux measurements. The diameters of these large artificial cells are 3.1 ± 0.1 mm (mean:t S.D.).

B.2. Lipid-polymer membrane artificial cells

Polymer membrane artificial cells are prepared by the drop method of B.1. followed by incorporation of the lipid component as follows.

B.2.1. *Materials*

1. As in B.1.1.
2. Lipid solution containing 1.4 g egg lecithin (Nutritional Bio-chemical Co.) and 0.86 g cholesterol (Sigma Co.) in 100 ml of tetradecane are prepared on the day of use.

B.2.2. *Procedure*

1. These are prepared using the procedure described above with the following modifications. After the removal of cyclohexane by aspiration and evaporation, 50 ml of lipid dissolved in tetradecane is added.
2. The artificial cells are kept in a lipid solution in the same open glass petri dish in a fume cabinet for one hour with occasional agitation. The petri dish is tilted to keep the artificial cells completely submerged in the lipid solution.
3. At the end of 1 h the lipid solution containing the artificial cells is carefully layered over 80 ml of the suspending aqueous solution in a 150 ml glass beaker. The artificial cells are transferred from the upper non-aqueous layer to the aqueous medium below by gentle stirring with a glass stirring rod, being careful not to rupture the artificial cells. As the artificial cells entered the aqueous medium, excess lipid tends to accumulate on the top of the artificial cells. As the excess lipid is removed by gentle stirring and washing with the aqueous solution, the artificial cells settle to the bottom of the beaker.

   (The lipid content of the artificial cells can be measured. The lipid is extracted from the artificial cells with chloroform. The chloroform extract is dried and analyzed for phospholipid by digestion and inorganic phosphate determination.)

B.3.　Lipid-polymer membrane artificial cells with macrocyclic carrier

Lipid-polymer membrane artificial cells are first prepared using the method described in B.2. above. These are then suspended in an aqueous medium containing 5×10^{-6} M valinomycin and 0.4% ethanol to allow the valinomycin to be incorporated into the lipid component of the lipid-polymer membrane. In flux studies, the control polyamide artificial cells and control lipid-polyamide artificial cells are also suspended in an aqueous medium with 0.4 ethanol but with no valinomycin.

B.4.　Incorporation of Na-K-ATPase to membrane of artificial cells

B.4.1.　*Materials*

1. Diamine solution is not the same as that in B.1.; instead it contains 1,6-hexanediamine (0.1 M), NaOH (0.2 M). Filter with Whatman #42 paper.
2. Sebacoyl chloride solution is also not the same as that in B.1.; instead, it is formed by adding 0.4 ml sebacoyl chloride to 30 ml of the mixed organic solvent immediately before use. (Glass syringe used.)
3. ATPase is obtained from blood bank human red blood cells using the method of Nakoa *et al.* (1963).

B.4.2.　*Procedure*

1. A 30 ml glass syringe fitted with an 18-gauge stainless steel needle and placed in a Harvard infusion pump, contains diamine solution (from B.4.1.) but with no hemoglobin.
2. Sebacoyl chloride solution (from B.4.1.).
3. The steps after this are the same as the procedure for polyamide artificial cells (B.4.2.) all the way to the step of suspension in the aqueous solution.
4. The following steps are for the incorporation of the ouabain-sensitive Na-K-ATPase into the artificial cell membranes.

5. The artificial cells are placed in a 3 ml beaker and all the suspending aqueous solution is removed. A solution consisting of 1 ml of diamine solution (from B.4.1.), 1 ml of the hemoglobin solution (from B.1.1.), and 0.2 ml of the ATPase solution containing 400 mg of protein (from B.4.1.) is added to the beaker containing the artificial cells.

6. Individual artificial cells with a thin layer of diamine, hemoglobin, and ATPase on the surface are picked up using a hollow glass rod of 5 mm internal diameter and gently dropped into a petri dish containing 100 ml of sebacoyl chloride solution (from B.4.1.). The polymerization is allowed to continue for 15 min. During this time, the Na+-K+-ATPase is incorporated in the membrane of each artificial cell.

7. Finally, the artificial cells are washed three times with 100 ml cyclohexane and the cyclohexane is decanted and evaporated after the final washing. To transfer the artificial cells from the organic to the aqueous medium, 10 ml of Tween 20 in 90 ml of distilled water are added. The artificial cells are washed 10 times with water, stored overnight in 500 ml of water at 4°C, and then washed another three times with 100 ml of water to remove any remaining traces of Tween 20. The ATPase-incorporated artificial cells are then analyzed for ATPase activity.

B.5. Standard alginate-polylysine-alginate artificial cells (tissues, cells, microorganisms)

B.5.1. *Materials*

1. Calcium-free perfusion solution:142 mM NaCl, 6.7 mM KCl, and 10 mM HEPES, pH 7.4.
2. Collagenase perfusion buffer: 67 mM NaCl, 6.7,mM KCl, 5 mM $CaCl_2$, 0.05% collagenase, and 100 mM HEPES, pH 7.5.
3. William's E medium (Gibco Laboratories; Burlington, ON).
4. Streptomycin and penicillin (Gibco).
5. Nylon monofilament mesh 74 μM (Cistron Corp.; Elmford, NY).
6. Buffered saline: 0.85% NaCl, 20 mM D-fructose, and 10 mM HEPES, pH 7.4.

7. Stock solution of sodium alginate: 4% sodium alginate, and 0.45% NaCl.
8. Iscove's Modified Dulbecco's Medium (IMDM) (GIBCOBRL, Life Technologies, NY).
9. Nylon filter 85 μm.
10. Hepatocytes: Obtained from Wistar rats as described under method in B.5.2.1.
11. Bone marrow stem cells: These are obtained from the bone marrow of Wistar rats as described in B.5.2.2.
12. Luria Bertani (LB) medium: 10 g/L bacto tryptone, 5 g/L bacto yeast extract, and 10 g/L sodium chloride. Adjust pH to 7.5 with 1 N NaOH.
13. Genetically engineered *E. coli* DH5. *E. coli* DH5 is a nonpathogenic bacterium (Section B.5.2.3).
14. Alginate solution: 2% sodium alginate, and 0.9% sodium chloride. Sodium alginate is Kelco Gel® low viscosity alginate, Keltone LV, MW 12,000-80,000 (Merck & Co.; Clark, NJ). Sterilize before use, either by filtration or by heat for 5 min.
15. Syringe pump, compact infusion pump model 975 (Harvard Apparatus; Mill, MA).
16. Poly-L-lysine, Mw 15,000–30,000: 0.05 % poly-L-lysine (Sigma), and 10 mM HEPES buffer saline, pH 7.2.
17. Citrate solution: 3% citrate, and 1:1 HEPES buffer saline, pH 7.2.
18. Calcium chloride solution: 1.4% calcium chloride, pH 7.2.
19. Poly-L-lysine, Mw 16,100: 0.05% poly-L-lysine (Sigma), and 10 mM HEPES buffer saline, pH 7.2.
20. CaCl$_2$ solution: 100 mM CaCl$_2$, 20 mM D-fructose, and 10 mM HEPES buffer, pH 7.4.
21. Hank's Balanced Salt Solution.
22. Poly-l-lysine-fructose solution: 0.05% poly-L-lysine, 0.85% NaCl, 20 mM D-fructose, and 10 mM HEPES buffer, pH 7.4.
23. Sodium alginate 0.2%: 0.2% sodium alginate, 0.85% NaCl, 20 mM D-fructose, and 10 mM HEPES buffer, pH 7.4.
24. Sodium citrate solution: 50 mM sodium citrate, 0.47% NaCl, and 20 mM D-fructose, pH 7.4.
25. Collagenase: type IV (Sigma).

26. Trypsin: Type I-S trypsin inhibitor (Sigma).
27. HEPES: (4-(2-hydroxyethyl)-1-piperazine ethane sulphonic acid) buffer (Boehringer Mannheim; Montreal, PQ).
28. Droplet generator 1: Contains 2 co-axially arranged jets: (i) the central jet consisted of a 26G stainless steel needle (Perfektum) (Popper & Sons, Inc.; New Hyde Park, NY), and (ii) a 16G surrounding air jet, through which the sample and air are respectively passed. To prevent the extruding sample from occluding the outlet of the surrounding air jet, the tip of the sample jet is constructed such that the tip projects 0.5 mm beyond the end of the air jet.
29. Droplet generator 2: It is a larger and slightly modified variant of the droplet generator 1. It is constructed with a 13G sample jet, and an 8G surrounding air jet. The ends of the jets are cut flush to each other. A 1.7×1.1 mm PTFE capillary tube (Pharmacia P-L Biochemicals; Montreal, PQ) is inserted into the sample jet until it protrudes approximately 15 mm from the outlet of the sample jet. The end of the capillary tubing is tapered to facilitate shearing by the flow of passing air from the air jet. The capillary tubing is approximately 3.2 m in length, and has the capacity to be filled with microspheres suspended in 2.5 ml of sodium alginate.
30. Commercial generators are now available. They are easier to use and more reproducible.

B.5.2. *Procedure*

Preparation of rat hepatocytes

1. Each rat is anesthetized with sodium pentobarbital and cannulated via the portal vein.
2. The thoracic vena cava is cut and the liver is perfused with a calcium-free perfusion buffer for 10 min at 40 ml/min.
3. Afterwards, the liver is perfused with the collagenase perfusion buffer for an additional 15 min at 25 ml/min.
4. The liver is then excised, placed in William's E medium supplemented with 100 μg/ml streptomycin and penicillin, and gently shaken to free loose liver cells from the liver tissue.

5. The cells are collected, filtered through a $74\,\mu m$ nylon monofilament mesh, and centrifuged to remove connective tissue debris, cell clumps, non-parenchymal cells, and damaged cells.
6. Isolated hepatocytes are prepared for encapsulation by first washing and suspending the cells with buffered saline.
7. The cells are then mixed with a 4% stock solution of sodium alginate, to make a cell suspension consisting of 20×10^6 cells/ml of 2% sodium alginate.

Preparation of rat bone marrow stem cells

Each rat is anesthetized with sodium pentobarbital, and both femurs are isolated. Iscove's Modified Dulbecco's Medium (IMDM) is used to flush out bone marrow cells from the femurs using a 5 ml syringe with a 22-gauge needle. The cell suspension is filtered through a nylon filter (85 μm). The bone marrow cells are then washed with IMDM and centrifuged at 50 g for 10 min at 4°C; this is repeated three times. After the last wash, the bone marrow cells (nucleated cells) are kept on ice until use.

Genetically-engineered E. coli DH5 cells and microorganism

1. Genetically-engineered bacteria *E. coli* DH5, containing the urease gene from *K. aerogenes,* is used. LB growth medium is used for primary cell cultivation. Incubation is carried out in 5 ml LB in 16 ml culture tubes at 37°C in an orbital shaker at 120 rpm. For the large-scale production of biomass, for microencapsulation purpose, a 250 ml Erlenmeyer flask containing 100 ml of the suitable medium is used.
2. Log phase bacterial cells are harvested by centrifuging at 10,000 g for 20 min at 4°C.
3. Discard the supernatant, and wash the cell biomass with sterile cold water 5 times to remove media components by centrifugation at 10,000 g for 10 min at 4°C.
4. Suspend the bacterial cells in an autoclaved sodium alginate ice cold solution.

5. The viscous alginate-bacterial suspension is pressed through a 23-gauge needle using a syringe pump.
6. Compressed air through a 16-gauge needle is used to shear the droplets coming from the tip of the 23-gauge needle.
7. The droplets are allowed to gel for 15 min in a gently stirred ice-cold solution of calcium chloride (1.4%).
8. After gelation in the calcium chloride, alginate gel beads are coated with poly-l-lysine for 10 min.
9. The beads are then washed with HEPES and coated with an alginate solution (0.1%) for 4–8 min.
10. The alginate-poly-L-lysine-alginate capsules are then washed in a 3% citrate bath to liquefy the gel in the artificial cells.

Encapsulation using the standard method

1. Hepatocytes, hepatocytes and bone marrow cells, or bacterial cells are suspended in an autoclaved 0.9% sodium alginate ice cold solution.
2. The viscous alginate suspension is pressed through a 23G stainless steel needle using a syringe pump. Sterile compressed air, through a 16G coaxial stainless steel needle, is used to shear the droplets coming out of the tip of the 23G needle.
3. The droplets are allowed to gel for 15 min in a gently stirred, heat sterilized and ice cold calcium chloride solution. Upon contact with the calcium chloride buffer, alginate gelation is immediate.
4. After gelation in the calcium chloride solution, alginate gel beads are reacted with poly-L-lysine (PLL), Mw 16,100 for 10 min The positively charged PLL forms a complex of semipermeable membrane.
5. The beads are then washed with HEPES, pH 7.2 and coated with an alginate solution (0.1%) for 4 min.
6. The alginate-poly-L-lysine-alginate capsules so formed are then washed in a 3% citrate bath to liquefy the gel in the artificial cells.
7. The APA artificial cells formed, which contains entrapped hepatocytes or bacterial cells, are stored at 4°C and used for experiments. The conditions are kept sterile during the process of microencapsulation.

B.5.3. *Comments*

1. Alginates are heteropolymer carboxylic acids, coupled by 1–4 glycosidic bonds of β-D-mannuronic (M) and α-L-gluronic acid unit (G). Alkali alginate are soluble in water, whereas alginic acids and the salts of polyvalent metal cations are insoluble. Thus, when a drop of sodium alginate solution enters a calcium chloride solution, rigid spherical gels are formed by ionoirotpic gelation.

2. All the solutions are kept in an ice-cold bath before use and during the process of bioencapsulation. The pH of the solutions is kept at 7.4 by buffering with HEPES. Except for sodium alginate, the solutions are sterilized by filtering through a sterile 0.2 μm Millipore filter.

3. Alginate concentration in the tested range, 1.00–2.25% (w/v), does not affect the bacterial cell viability or cell growth. The quality of artificial cells improves with increasing alginate concentration from 1% to 1.75% (w/v). The use of 2% (w/v) alginate resulted in perfectly spherical shaped and sturdy artificial cells, with the maximum number of encapsulated bacterial cells. An increase in liquid flow rate of the alginate-cell or bacterial suspension through the syringe pump from 0.00264 to 0.0369 ml/min resulted in increase in artificial cell diameter. The flow rate in the range of 0.00724 to 0.278 ml/min resulted in good spherical artificial cells. At an air flow rate of 2 l/min., the artificial cells have an average of 500 ± 45 μm diameter. At the air flow rates increase to above 3 l/min, the artificial cells become irregular in shape. These results indicate that alginate concentration, air flow rate, and liquid flow rate are critical for obtaining artificial cells of desired characteristics and permselectivity (Prakash and Chang, 1996). We find that the following composition is most suitable for our purpose: 2% (w/v) alginate, 0.0724 ml/min liquid flow rate, 2 l/min. air flow rate. Artificial cells prepared in this way are permeable to albumin, but impermeable to molecules with higher molecular weights (Coromilli and Chang, 1993). Thus, larger molecular weight hepato-stimulating factors (Kashani and Chang, 1991) and globulin (Coromilli and Chang, 1993) cannot cross the membrane of the standard artificial cells.

B.6.   Two-step method for alginate-polylysine-alginate artificial cells (tissues, cells, microorganisms)

B.6.1.   *Procedure*

The standard method described above is not optimal for encapsulating high concentrations of cells or microorganisms. Cells or microorganisms may be trapped in the membrane matrix. This can weaken the membrane. If the cells are exposed to the surface, this may also result in loss of immunoisolation and rejection. As a result, a two-step method has been developed to prevent this problem (Wong and Chang, 1991; Chang and Wong, 1992).

1. The hepatocytes or hepatocytes and bone marrow stem cells suspended in sodium alginate are entrapped within small solid calcium alginate microspheres. This is done by filling a 5 ml syringe with the cell suspension, and extruding the sample with a syringe infusion pump through the sample jet of the first droplet generator. The droplets formed at the end of the sample jet are allowed to fall drop-wise into a Pyrex dish (125.65 mm) containing 300 ml $CaCl_2$ solution. Every 5 min the cells in the syringe are resuspended by gentle inversion of the syringe to minimize the effect of cells sedimenting in the alginate solution. To prepare smaller droplets, the air flow and infusion rate through the droplet generator are 2–3 l/min and 0.28–0.39 ml/min, respectively; the clearance height between the end of the sample jet and the surface of the calcium solution is set approximately at 20 cm. A strainer cup is fitted inside the dish to collect the droplets, and to facilitate the removal of the formed microspheres.

2. The microspheres are allowed to cure for approximately 15 min, after which they are removed and temporarily stored in Hank's Balanced Salt Solution supplemented with 10%, 100 mM $CaCl_2$.

3. 1.0 ml of formed microspheres are collected and washed three times with buffered saline.

4. The final saline washing is aspirated and 1 ml of 1.2–1.6% sodium alginate is added to 1.0 ml of washed microspheres. The sodium

alginate is prepared by diluting the 4% stock solution with buffered saline. With a 5 ml syringe, the length of the PTFE capillary tubing is filled with the sodium alginate and a suspension of microspheres. The tapered end of the capillary tubing is inserted through the top of the sample jet of the second droplet generator until the tip of the tubing extended approximately 15 mm beyond the end of the sample jet. In order to prepare larger droplets that contain the smaller alginate microspheres, the air flow and extrusion rate through the modified droplet generator should be 7–9 l/min and 0.28–0.39 ml/min, respectively. The tip of the capillary tubing is set approximately 20 cm above the surface of the calcium solution. With a 5 ml syringe still attached to the other end of the tubing, the microspheres suspension in the tubing is extruded with a Harvard infusion pump. Similarly, the drops formed at the end of the sample jet are allowed to fall dropwise into a Pyrex dish containing a strainer cup and filled with 300 ml of 100 mM $CaCl_2$.

5. The spheres are allowed to cure in the calcium solution for approximately 15 min, after which they are removed and washed with buffered saline.

6. The alginic acid matrix on the surface of the sphere is stabilized with poly-l-lysine by immersing 5 ml (settled volume) of macrospheres in 80 ml of 0.05% poly-l-lysine-fructose solution for 10 min.

7. The spheres are then drained, washed with buffered saline, and immersed into 200 ml of 0.2% sodium alginate for 10 min to apply an external layer of alginate.

8. After 10 min, the spheres are collected and immersed in 200 ml 50 mM sodium citrate solution to solubilize the intracapsular calcium alginate. This may require up to 30 min with frequent changes of the sodium citrate solution.

B.6.2. *Comments*

The two-step method prevents the entrapment of small cells in the membrane matrix. Artificial cells prepared in this way, when implanted, are much more stable and with reduced rejection (Wong and Chang, 1991; Chang and Wong, 1992).

B.7. Macroporous agar membrane artificial cells

When using cells or microorganisms to act on macromolecules, the above methods cannot be used. Thus, in using microorganisms to act on cholesterol bound to lipoprotein, the microorganisms have to be encapsulated in macroporous artificial cells (Garofalo and Chang, 1991).

B.7.1. *Materials*

1. A solution of 2% agar and 2% sodium alginate is autoclaved for 15 min and cooled to 45°C to 50°C.
2. *Pseudomonas pictorum* is cultured in nutrient broth at 25°C, followed by harvesting and resuspension in a cholesterol medium. After the suspension is cultured for 15 days at 25°C, it is used as an inoculum for biomass production. The culture is grown in bovine calf serum at 37°C for 36 h, and then harvested. This is used to prepare bacterial suspensions for immobilization. The concentration is about 0.4 mg of dry cell/ml (*see comments*) .

B.7.2. *Procedure*

3. *P. pictorum* suspended in 0.4 ml of 0.9% NaCl is added dropwise to 3.6 ml of agar alginate solution at 45°C, with stirring carried out vigorously.
4. 3 ml of the mixture obtained is kept at 45°C while it is being extruded through the syringe. The extruded drops are collected into cold (4°C) 2% calcium chloride and allow to harden. These agar-alginate beads are about 2 mm in diameter.
5. After 15 min, the supernatant is discarded and the beads are resuspended in 2% sodium citrate for 15 min.
6. Then, they are washed and stored in 0.9% saline at 4°C.

When testing for immobilized bacterial activity, 1 ml of beads/artificial cells is placed in a sterile 50 ml flask. 5 ml of serum are added and a foam plug is fitted. Samples are withdrawn at specified intervals. When empty beads or artificial cells are prepared, the bacterial suspension is replaced by saline, and all the other steps are kept the same.

B.7.3. *Comments*

1. Temperature is a very critical parameter in the immobilization of *P. pictorum* (Garofalo and Chang, 1991). A low temperature produces gelation of the polymer in the syringe or conduits. A high temperature prevents gelation but increases the mortality rate of *P. pictorum*. Exposing *P. pictorum* to 55°C for 10 min or more can completely inhibit enzymatic activity. However, up to 20 min of exposure to 45°C does not significantly inhibit cholesterol activity. Open pore agar beads stored at 4°C did not show any sign of deterioration. The beads retain its enzymatic activity even after 9 mth of storage.

References

Abshire, T.C., Pollock, B.H., Billett, A.L., *et al.* (2000) Weekly polyethylene glycol conjugated L-asparaginase compared with biweekly dosing produces superior induction remission rates in childhood relapsed acute lymphoblastic leukemia: a Pediatric Oncology Group study. *Blood* **96**:1709–1715.

Abuchowski, A., Kazo, G.M., Verhoest, C.R. Jr., *et al.* (1984) Cancer therapy with chemically modified enzymes. I. Antitumor properties of polyethylene glycol-asparaginase conjugates. *Cancer Biochem Biophys* **7**:175–186.

Aebischer, P., *et al.* (1996) Gene therapy for amyotrophic lateral sclerosis (ALS) using a polymer encapsulated xenogenic cell line engineered to secrete hCNTF. *Hum Gene Ther* **1**:851–860.

Aebischer, P., *et al.* (1996) Intrathecal delivery of CNTF using encapsulated genetically modified xenogeneic cells in amyotrophic lateral sclerosis patients. *Nat Med* **2**:696–699.

Agisaka, K., and Iwashita, Y. (1980) Modification of human hemoglobin with polyethylene glycol: a new candidate for blood substitute. *Biochem Biophys Res Commun* **97**:1076.

Agishi, T., Yamashita, N., and Ota, K. (1980). Clinical results of direct charcoal hemoperfusion for endogenous and exogenous intoxication. In: Sideman, S., and Chang, T.M.S. (eds.), *Hemoperfusion: Kidney and Liver Support and Detoxification*, Part I, pp. 255–263. Hemisphere, Washington, DC.

Agishi, T., Funakoshi, Y., Honda, H., *et al.* (1988) (Pyridoxalated hemoglobin) — (polyoxyethylene) conjugate solution as blood substitute for normothermic whole body rinse-out. *Biomater Artif Cells Artif Organs* **16**:261–270.

Agishi, T., K., Sonda, K., Nakajima, I., *et al.* (1992) Modified hemoglobin solution as possible perfusate relevant to organ transplantation. *Biomater Artif Cells Immobil Biotechnol* **20**:539–544.

Alayash, A.I. (2004) Oxygen therapeutics: can we tame hemoglobin. *Nat Rev Drug Discov* **3**:152–159.

Alayash, A.I., and Fratantoni, J.C. (1992) Effects of hypothermic conditions on the oxygen carrying capacity of crosslinked hemoglobins. *Biomater Artif Cells Immobil Biotechnol* **20**:259–262.

Alayash, A.I., Brockner Ryan, B.A., and Fratantoni, J.C. (1992) Hemoglobin-based oxygen carriers: structural alterations that affect free radical formation. *Biomater Artif Cells Immobil Biotechnol* **20**:277–282.

Al-Hendy, A., Hortelano, G., Tannenbaum, G.S., *et al.* (1996) Growth retardation — an unexpected outcome from growth hormone gene therapy in normal mice with microencapsulated myoblasts. *Hum Gene Ther* **7**:61–70.

Alison, M.R., *et al.* (2000) Cell differentiation: hepatocytes from non-hepatic adult stem cells. *Nature* **406**:257.

Allen, J.W., Hassanein, T., and Bhatia, S.N. (2001) Advances in bioartificial liver devices. *Hepatology* **34**:447–455.

Amano, I., Kano, H., and Takahira, H., *et al.* (1978). Hepatic assist system using bead-type charcoal. In: Chang, T.M.S. (ed.), *Artificial Kidney, Artificial Liver and Artificial Cells*. pp. 89–98. Plenum Press, New York.

Amberson, W.R. (1937). Blood substitute. *Biol Rev* **12**:48.

Amend, J., Ou, C., Ryan-MacFarlane, C., Anderson, P.J., *et al.* (1996) Systemic responses to SFHS-infusion in hemorrhaged dogs. *Artif. Cells Blood Substit Immobil Biotechnol* **24**:19–34.

Anderson, A., and Guroff, G. (1974) Phenylketonuria. *Comp Pathol Bull* **6**:1–31.

Anderson, P.J., Ning, J., and Biro, G.P. (1992) Clearance of differentially labelled infused hemoglobin and polymerized hemoglobin from dog plasma and accumulation in urine and selected tissues. *Biomater Artif Cells Immobil Biotechnol* **20**:781–788.

Arbeloa, M., Neufeld, R.J., and Chang, T.M.S. (1986) Analysis of microencapsulated urease, fluidized bed reactor under steady and transient state operating conditions. *J Memb Sci* **29**:321–331.

Asselin, B.L., Whitin, J.C., Coppola, D.J., *et al.* (1993) Comparative pharmacokinetic studies of three asparaginase preparations. *J Clin Oncol* **11**:1780–1786.

Avramis, V.I., Sencer, S., Periclou A.P., *et al.* (2002) A randomized comparison of native *Escherichia coli* asparaginase and polyethylene glycol conjugated asparaginase for treatment of children with newly diagnosed standard-risk acute lymphoblastic leukemia: a Children's Cancer Group study. *Blood* **99**:1986–1994.

Awasthi, V.D., Garcia, D., Klipper, R., *et al.* (2004) Neutral and anionic LEH: effect of postinserted poly(ethylene glycol)-distearolyphosphatidyletheanolamine on distribution and circulation kinetics. *J Pharmacol Exp Ther* **309**:241–248.

Bachoud-Levi, A.C., *et al.* (2000) Neuroprotective gene therapy for Huntington's disease using a polymer encapsulated BHK cell line engineered to secrete human CNTF. *Hum Gene Ther* **11**:1723–1729.

Bakker, J.C., Berbers, W.A.M., Bleeker, W.K., *et al.* (1992) Characteristics of crosslinked and polymerized hemoglobin solutions. *Biomater Artif Cells Immobil Biotechnol* **20**:233–242.

Bakker, J.C., Bleaker, W.K., and Van der Plas, J. (1988). Hemoglobin interdimerically cross-linked with NEPLP. *Biomater Artif Cells Artif Organs* **16**:635–636.

Bakker, J.C., and Bleaker, W.K. (1994) Blood substitutes based on modified hemoglobin. *Vox Sang* **67**(Suppl.):139–142.

Banghan, A.D., *et al.* (1965) Diffusion of univalent ions across the lamellae of swollen phospholipids. *J Mol Biol* **13**:238–252.

Baron, B.J., and Scalea, T.M. (1996) Acute blood loss. *Emerg Med Clin North Am* **14**:35–55.

Barre, P., Gonda, A., and Chang, T.M.S. (1986) Routine clinical applications of hemoperfusion-hemodialysis in uremia. *Int J Artif Organs* **9**:305–308.

Barre, P., and Chang, T.M.S. (1984) Application of coated charcoal hemoperfusion to clinical practice. *Renal Fam J* **6**:20–25.

Barve, A., Sen, A.P., and Gulati, A. (1997) Dose response effect of diaspirin crosslinked hemoglobin (DCLHb) on systemic hemodynamics and regional blood circulation in rats. *Artif Cells Blood Substit Immobil Biotechnol* **25**:75–84.

Bartels, O., Neidhardt, B., and Neidhardt, M., *et al.* (1977). Untersuchungen und entahrungen mit der kohlehaemoperfusion bei leberkoma (investigation and experience with carbon hemoperfusion for liver coma). In: Demling, L., and Bartels, O. (eds.), *Entgiftung mit Hamoperfusion (Detoxification by Hemoperfusion)*. pp. 110. Bundemagel-Verlag, Freiburg.

Barwe, S.P., Kim, S., Rajasekarean, S.A., *et al.* (2007). Janus model of the Na, K-ATPase beta-subunit transmembrane domain: distinct faces mediate alpha/beta assembly and beta-beta homo-oligomerization. *J Mol Biol* **19**:706–714.

Basic, D., Vacek, I., and Sun, A.M. (1996) Microencapsulation and transplantation of genetically engineered cells: a new approach to somatic gene therapy. *Artif Cells Blood Substit Immobil Biotechnol* **24**:219–255.

Bayley, H. (1997) Building doors to cells. *Sci Am* **277**:62–67.

Bayley, H. (1999) Designed membrane channels and pores. *Curr Opin Biotechnol* **10**:94–103.

Beach, M.C., Morley, J., Spiryda, L., and Weinstock, S.B. (1988) Effects of liposome encapsulated hemoglobin on the reticuloendothelial system. *Biomater Artif Cells Artif. Organs* **16**:635–636.

Beissinger, R.L., Farmer, M.C., and Gossage, J.L. (1986) *Trans Am Soc Artif Intern Organs* **32**:58.

Bekyarova, G., Yankova, T., and Galunska, B. (1996) Increased antioxidant capacity, suppression of free radical damage and erythrocyte aggregability after combined application of alpha-tocopherol and FC-43 perfluorocarbon emulsion in early postburn period in rats. *Artif Cells Blood Substit Immobil Biotechnol* **24**:629–641.

Benesch, R., Benesch, R.E., Yung, S., and Edalji, R. (1975) Hemoglobin covalently bridged across the polyphosphate binding site. *Biochem Biophys Res Commun* **63**:1123.

Bensinger, W., Baker, D.A., and Buckner, C.D., *et al.* (1981) Immunoadsorption for removal of A and B blood-group antibodies. *New Engl J Med* **314**:160–162.

Better, O.S., Brunner, G., Chang, T.M.S., *et al.* (1979) Controlled trials of hemoperfusion for intoxication. *Ann Intern Med* **91**:925.

Biberstein, M.P., Ward, D.M., and Ziegler, M.G. (1983) Use of betablockade and hemoperfusion for acute theophylline poisoning. *Western J Med* **141**:485–490.

Biessels, P.T.M., Berbers, G.A.M., Broeders, G.C.J.M., *et al.* (1992) Detection of membrane fragments in hemoglobin solutions. *Biomater Artif Cells Immobil Biotechnol* **20**:439–442.

Binette, T.M., Dufour, J.M., and Korbutt, G.S. (2001) *In vitro* maturation of neonatal porcine islets: a novel model for the study of islet development and xenotransplantation. *Ann NY Acad Sci* **944**:47–61.

Biro, G.P. (1985) Fluorocarbons in the resuscitation of hemorrhage. *Perfluorochem Oxygen Transport* **23**:143.

Biro, G.P. (1993) Perfluorocarbon-based red blood cell substitutes. *Transfus Med Rev* **7**:84–95.

Biro, G.P. (1994) Central hemodynamics and blood flow distribution during infusion of perflubron emulsion or its vehicle. *Artif Cells Blood Substit Immobil Biotechnol* **22**:1343–1353.

Biro, G.P., Ou, C., Ryan-MacFarlane, C., and Anderson, P.J. (1995) Oxyradicaly generation after resuscitation of hemorrhagic shock with blood or stromafree hemoglobin. *Artif Cells Blood Substit Immobil Biotechnol* **23**:631–645.

Bleeker, W., Agterberg J., La Hey, E., *et al.* (1996) Hemorrhagic disorders after administration of gluteraldehyde-polymerized hemoglobin. In: Winslow, R.M., Vandegriff, K.D., and Intaglietta, M. (eds.), *Blood Substitutes: New Challenges.* pp. 112–123. Birkhäuser, Boston.

Bleeker, W.K., Zappeij, L.M., den Boer, P.J., *et al.* (1995) Evaluation of the immunogenicity of polymerized hemoglobin solutions in a rabbit model. *Artif. Cells, Blood Substit Immobil Biotechnol* **23**:461–468.

Bloch, J., Bachoud-Lévi, A.C., Déglon, N., *et al.* (2004) Neuroprotective gene therapy for Huntington's disease, using polymer-encapsulated cells engineered to secrete human ciliary neurotrophic factor: results of a phase I study. *Hum Gene Ther* **15**:968–975.

Blume, V., Helmsteadt, D., Sybrecht, G., *et al.* (1976a) Haperfusions therpie des akuten Leberversagens. *Dtsch Med Wochenschr* **14**:559.

Bolin, R.B., Geyer, R.P., and Nemo, G.J. (eds.) (1983) *Advances in Blood Substitute Research.* Alan R. Liss Inc., New York.

Bonhard, K. (1988). Hemoglobin preparations for perfusion and infusion problems of large-scale production. *Biomater Artif Cells Artif Organs* **16**:85–92.

Bonomini, V., and Chang, T.M.S. (eds.) (1982) *Hemoperfusion* (Contributions to Nephrology Series). S. Karger AG, Basel, Switzerland.

Borden, E.C. (ed.) (2002) *Melanoma: Biologically Targeted Therapeutics.* Humana Press, Totowa, New Jersey.

Bourget, L., and Chang, T.M.S. (1985) Phenylalanine ammonia-lyase immobilized in semipermeable microcapsules for enzyme replacement in phenylketonuria. *FEBS Lett* **180**:5–8.

Bourget, L., and Chang, T.M.S. (1984) Artificial cell-microencapsulated phenylalanine ammonia-lyase. *Appl Biochem Biotechnol* **10**:57–59.

Bourget, L., and Chang, T.M.S. (1986) Phenylalanine ammonia-lyase immobilized in microcapsules for the depleture of phenylalanine in plasma in phenylketonuric rat model. *Biochim Biophys Acta* **883**:432–438.

Bourget, L., and Chang, T.M.S. (1989) Effects of oral administration of artificial cells immobilized phenylalanine ammonia-lyase on intestinal amino acids of phenylketonuric rats. *J Biomater Artif Cells Artif Organs* **17**:161–182.

Bowersox, J.C., and Hess, J.R. (1994) Trauma and military applications of blood substitutes. *Artif Cells Blood Substit Immobil Biotechnol* **22**:145–159.

Bowes, M.P., Burhop, K.E., Zivin, J.A., and Abraham, D.J. (1994) Diaspirin cross-linked hemoglobin improves neurological outcome following reversible but not irreversible CNS ischemia in rabbits. *Stroke* **25**:2253–2257.

Braun, R.D., Linsenmeier, R.A., and Goldstick, T.K. (1992) New perfluorocarbon emulsion improves tissue oxygenation in retina. *J Appl Physiol* **72**:1960–1968.

Bruni, S., and Chang, T.M.S. (1995) Kinetics of UDP-glucuronosyltransferase in bilirubin conjugation by encapsulated hepatocytes for transplantation into Gunn rats. *J Artif Organs* **19**:449–457.

Bruni, S., and Chang, T.M.S. (1989) Hepatocytes immobilized by microencapsulation in artificial cells: effects on hyperbiliru-binemia in Gunn rats. *J Biomater Artif Cells Artif Organs* **17**:403–412.

Bruni, S., and Chang, T.M.S. (1999) Comparision of bilirubin conjugation in encapsulated hepatocytes, hepatocyte homogenate and intact hepatocytes. *Artif Cells Blood Substit Immobil Biotechnol* **27**:343–356.

Bruni, S., and Chang, T.M.S. (1999) Kinetic studies of hepatocyte UDP glucuronosyltransferase: evidence of an allosteric enzyme. *Artif Cells Blood Substit Immobil Biotechnol* **27**:357–366.

Bruni, S., and Chang, T.M.S. (1995) Effects of donor strains and age of the recipient in the use of microencapsulated hepatocytes to control hyperbilirubinemia in the Gunn rat. *Int J Artif Organs* **18**:332–339.

Bruni, S., and Chang, T.M.S. (1991) Encapsulated hepatocytes for controlling hyperbilirubinemia in Gunn rats. *Int J Artif Organs* **14**:239–241.

Bucci, E., Fronticelli, C., Razynska, A., *et al.* (1992) Hemoglobin tetramers stabilized with polyaspirins. In: Chang, T.M.S. (ed.), *Blood Substitutes and Oxygen Carrier.* pp. 76–85. Marcel Dekker, New York.

Bucci, E., and Fronticelli, B. (1992) Hemoglobin tetramers stabilized with polyaspirins. *Biomater Artif Cells Immobil Biotechnol* **20**:243–252.

Bucci, E., Razynska, A., Kwansa, H., *et al.* (1996) Production and characteristics of an infusible oxygen-carrying fluid based on hemoglobin intramolecularly cross-linked with sebacic acid. *J Lab Clin Med* **128**:146–153.

Buehler, P.W., Mehendale, S., Wang, H., *et al.* (2000) Resuscitative effects of polynitroxylated alpha-alpha-cross-linked Hb following severe hemorrhage in the rat. *Free Radic Biol Med* **29**:764–774.

Bunn, H.F., and Jandl, J.H. (1968) The renal handling of hemoglobin. *Trans Assoc Am Phys* **81**:147.

Burhop, K.E., Farrell, T.L., Nigro, C., *et al.* (1992) Effects of intravenous infusions of diaspririn cross-linked hemoglobin (DCLHb) on sheep. *Biomater Artif Cells Immobil Biotechnol* **20**:581–586.

Burhop, K.E., and Estep, T.E. (2001) Hb induced myocardial lesions. *Artif Cells Blood Substit Immobil Biotechnol* **29**:101–106.

Burkard, M.E., and Van Liew, I.I.D. (1994) Oxygen transport to tissue by persistent bubbles: theory and simulation. *J Appl Physiol* **77**:2874–2978.

Calafiore, R., *et al.* (1999) Transplantation of minimal volume microcapsules in diabetic high mammalians. *Ann NY Acad Sci* **875**:219–232.

Campbell, J., and Chang, T.M.S. (1975) Enzymatic recycling of coenzymes by a multienzyme system immobilized within semipermeable collodion microcapsules. *Biochim Biophys Acta* **397**:101–109.

Campbell, J., and Chang, T.M.S. (1976) The recycling of NAD+ (free and immobilized) within semipermeable aqueous microcapsules containing a multi-enzyme system. *Biochem Biophys Res Commun* **69**:562–569.

Campbell, J., and Chang, T.M.S. (1977) Immobilized multienzyme systems and coenzyme requirements: perspectives in biomedical applications. In: Chang, T.M.S. (eds.), *Biomedical Applications of Immobilized Enzymes and Proteins*, Vol 2, pp. 281–302. Plenum Press, New York.

Campbell, J., and Chang, T.M.S. (1978) Microencapsulated multi-enzyme systems as vehicles for the cyclic regeneration of free and immobilized coenzymes. *Enzyme Eng* **3**:371–377.

Campbell, J., Chawla, A.S., and Chang, T.M.S. (1977) The use of immobilized enzyme-membrane sandwich reactors in automated analysis. *Anal Biochem* **83**:330–335.

Casciani, C., Splendiani, G., and Chang, T.M.S. (1993) Hemoperfusion, sorbent and immobilized bioreactants. *Biomater Artif Cells Immobil Biotechnol* (special issue).

Cattaneo, M., and Chang, T.M.S. (1991) The potential of microencapsulated ureasezeolite oral sorbent for the removal of urea in uraemia. *J Am Soc Artif Intern Organs* **37**:80–87.

Cerny, L.C., Barnes, B., Fisher, L., *et al.* (1996) A starch-hemoglobin resuscitative compound. *Artif Cells Blood Substit Immobil Biotechnol* **24**:153–160.

Chan, P.H., Yang, G.Y., Chen, S.F., *et al.* (1991) Cold-induced brain edema and infarction are reduced in transgenic mice overexpressing CuZn-superoxide dismutase. *Ann Neurol* **29**:482–486.

Chang, E.J., Lee, T.H., Mun, K.C., *et al.* (2004a) Effects of polyhemoglobin-antioxidant enzyme complex on ischemia-reperfusion in kidney. *Transplant Proc* **36**: 1952–1954.

Chang, E.J., Lee, S.H., Mun, K.C., *et al.* (2004b) Effect of artificial cells on hepatic function after ischemia-reperfusion injury in liver. *Transplant Proc* **36**:1959–1961.

Chang, T.M.S. (1957) Hemoglobin corpuscles. Report of a research project for Honours Physiology, Medical Library, McGill University. Also reprinted 1988 as part of "30th Anniversary in Artificial Red Blood Cells Research." *Biomater Artif Cells Artif Organs* **16**:1–9.

Chang, T.M.S. (1964) Semipermeable microcapsules. *Science* **146**:524–525.

Chang, T.M.S. (1965) Semipermeable aqueous microcapsules. Ph.D. thesis, McGill University.

Chang, T.M.S., Macintosh, F.C., and Mason, S.G. (1966) Semipermeable aqueous microcapsules: I. Preparation and properties. *Can J Physiol Pharmacol* **44**: 115–128.

Chang, T.M.S. (1966) Semipermeable aqueous microcapsules ("artificial cells"): with emphasis on experiments in an extracorporeal shunt system. *Trans Am Soc Artif Intern Organs* **12**:13–19.

Chang, T.M.S., Johnson, L.J., and Ransome, O.J. (1967) Semipermeable aqueous microcapsules: IV. Nonthrombogenic microcapsules with heparin-complexed membranes. *Can J Physiol Pharmacol* **45**:705–715.

Chang, T.M.S. (1967) Microcapsules as artificial cells. *Science* **3**:62–67.

Chang, T.M.S., and Poznansky, M.J. (1968a) Semipermeable aqueous microcapsules (artificial cells): V. Permeability characteristics. *Biomed Mater Res* **2**:187–199.

Chang, T.M.S., and Poznansky, M.J. (1968b) Semipermeable microcapsules containing catalase for enzyme replacement in acatalsaemic mice. *Nature* **218**:242–245.

Chang, T.M.S., Pont, A., Johnson, L.J., and Malave, N. (1968) Response to intermittent extracorporeal perfusion through shunts containing semipermeable microcapsules. *Trans Am Soc Artif Intern Organs* **14**:163–168.

Chang, T.M.S. (1969a) Removal of endogenous and exogenous toxins by a microencapsulated absorbent. *Can J Physiol Pharmacol* **47**:1043–1045.

Chang, T.M.S. (1969b) Artificial cells made to order. *New Sci* **4**:18–19.

Chang, T.M.S. (1969c) Clinical potential of enzyme technology (1st incentive lecture). *Sci Tools* **16**:33–39.

Chang, T.M.S. (1969d) Lipid-coated spherical ultrathin membranes of polymer or cross-linked protein as possible cell membrane models. *Fed Proc* **28**:461.

Chang, T.M.S., and Malave, N. (1970) The development and first clinical use of semipermeable microcapsules (artificial cells) as a compact artificial kidney. *Trans Am Soc Artif Intern Organs* **16**:141–148.

Chang, T.M.S. (1970) Nonthrombogenic microcapsules. US Patent, 3, 522, 346.

Chang, T.M.S. (1971a) The *in vivo* effects of semipermeable microcapsules containing L-asparaginase on 6C3HED lymphosarcoma. *Nature* **229**:117–118.

Chang, T.M.S. (1971b) Stabilisation of enzymes by microencapsulation with a concentrated protein solution or by microencapsulation followed by cross-linking with glutaraldehyde. *Biochem Biophys Res Commun* **44**:1531–1536.

Chang, T.M.S. (1971c) Nonthrombogenic microcapsules. Canadian Patent, 876, 100.

Chang, T.M.S. (1971d) A new concept in artificial kidneys using semipermeable microcapsules. *Chemeca '70*. pp. 48–54. Butterworths, Australia.

Chang, T.M.S., Gonda, A., Dirks, J.H., and Malave, N. (1971a) Clinical evaluation of chronic, intermittent, and short term hemoperfusions in patients with chronic renal failure using semipermeable microcapsules (artificial cells) formed from membrane-coated activated charcoal. *Trans Am Soc Artif Intern Organs* **17**:246–252.

Chang, T.M.S., MacIntosh, F.C., and Mason, S.G. (1971b) Encapsulated hydrophilic compositions and methods of making them. Canadian Patent, 873, 815.

Chang, T.M.S. (1972a) *Artificial Cells*. Charles C. Thomas, Springfield, IL (out of print but available for free online viewing at www.artcell.mcgill.ca).

Chang, T.M.S. (1972b) Haemoperfusions over microencapsulated adsorbent in a patient with hepatic coma. *Lancet* **2**:1371–1372.

Chang, T.M.S. (1972c) A new approach to separation using semipermeable microcapsules (artificial cells): combined dialysis, catalysis, and absorption. In: Li, N. (ed.), *Recent Developments in Separation Science*. pp. 203–216. Chemical Rubber Co., Cleveland, OH.

Chang, T.M.S. (1972d) *In vitro* and *in vivo* kinetics of enzymes immobilized by microencapsulation. *Biotechnol Bioeng* **14**:520–525 (symposium volume **3**:395–399).

Chang, T.M.S. (1972e) Microcapsule artificial kidney and medium molecular weight clearance. *Proc Eur Dial Transplant Assoc* **9**:568–574.

Chang, T.M.S. (1972f) Effects of local applications of microencapsulated catalase on the response of oral lesions to hydrogen peroxide in acatalasemia. *J Den Res* **51**:319–321.

Chang, T.M.S., Gonda, A., Dirks, J.H., *et al.* (1972g) ACAC microcapsule artificial kidney for the long term and short term management of eleven patients with chronic renal failure. *Trans Am Soc Artif Intern Organs* **18**:465–472.

Chang, T.M.S., Coffey, J.F., Barre, P., *et al.* (1973a) Microcapsule artificial kidney: treatment of patients with acute drug intoxication. *Can Med Assoc J* **108**:429–433.

Chang, T.M.S., Coffey, J.F., Lister, C., *et al.* (1973b) Methaqualone, methyprylon, and glutethimide clearance by the ACAC microcapsule artificial kidney: *in vitro* and in patients with acute intoxication. *Trans Am Soc Artif Intern Organs* **19**:87–91.

Chang, T.M.S. (1973a) Blood compatible microcapsules containing detoxicants. US Patent, 3, 725, 113.

Chang, T.M.S. (1973b) The use of semipermeable microcapsules (artificial cells) as microsystems for separation. *Am Chem Soc* **33**:581–583.

Chang, T.M.S. (1973c) Immobilization of enzymes, adsorbents, or both within semipermeable microcapsules (artificial cells) for clinical and experimental treatment of metabolite-related disorders. In: Bergsma, D. (ed.), *Enzyme Therapy for Congenital Diseases*, Birth Defects: Original Article Series, Vol. 9, No. 2, pp. 66–76. Williams & Wilkins Co., Baltimore, MD.

Chang, T.M.S. (1973d) L-Asparaginase immobilized within semipermeable micro-capsules: *in vitro* and *in vivo* stability. *Enzyme* **14**:95–104.

Chang, T.M.S. (1973e) Recent trends in biocompatible materials and micro-capsular absorbents for kidney function replacement. In: Kenedi, R.M. (ed.), *Perspectives in Biomedical Engineering.* pp. 39–44. University Park Press (MacMillan), Baltimore, MD.

Chang, T.M.S. (1973f) Biomedical applications of artificial cells. *Biomed Eng* **8**:334–339.

Chang, T.M.S., and Migchelsen, M. 1973. Characterization of possible "toxic" metabolites in uremia and hepatic coma based on the clearance spectrum for larger molecules by the ACAC microcapsule artificial kidney. *Trans Am Soc Artif Intern Organs* **19**:314–319.

Chang, T.M.S., Migchelsen, M., Coffey, J.F. and Stark, A. (1974). Serum middle molecule levels in uremia during long term intermittent hemoperfusions with the ACAC (coated charcoal) microcapsule artificial kidney. *Trans Am Soc Artif Intern Organs* **20**:364–371.

Chang, T.M.S. (1974a) Enzymes immobilized by microencapsulation: preparation and biomedical applications. In: Salmona, M., Saronio, C., and Garattini, S. (eds.), *Insolubilized Enzymes.* pp. 15–27. Raven Press, New York.

Chang, T.M.S. (1974b) Platelet-surface interaction: effect of albumin coating or heparin complexing on thrombogenic surfaces. *Can J Physiol Pharmacol* **52**:275–285.

Chang, T.M.S. (1974c) A comparison of semipermeable microcapsules and standard dialysers for use in separation. *J Sep Purification Meth* **3**:245–262.

Chang, T.M.S. (1974d) A spherical ultrathin polymeric membrane system containing enzymes. *Polymer Preprints* **15**:381–382.

Chang, T.M.S. (1974e) Artificial cells and microcapsules: comparison of structural and functional differences. In: Vandager, J.E. (ed.), *Microencapsulation: Processes and Applications.* pp. 95–102. Plenum Press, New York.

Chang, T.M.S. (1974f) Effects of different routes of *in vivo* administration of microencapsulated enzymes. *Enzyme Eng* **2**:419–424.

Chang, T.M.S. (1974g) Future prospectives: pediatric hemodialysis based on artificial cells. *Clin Proc Child Hosp Nat Med Cent* **30**:304–311.

Chang, T.M.S. (1974h) Performance characteristics of the microcapsule artificial kidney. In: Whelpton, D. (ed.), *Renal Dialysis.* pp. 135–147. Sector Publisher, London.

Chang, T.M.S. (1975a) Rationale for the use of the ACAC microcapsule artificial kidney for the treatment of patients with chronic renal failure. *Ateneo Parmense Acta Biomed* **46**:353–368.

Chang, T.M.S. (1975b) Artificial cells as carriers for biologically active materials in therapy. In: Mattila, M.J. (ed.), *Clinical Pharmacology.* pp. 81–90. Proceedings of the 6th International Congress of Pharmacology, Finland, Vol 5.

Chang, T.M.S. (1975c) Artificial cells. *Chem Tech* **5**:80–85.

Chang, T.M.S. (1975d) Artificial kidney and artificial liver. *Lancet* **4**:1451–1452.

Chang, T.M.S. (1975e) Experience with treatment of acute liver failure patients by hemoperfusion over biocompatible microencapsulated (coated) charcoal. In: Williams, R., and Murray-Lyon L.M. (eds.), *Artificial Liver Support.* pp. 229. Pitman Medical Publishers, London.

Chang, T.M.S. (1975f) Biomedical research for the treatment of drug abuse and related problems. In: Maykut, M.O. (ed.), *Biomedical Research in Narcotic Abuse Problems.* pp. 69–78. Health and Welfare Canada.

Chang, T.M.S. (1975g) Microencapsulated adsorbent hemoperfusion for uremia, intoxication and hepatic failure. *Kidney Int* **7**:S387–S392.

Chang, T.M.S. (1975h) The one shot vaccine in socio-economic and ethical implications of enzyme engineering. In: Heden, C.G. (ed.), *International Federation of Institutes for Advanced Studies.* pp. 17–28. Stockholm, Sweden.

Chang, T.M.S. (1975i) A microcapsule artificial kidney: a step toward miniaturization. *J Dial Transplant* **4**:23–25.

Chang, T.M.S. (1975j) Experience with the treatment of acute liver failure patients by haemoperfusion over biocompatible microencapsulated (coated) charcoal. In: Williams, R. (eds.), *Artificial Support Systems for Acute Hepatic Failure.* pp. 229–233. Whitefriars Press, London.

Chang, T.M.S. (1975k) Immobilized enzymes and their biomedical applications. In: Weetall, H.H. (ed.), *Immobilized Enzymes, Antigens, Antibodies, and Peptides.* pp. 245–292. Marcel Dekker, Inc., New York.

Chang, T.M.S., Chirito, E., Barre, P., *et al.* (1975) Clinical performance characteristics of a new combined system for simultaneous hemoperfusion hemodialysis-ultrafiltration in series. *Trans Am Soc Artif Intern Organs* **21**:502–508.

Chang, T.M.S. (1976a) Biodegradable semipermeable microcapsules containing enzymes, hormones, vaccines, and other biologicals. *J Bioeng* **1**:25–32.

Chang, T.M.S. (1976b) Enzymes immobilized by microencapsulation within spherical ultra-thin polymeric membranes. *J Macromol Sci Chem* **A10**:245–258.

Chang, T.M.S. (1976c) Hemoperfusion alone and in series with ultrafiltration or dialysis for uremia, poisoning and liver failure. *Kidney Int* **10**:S305–S311.

Chang, T.M.S. (1976d) Methods for the therapeutic applications of immobilized enzymes. *Meth Enzymol* **XLIV**:676–698.

Chang, T.M.S. (1976e) Microencapsulation of enzymes and biologicals. *Meth Enzymol* **XLIV**:201–217.

Chang, T.M.S. (1976f) Semipermeable microcapsules as artificial cells: clinical applications and perspectives. In: Nixon, J. (ed.), *Microencapsulation*. pp. 57–65. Marcel Dekker, Inc., New York.

Chang, T.M.S. (1976g) Microcapsule artificial kidney: including updated preparative procedures and properties. *Kidney Int* **10**:S218–S224.

Chang, T.M.S. (1976h) Blood compatible microcapsules containing detoxicants. Canadian Patent, 982, 941.

Chang, T.M.S. (1976) Hemoperfusion alone and in series with ultrafiltration or dialysis for uremia, poisoning and liver failure. *Kidney Int* **10** (Suppl. 7):S305.

Chirito, E., Reiter, B., Lister, C., and Chang, T.M.S. (1977a) Artificial liver: the effect of ACAC microencapsulated charcoal hemoperfusion on fulminant hepatic failure. *Artif Organs* **1**:76–83.

Chang, T.M.S. (1977a) Encapsulation of enzymes, cell contents, cells, vaccines, antigens, antiserum, cofactors, hormones, and proteins. In: Chang, T.M.S. (ed.), *Biomedical Applications of Immobilized Enzymes and Proteins,* Vol 1, pp. 69–90. Plenum Press, New York.

Chang, T.M.S. (1977b) Protective effects of microencapsulation (coating) on platelet depletion and particulate embolism in the clinical applications of charcoal haemoperfusion. In: Kenedi, R.M., Courtney, J.M., Gaylor, J.D.S. and Gilchrist, T. (eds.), *Artificial Organs*. pp. 164–177. MacMillan Press Ltd., London, UK.

Chang, T.M.S., Chirito, E., and Barre, P., *et al.* (1977b) Clinical evaluation of the clearance profiles of a portable, compact, dialysate-free system incorporating microencapsulated charcoal hemoperfusion for blood purification with ultrafiltration for fluid removal. *J Dial* **1**:239–259.

Chang, T.M.S. (ed.) (1977c) *Biomedical Applications of Immobilized Enzymes & Proteins,* Vol 1. Plenum Publishing Corporation, New York.

Chang, T.M.S. (ed.) (1977d) *Biomedical Applications of Immobilized Enzymes & Proteins,* Vol 2. Plenum Publishing Corporation, New York.

Chang, T.M.S. (1977e) Artificial kidney, artificial liver, and detoxifiers based on artificial cells, immobilized proteins, and immobilized enzymes. In: Chang, T.M.S. (ed.), *Biomedical Applications of Immobilized Enzymes and Proteins,* Vol 1. pp. 281–295. Plenum Press, New York.

Chang, T.M.S. (1977f) Criteria, evaluation, and perspectives of various microencapsulated charcoal hemoperfusion systems. *J Dial Transplant* **6**: 50–53.

Chang, T.M.S. (1977g) Enzyme immobilized within spherical ultrathin polymeric membranes. *Polymer Grafts in Biochemistry-Bioactive Material* (Preprint). *American-Japanese Chemical Societies Joint Meeting,* Hawaii.

Chang, T.M.S. (1977h) Experimental therapy using semipermeable micro-capsules containing enzymes and other biologically active material. In: Chang, T.M.S. (ed.), *Biomedical Applications of Immobilized Enzymes and Proteins,* Vol 1, pp. 147–162. Plenum Press, New York.

Chang, T.M.S. (1977i) Introduction. In: Chang, T.M.S. (ed.), *Biomedical Applications of Immobilized Enzymes and Proteins,* Vol 1. Plenum Press, New York.

Chang, T.M.S. (1977j) Novel approaches in artificial kidney, artificial liver, and detoxification. *J Dialysis Transplant* **8**:20–26.

Chang, T.M.S. (1977k) Rationale and strategies for the therapeutic applications of immobilized enzymes. In: Chang, T.M.S. (ed.), *Biomedical Applications of Immobilized Enzymes and Proteins,* Vol 1, pp. 93–104. Plenum Press, New York.

Chang, T.M.S. (1978a) A 1978 perspective of hemoperfusion. *Artif Organs* **2**:359–362.

Chang, T.M.S. (1978b) Artificial cells and charcoal hemoperfusion *Int J Artif Organs* **1**:111–112.

Chang, T.M.S. (1978c) Artificial cells for artificial kidney, artificial liver and detoxification. In: Chang, T.M.S. (ed.), *Artificial Kidney, Artificial Liver and Artificial Cells.* pp. 57–77. Plenum Press, New York.

Chang, T.M.S. (1978d) Biomedical application of microencapsulated enzyme. *Hind Antibiot Bull* **2**:5–9.

Chang, T.M.S. (1978e) Future of haemodialysis. In: Davidson, A. (ed.), *Dialysis '77.* pp. 269–277. Pitman Medical Publisher, London, UK.

Chang, T.M.S. (1978f) Hemoperfusion in chronic schizophrenia. *Int J Artif Organs* **1**:253–255.

Chang, T.M.S. (1978g) Introductory chapter. In: Chang, T.M.S. (ed.), *Artificial Kidney, Artificial Liver and Artificial Cells.* pp. 3–8. Plenum Press, New York.

Chang, T.M.S. (1978h) Microcapsule artificial kidney in replacement of renal function: with emphasis on adsorbent hemoperfusion. In: Drukker, W., Parsons, F.M. and Maher, J.F. (eds.), *Replacement of Renal Function by Dialysis.* pp. 217–231. Martinus Nijhoff Publishers, The Hague.

Chang, T.M.S. (1978i) Microcapsule artificial kidney. *Proceedings of the Canadian Clinical Engineering Conference.* pp. 134–137. National Research Council, Ottawa, Canada.

Chang, T.M.S. (1978j) Artificial cells in 1977. *Int J Artif Organs* **1**:35–43.

Chang, T.M.S. (1978k) Biomedical applications of microencapsulation. In: Kondo, T. (ed.), *Microencapsulation.* pp. 369–376. Tokyo, Japan.

Chang, T.M.S. (ed.) (1978l) *Artificial Kidney, Artificial Liver, and Artificial Cells,* Plenum Publishing Corporation, New York.

Chang, T.M.S. (1978m) Adsorbent hemoperfusion — general discussion. In: Chang, T.M.S. (ed.), *Artificial Kidney, Artificial Liver and Artificial Cells.* pp. 301–304. Plenum Press, New York.

Chang, T.M.S., Lister, C., Chirito, E., *et al.* (1978) Effects of hemoperfusion rate and time of initiation of ACAC charcoal hemoperfusion on the survival of fulminant hepatic failure rats. *Trans Am Soc Artif Intern Organs* **24**:243–245.

Chang, T.M.S., and Kuntarian, N. (1978). Galactose conversion using a microcapsule immobilized multienzyme cofactor recycling system. *Enzyme Eng* **4**: 193–197.

Chang, T.M.S., and Malouf, C. (1978) Artificial cells microencapsulated multienzyme system for converting urea and ammonia to amino acid using alpha-ketoglutarate and glucose as substrate. *Trans Am Soc Artif Intern Organs* **24**:18–20.

Chang, T.M.S., Chirito, E., Barre, P., et al. (1979a) Long-term clinical assessment of combined ACAC hemoperfusion-ultrafiltration in uremia. *Artif Organs* **3**:127–131.

Chang, T.M.S., Malouf, C., and Resurreccion, E. (1979b) Artificial cells containing multienzyme systems for the sequential conversion of urea into ammonia, glutamate, then alanine. *Artif Organs* **3**(Suppl.):284–287.

Chang, T.M.S., and Malouf, C. (1979) Effects of glucose dehydrogenase in converting urea and ammonia into amino acid using artificial cells. *Artif Organs* **3**:38–41.

Chang, T.M.S. (1979a) A visitor's impression of uremia therapy in China. *Artif Organs* **3**(Suppl.):34–35.

Chang, T.M.S. (1979b) Artificial cells as drug carriers in biology and medicine. In: Gregoriadis, G. (ed.), *Drug Carriers in Biology and Medicine*. pp. 271–285. Academic Press, Inc., New York.

Chang, T.M.S. (1979c) Assessments of clinical trials of charcoal hemoperfusion in uremic patients. *Clin Nephrol* **11**:111–119.

Chang, T.M.S. (1979d) Hemoperfusion: clearance, blood compatibility and safety. *Int J Artif Organs* **2**:276–277.

Chang, T.M.S. (1980a) Artificial cells: the use of hybrid systems. *Artif Organs* **4**: 264–271.

Chang, T.M.S. (1980b) Artificial red blood cells. *Trans Am Soc Artif Intern Organs* **26**:354–357.

Chang, T.M.S. (1980c) Biomedical applications of artificial cells. In: Goldberg, E.P., and Nakajima, A. (eds.), *Biomedical Polymers*. pp. 171–187. Academic Press, New York.

Chang, T.M.S. (1980d) Blood compatible coating of synthetic immunoadsorbents. *Trans Am Soc Artif Intern Organs* **26**:546–549.

Chang, T.M.S. (1980e) Clinical experience with ACAC coated charcoal hemoperfusion in acute intoxication. *Clin Toxicol* **17**:529–542.

Chang, T.M.S. (1980f) Guest editorial: hybrid artificial organs. *Artif Organs* **4**:258.

Chang, T.M.S. (1980g) New approaches using immobilized enzymes for the removal of urea and ammonia. *Enzyme Eng* **5**:225–229.

Chang, T.M.S. (1980h) Present status and future prospectives of artificial cells in hemoperfusion. In: Sideman, S., and Chang, T.M.S. (eds.), *Hemoperfusion: Artificial Kidney and Liver Support and Detoxification*. pp. 93–104. Hemisphere Publishing Corp., Washington, DC.

Chang, T.M.S. (1980i) Use of immobilized enzymes in medical practice. In: Danehy, J.P., and Wolnak, B. (eds.), *Enzymes: The Interface Between Technology and Economics*. pp. 123–130. Marcel Dekker Inc., New York.

Chang, T.M.S. (1980j) Animal model studies: enzyme entrapment, neural delivery, and transplantation. In: Desnick, R.J. (ed.), *Enzyme Therapy in Genetic Diseases:*

2. *Birth Defects.* Original Article Series, Vol. 15. No. 1, pp. 231–235. Alan R. Liss Inc., New York.

Chang, T.M.S. (1980k) Microencapsulation of specific sorbents for medical applications. In: Giordanco, C. (ed.), *Sorbents and Their Clinical Applications.* pp. 195–217. Academic Press, Inc., New York.

Chang, T.M.S. (1980l) Guest editor of symposium volume on "Hybrid Artificial Organs". *Artif Organs* **4**:264–301.

Chang, T.M.S. (1980m) Hemoperfusion: closing remarks. In: Sideman, S., and Chang, T.M.S. (eds.), *Hemoperfusion: Artificial Kidney and Liver Support and Detoxification.* pp. 403–405. Hemisphere Publishing Corp., Washington, DC.

Chang, T.M.S. (1978n) Adsorbent hemoperfusion — general discussion. In: Chang, T.M.S. (ed.), *Artificial Kidney, Artificial Liver and Artificial Cells.* pp. 301–304. Plenum Press, New York.

Chang, T.M.S., Lister, C., Chirito, E., *et al.* (1978) Effects of hemoperfusion rate and time of initiation of ACAC charcoal hemoperfusion on the survival of fulminant hepatic failure rats. *Trans Am Soc Artif Organs* **24**:243–245.

Chang, T.M.S., Espinosa-Melendez, E., Francoeur, T.E., *et al.* (1980a) Albumin-collodion activated coated charcoal hemoperfusion in the treatment of severe theophylline intoxication in a 3-year old patient. *Pediatrics* **65**:811–814.

Chang, T.M.S., and Lister, C. (1980) Analysis of possible toxins in hepatic coma including the removal of mercaptan by albumin-collodion charcoal. *Int J Artif Organs* **3**:108–112.

Chang, T.M.S., Winchester, J., Rosenbaum, J., *et al.* (1980b) Adsorbent hemoperfusion for blood purification. *Trans Am Soc Artif Intern Organs* **26**:593–597.

Chang, T.M.S. (1981a) Lecture series on artificial cells. Monograph in Chinese of a 12 hour lecture series by Chang, T.M.S at the Chinese Academy of Sciences, Beijing, China. Monograph prepared and translated into Chinese by Wang, H.Z., Zheng, W.H., Miao, E.H., *et al.* Chinese Medical Association, Hopei, Shijiazhuang.

Chang, T.M.S. (1981b) Hemoperfusion, exchange transfusion, cross circulation, liver perfusion, hormones and immobilized enzymes. In: Brunner, G., and Schmidt, F.W. (eds.), *Artificial Liver Support.* pp. 126–133. Springer-Verlag, Berlin, Germany.

Chang, T.M.S. (1981c) Artificial cells. *Proceedings of the World Congress of Chemical Engineering.*

Chang, T.M.S. (1981d) Biocompatibility and experimental therapy of immobilized enzymes and proteins. In: Williams, D.F. (ed.), *Systemic Aspects of Biocompatibility.* Vol 2, pp. 181–188.

Chang, T.M.S. (1981e) Controversies and issues in hemoperfusion. *Artif Organs* **5**(Suppl.):10–15.

Chang, T.M.S. (1981f) Current status of sorbent microencapsulation. *Advances in Basic and Clinical Nephrology, Proceedings of the 8th International Congress of Nephrology.* pp. 400–406. S. Karger Publisher, Basel, Switzerland.

Chang, T.M.S., and Shu, C.D. (1981) Artificial cells for hemoperfusion. *Chinese J Organs Transplant, Zhonghua Qiguan Yizhi Zazhi* **2**:50–55.

Chang, T.M.S., and Lister, C. (1981) Middle molecules in hepatic coma and uremia. *Artif Organs* **4**(Suppl.):169–172.

Chang, T.M.S., Lacaille, Y., Picart, X., *et al.* (1981) Composite artificial kidney: a single unit combining hemodialysis and hemoperfusion *Artif. Organs* **5**(Suppl.): 200–203.

Chang, T.M.S. (1982a) Earlier hemoperfusion in fulminant failure. *Lancet* **2**:1039.

Chang, T.M.S. (1982b) Artificial cells encapsulated enzymes and adsorbents in congenital metabolic disorders. In: Cockburn, F., and Gitzelman, R. (eds.), *Inborn Errors of Metabolism in Humans.* pp. 131–138. Alan R. Liss Inc., New York.

Chang, T.M.S. (1982c) Blood compatible adsorbent hemoperfusion for extracorporeal blood treatment. In: Criban, T.C.J., Vissar, J., and Nivard, R.J.F. (eds.), *Theoretical Aspects, Industrial and Biomedical Applications of Affinity Chromatography and Related Techniques.* pp. 357–364. Elsevier Scientific Publishing Co., Amsterdam, Netherlands.

Chang, T.M.S. (1982d) Hemoperfusion and artificial cells. *Contrib Nephrol* **29**:3–6.

Chang, T.M.S. (1982e) Hemoperfusion in 1981. *Contrib Nephrol* **29**:11–22.

Chang, T.M.S. (1982f) Microcapsules as artificial cells. In: Langer, R. (ed.), *Microcapsules and Microcarriers in Biotechnology.* Massachusetts Institute of Technology, Massachusetts, USA.

Chang, T.M.S. (1982g) Plasma perfused over immobilized protein A for breast cancer. *New Engl J Med* **306**:936.

Chang, T.M.S., Barre, P., Kuruvilla, S., *et al.* (1982a) Hemoperfusion-hemodialysis in a single unit: composite artificial kidney. *J Artif Support Sys* **1**:63–67.

Chang, T.M.S., Barre, P., Kuruvilla, S., *et al.* (1982b) Phase one clinical trial of a new composite artificial kidney: a single unit combining dialysis with hemoperfusion. *Trans Am Soc Artif Inter Organs* **28**:43–48.

Chang, T.M.S., Shu, C.D., Yu, Y.T., and Grunwald, J. (1982c) Artificial cells immobilized enzymes for metabolism disorders. In: Crawford, M., Gibbs, D., and Watts, R.W.E. (eds.), *Advances in the Treatment of Inborn Errors of Metabolism.* pp. 175–184. John Wiley & Sons Ltd., UK.

Chang, T.M.S., Yu, Y.T., and Grunwald, J. (1982d) Artificial cells, immobilized multienzyme systems and cofactors. *Enzyme Eng* **6**:451–561.

Chang, T.M.S. (1983a) A multifaceted approach to artificial liver support. In: Starzl, T.E., Chang T.M.S., Williams, R., and Nose, Y. (eds.), *Liver Support/Transplants and Artificial Organs,* Panel Conference, *Trans Am Soc Artif Intern Organs* **29**: 795–799.

Chang, T.M.S. (1983b) Artificial cells. *Proceedings of the Symposium on Plastics in Artificial Organs.* American Chemical Society, Seattle, USA.

Chang, T.M.S. (1983c) Membrane biotechnology. *Artificial Cells, Proceedings, Membrane Technology Conference,* Oregon, USA.

Chang, T.M.S. (1983d) Past, present and future of hemoperfusion. In: Piskin, E., and Chang, T.M.S. (eds.), *Past, Present and Future of Artificial Organs.* pp. 109–127. Meteksan Pub. Co., Ankara, Turkey.

Chang, T.M.S. (1983e) Artificial liver support: requirements and approaches. *Int J Artif Organs* **6**:178–182.

Chang, T.M.S. (1983f) Past, present and future of artificial cells. In: Piskin, E., and Chang, T.M.S. (eds.), *Past, Present and Future of Artificial Organs.* pp. 10–24. Meteksan Publ. Co., Ankara, Turkey.

Chang, T.M.S., and Barre, P. (1983) Effect of desferrioxamine on removal of aluminium and iron by coated charcoal haemoperfusion and haemodialysis. *Lancet* Nov:1051–1053.

Chang, T.M.S., Barre, P., Kuruvilla, S., *et al.* (1983) Laboratory and clinical assessments of hemoperfusion and hemodialysis in a single unit: composite artificial kidney. In: Piskin, E., and Chang, T.M.S. (eds.), *Past, Present and Future of Artificial Organs.* pp. 196–200. Meteksan Pub. Co., Ankara, Turkey.

Chang, T.M.S., and Zheng, W.H. (1983) Hemoperfusion with collodion coated Chinese petroleum based spherical charcoal. *Chinese J Biomed Eng* **2**:170–177.

Chang, T.M.S., Espinosa-Melendez, E., Lal, S., *et al.* (1983) B endorphin and other middle molecules in chronic schizophrenia and removal by adsorbent hemoperfusion. In: Ernst K., and Seidel, K. (eds.), *Detoxifications in Psychiatric Disorders.* pp. 69–77. Veb Verlag Volk und Gesundheit, Berlin.

Chang, T.M.S. (1984a) Artificial cells. In: Lim, F. (ed.), *Biomedical Applications of Microencapsulation.* pp. 85–103. CRC Press, Boca Raton, Florida, USA.

Chang, T.M.S. (1984b) Microencapsulation: an introduction. *Appl Biochem Biotechnol* **10**:3–4.

Chang, T.M.S. (1984c) Artificial cells in hemoperfusion and other applications. *Nephron* **36**:161–168. (Special festschrift issue for Professor Kolff.)

Chang, T.M.S. (1984d) Artificial cells in medicine and biotechnology. *Appl Biochem Biotechnol* **10**:5–24.

Chang, T.M.S. (1984e) Artificial cells. In: Paul, J.P., Gaylor, J.D.S., Courtney, J.M., and Gilchrist, T. (eds.), *Biomaterial in Artificial Organs.* pp. 103–111. McMillan Press, London.

Chang, T.M.S. (1984f) Artificial cells. In: Gebelein, C.G. (ed.), *Polymeric Materials and Artificial Organs.* pp. 171–180. American Chemical Society, Washington, DC.

Chang, T.M.S. (1984g) Coated charcoal hemoperfusion. *Life Support Syst* **2**: 99–106.

Chang, T.M.S. (1984h) Liver support systems. In: Bartlett R.H. (ed.), *Life Support Systems in Intensive Care.* pp. 461–485. Year Book Publisher.

Chang, T.M.S. (ed.) (1984i) Microencapsulation and artificial cells. Humana Press, Clifton, New Jersey, USA.

Chang, T.M.S., Lister, C., and Varma, R. (1984a) Kinetics of coated charcoal hemoperfusion in the removal of Desferal for the elimination of aluminum. In: Atsumi, K., Maekawa, M., and Ota, K. (eds.), *Progress in Artificial Organs.* pp. 602–605. ISAO Press, Cleveland, USA.

Chang, T.M.S., Odaka, M., Schreiner, G.E., *et al.* (1984b) Adsorbents in hemoperfusion. In: Atsumi, K., Maekawa, M., and Ota, K. (eds.), *Progress in Artificial Organs.* pp. 624–627. ISAO Press, Cleveland, USA.

Chang, T.M.S. (1985a) Artificial cells with regenerating multienzyme systems. *Meth Enzymol* **112**:195–203.

Chang, T.M.S. (1985b) Biomedical applications of artificial cells immobilized enzymes, cells and other biologically active materials. In: Laskin, A.I. (ed.), *Enzymes and Immobilized Cells in Biotechnology.* pp. 263–281. Addison-Wesley Publishing Co., Reading, Massachusetts, USA.

Chang, T.M.S. (1985c) Biotechnology of artificial cells including its application in artificial organs. "The Principles, Applications and Egulations of Biotechnology in Industry, Agriculture and Medicine". In: Cooney, C.L., and Humphrey, A. (eds.), *Comprehensive Biotechnology.* pp. 53–72. Pergamon Press, New York.

Chang, T.M.S. (1985d) The future potentials of blood substitutes in transfusion. *Transplantation Today* Sept:40–45.

Chang, T.M.S., Barre, P., and Kuruvilla, S. (1985) Long-term reduced time hemoperfusion-hemodialysis compared to standard dialysis: a preliminary crossover analysis. *Trans Am Soc Artif Intern Organs* **31**:572–576.

Chang, T.M.S., and Ho, B.L. (eds.) (1985) Hemoperfusion and artificial organs. China Academic Publishers, Beijing, China.

Chang, T.M.S. (1986a) Application of artificial cells in hemoperfusion, blood substitutes, immunosorption, artificial pancreas, and enzyme replacement. In: Nose, Y., Kjellstrand, C., and Ivanovich, P. (eds.), *Progress in Artificial Organs.* pp. 232–238. ISAO Press, Cleveland, USA.

Chang, T.M.S. (1986b) Experimental artificial liver support with emphasis on fulminant hepatic failure: concepts and reviews. *Seminars in Liver Diseases Series on Fulminant Hepatic Failure* **6**:148–159.

Chang, T.M.S. (1986c) Experimental evaluation of artificial liver support. In: Williams, R. (ed.), *Liver Failure.* pp. 191–206. Clinics in Critical Medicine Series, Churchill Livingstone Inc., UK.

Chang, T.M.S. (1986d) Hemoperfusion in poisoning, fulminant hepatic failure, aluminium and iron removal, uremia and immunosorption. In: Nose, Y., Kjellstrand, C., Ivanovich, P. (eds.), *Progress in Artificial Organs.* pp. 595–600. ISAO Press, Cleveland, USA.

Chang, T.M.S., and Trevino Becerra, A. (eds.) (1986) Hemoperfusion. *Int J Artif Organs* (special issue), **9**:279–368.

Chang, T.M.S. (1987a) Modified hemoglobin as red blood cell substitutes. *J Biomater Artif Cells Artif Organs* **15**:323–328.

Chang, T.M.S. (1987b) Recycling of NAD(P) by multienzyme systems immobilised by microencapsulation in artificial cells. *Meth Enzymol* **136**:67–82.

Chang, V., Battistin, R., Rodrigue, A., and Chang, T.M.S. (1987a) Flow analysis for optimal configuration of hemoperfusion device. *Int J Artif Organs* **10**:115–120.

Chang, T.M.S., and Varma, R. (1987) Pyridoxalated heterogenous and homologous polyhemoglobin and hemoglobin: systemic effects of replacement transfusion in rats previously receiving immunising doses. *J Biomater Artif Cells Artif Organs* **15**: 443–452.

Chang, T.M.S., Farmer, M., Geyer, R.P., and Moss, G. (1987b) Blood substitutes based on modified hemoglobin and fluorochemicals. *TASAIO J* **33**.

Chang, T.M.S., and Nicolaev, N. (eds.) (1987) Hemoperfusion, sorbent and immobilized bioreactants. *Int J Biomater Artif Cells Artif Organs* (special issue) **15**:1–321.

Chang, T.M.S. (1988a) Artificial cells with ultrathin lipid-polymer or lipid-protein membranes In: Gaber, B.P., Schnur, J.M., and Chapman, D. (eds.), *New Technological Applications of Phospholipid Bilayers, Thin Films and Vesicles.* pp. 215–223. Plenum Press, New York.

Chang, T.M.S. (1988b) Medical application of artificial cells in transfusion, phenylketonuria, essential amino acid production and liver failure. *Ann N Y Acad Sci* **542**:507–514.

Chang, T.M.S. (1988c) 30th anniversary of artificial red blood cell research. *J Biomater Artif Cells Artif Organs* **16**:1–9.

Chang, T.M.S. (1988d) Blood substitutes prepared from modified hemoglobin. In: Lowe, K.C. (ed.), *Blood Substitutes: Preparation, Physiology and Medical Application.* pp. 87–93.

Chang, T.M.S. (1988e) Methods in the medical applications of immobilized proteins, enzymes and cells. *Meth Enzymol* **137**:444–457.

Chang, T.M.S. (1988f) The role of biotechnology in bioartificial or hybrid artificial cells and organs: editorial. *J Biomater Artif Cells Artif Organs* **16**:704–705.

Chang, T.M.S. (1988g) Red blood cells substitutes: microencapsulated hemoglobin and cross-linked hemoglobin including pyridoxylated polyhemoglobin and conjugated hemoglobin. *J Biomater Artif Cells Artif Organs* **16**:11–29.

Chang, T.M.S., and Varma, R. (1988) Immunological and systemic effects of transfusions in rats using pyridoxylated hemoglobin and polyhemoglobin from homologous and heterogenous sources. *J Biomater Artif Cells Artif Organs* **16**: 205–215.

Chang, T.M.S., Langer, R., Sparks, R., *et al.* (1988) Drug delivery system in biotechnology. *J Artif Organs* **12**:248–251.

Chang, T.M.S., and Lister, C. (1988/1989) Plasma intestinal concentration patterns suggestive of entero-portal recirculation of amino acids: effects of oral administration of asparaginase, glutaminase and tyrosinase immobilized by microencapsulation in artificial cells. *J Biomater Artif Cells Artif Organs* **16**:915–926.

Chang, T.M.S. (1989a) Preparation and characterization of xanthine oxidase immobilized by microencapsulation in artificial cells for the removal of hypoxanthine. *Biomater Artif Cells Artif Organs* **17**:611–616.

Chang, T.M.S. (1989b) The use of modified hemoglobin as an oxygen carrying blood substitute. *Transfu Med Rev* **3**:213–218.

Chang, T.M.S. (1989c) Hybrid artificial cells encapsulation and immobilisation of cell culture, organelles and multienzyme system. In: Baquey, C., and Dupuy, B. (eds.), *Hybrid Artif Organs* **177**:239–246. (colloque inserm).

Chang, T.M.S., Barre, P., Lister, C., and Kuruvilla, C. (1989a) Artificial cells in medical applications with emphasis on hemoperfusion for aluminium removal and crossover control clinical trial on hemoperfusion-hemodialysis. *Contrib Nephrol* **70**:237–249.

Chang, T.M.S., and Geyer, R. (eds.) (1989) *Blood Substitutes.* Marcel Dekker Publisher, USA.

Chang, T.M.S., Bourget, L., and Lister, C. (1989b) Novel method of amino acid removal by immobilized bioreactant based on new findings of enterorecirculation of amino acids. Canadian Patent 1992.

Chang, T.M.S. (1990a) Biotechnological and medical application of immobilization of hepatocytes, microorganisms and enzyme system by microencapsulation in artificial cells. *Enzyme Eng* **10**:109–115.

Chang, T.M.S. (1990b) Modified hemoglobin: endotoxin and safety studies. *J Biomater Artif Cells Artif Organs* **18**:vii–viii.

Chang, T.M.S., and Lister, C. (1990) A screening test of modified hemoglobin blood substitute before clinical use in patients based on complement activation of human plasma. *J Biomater Artif Cells Artif Organs* **18**:693–702.

Chang, T.M.S., and Daka, J. (1990) A novel method for bilirubin removal. Canadian Patent granted.

Chang, T.M.S., Bourget, L., and Lister, C. (1991) Novel method of amino acid removal by immobilized bioreactant based on new findings of enterorecirculation of amino acids. USA Patent No. 5, 147, 641, (issued Sept. 15, 1992).

Chang, T.M.S., and Odaka, M. (eds.) (1991) Hemoperfusion, sorbent and immobilized bioreactants. *Biomater Artif Cells Immobil Biotechnol* (special issue) **19**:1–298.

Chang, T.M.S. (1991a) Artificial cells. In: Dulbecco, R. (ed.), *Encyclopedia of Human Biology*, Vol. 1, pp. 377–383. Academic Press, Inc., San Diego California.

Chang, T.M.S. (1991b) Therapeutic application of immobilized protein and cells. In: Taylor, R.F. (ed.), *Immobilized Proteins of Bioprocess Technology Series*. pp. 305–318. Marcel Dekker, Inc., New York.

Chang, T.M.S. (ed.) (1992a) *Blood Substitutes and Oxygen Carriers*. Marcel Dekker Publisher, USA. (Book review "The authors of the papers represent a list of 'Who's Who' in the field." *Hematol Pathol*).

Chang, T.M.S. (1992b) Artificial cells for artificial liver support. In: Brunner, G. (ed.), *Artificial Liver Support*. pp. 261–270. Springer-Verlag, Berlin.

Chang, T.M.S. (1992c) Living cells and microorganism immobilized by microencapsulation inside artificial cells. In: Goosen, M. (ed.), *Fundamentals of Animal Cell Encapsulation and Immobilization*. pp. 183–196. CRC Press, Boca Raton, USA.

Chang, T.M.S. (1992d) Artificial cells in immobilization biotechnology. *Biomater Artif Cells Immobil Biotechnol* **20**:1121–1143.

Chang, T.M.S. (1992e) Artificial cells — 35 years. *J Artif Organs* **16**:8–12.

Chang, T.M.S. (1992f) Artificial liver support based on artificial cells with emphasis on encapsulated hepatocytes. *J Artif Organs* **16**:71–74.

Chang, T.M.S. (1992g) Blood substitutes based on modified hemoglobin prepared by encapsulation or crosslinking. *Biomater Artif Cells Immobil Biotechnol* **20**:154–174.

Chang, T.M.S. (1992h) Hybrid artificial cells microencapsulation of living cells. *ASAIO J* **38**:128–130.

Chang, T.M.S. (1992i) Recent advances in artificial cells with emphasis on biotechnological approaches based on microencapsulation. In: Donbrow, M. (ed.), *Microcapsules and Nanoparticles in Medicine and Pharmacy*. pp. 323–339. CRC Press Inc., Boca Raton, USA.

Chang, T.M.S., and Lister, C. (1992) An *in vitro* screening test for modified hemoglobin to bridge the gap between animal safety studies and clinical use in patients. *Biomater Artif Cells Immobil Biotechnol* **20**:481–487.

Chang, T.M.S., and Varma, R. (1992) Effect of a single replacement of Ringer lactate, hypertonic saline/dextran, 7g% albumin, stroma-free hemoglobin, o-Raffinose polyhemoglobin, or whole blood, on the long term survival of unanesthetized rats with lethal hemorrhagic shock after 67% acute blood loss. *Biomater, Artif Cells Immobil Biotechnol* **20**:433–440.

Chang, T.M.S., and Wong, H. (1992) A novel method for cell encapsulation in artificial cells. USA Patent No. 5, 084, 350, (issued Jan. 28, 1992).

Chang, T.M.S., and Yu, W.P. (1992) Biodegradable polymer membrane containing hemoglobin as potential blood substitutes. British Provisional Patent No. 9219426.5, (issued Sept. 14, 1992).

Chang, T.M.S., Lister, C., Nishiya, T., and Varma, R. (1992a) Immunological effects of hemoglobin, microencapsulated hemoglobin, polyhemoglobin and conjugated hemoglobin using different immunizing schedules. *Biomater Artif Cells Immobil Biotechnol* **20**:519–526.

Chang, T.M.S., Mobed, M., Nishiya, T. (1992b) Purification and characterization of liposomes encapsulating hemoglobin as potential blood substitutes. *Biomater Artif Cells Immobil Biotechnol* **20**:53–70.

Chang, T.M.S. (1993b) Safety studies of modified hemoglobin as oxygen carrying blood substitute. *J Hematol Pathol* **7**:47–52.

Chang, T.M.S. (1993c) Bioencapsulation in biotechnology. In: Oncelet, D., and Lievremont M. (guest eds.), *Biomater Artif Cells Immobil Biotech* **21**:291–298.

Chang, T.M.S., and Lister, C. (1993a) Use of finger-prick human blood samples as a more convenient way for *in vitro* screening of modified hemoglobin blood substitutes for complement activation: a preliminary report. *Biomater Artif Cells Immobil Biotechnol* **21**:685–690.

Chang, T.M.S., and Lister, C. (1993b) Screening test for modified hemoglobin blood substitute before use in human. USA Patent No. 5, 200, 323.

Chang, T.M.S. (1994a) Artificial cell including blood substitutes and biomicro-encapsulation: from ideas to applications. *J Artif Cells Blood Substit Immobil Biotechnol* **22**:vii–xiv.

Chang, T.M.S. (1994b) Efficacy of rbc substitutes *J Artif Cells Blood Substit Immobil Biotechnol* **22**:ii–iii.

Chang, T.M.S., Reiss, J., Winslow, R. (eds.) (1994) Blood substitutes: general. *Artif Cells, Blood Substit Immobil Biotechnol* (special issue) **22**:123–360.

Chang, T.M.S., and Lister, C. (1994) Plasma and blood for safety studies in research, development, industrial production and preclinical analysis. In: Chang, T.M.S., Winslow, R., and Reiss, J. (guest eds.), *J Artif Cells Blood Substit Immobil Biotechnol* **22**:171–180.

Chang, T.M.S., and Varma, R. (1994) Assessment of blood substitute: 1. Efficacy studies in anesthetized and conscious rats with loss of 1/3, 1/2 and 2/3 total blood volume. In: Chang, T.M.S., Winslow, R., and Reiss, J. (guest eds.), *J Artif Cells Blood Substit Immobil Biotechnol* **22**:159–169.

Chang, T.M.S. (1995) Modified hemogolobin as red blood cell substitutes. In: Shek, P.N. (ed.), *Liposome in Biomedical Application.* pp. 209–216. Harwood Academic Publisher GmbH USA.

Chang, T.M.S. (1995a) Artificial cells with emphasis on bioencapsulation in biotechnology. *Biotechnol Ann Rev* **1**:267–296.

Chang, T.M.S. (1995b) Crosslinked hemoglobin being well into clinical trials, increasing research efforts are now on a second generation red blood cell substitute based on encapsulated hemoglobin. *Artif Cells Blood Substit Immobil Biotechnol* **23**:257–263.

Chang, T.M.S. (1995c) Kunstliche Zellen und kunstlicher Blutersatz (Artificial Cells & Blood Substitutes). In: Bucherl, E.S. (ed.), *Kunstliche Organe (Artificial Organs).* pp. 32–40. Wissenschaftliche Buchgesellschaft, Darmstadt, Germany.

Chang, T.M.S. (1995d) Present status and furture perspectives of modified hemoglobin as red blood cell substitutes. In: Tsuchida, E. (ed.), *Red Blood Cells Substitutes.* pp. 21–34. John Wiley & Sons, New York.

Chang, T.M.S., and Weinstock, S. (eds.) (1995) Blood substitutes. *Artif Cells Blood Substit Immobil Biotechnol* (special issue) **23**:257–459.

Chang, T.M.S., Bourget, L., and Lister, C. (1995) A new theory of enterorecirculation of amino acids and its use for depleting unwanted amino acids using oral enzyme-artificial cells, as in removing phenylalanine in phenylketonuria. *Artif Cells Blood Substit Immobil Biotechnol* **25**:1–23.

Chang, T.M.S. (1996) Editorial: past, present and future perspectives on the 40th anniversary of hemoglobin based red blood cell substitutes. *Artif Cells Blood Substit Immobil Biotechnol* **24**:ixxxvi.

Chang, T.M.S., and Prakash, S. (1996) Artificial cells for bioencapsulation of cells and genetically engineered *E. coli*: for cell therapy, gene therapy and removal of urea and ammonia. Chapter 75 "Expression and Detection of Recombinant Genes". *Methods in Molecular Biology.*

Chang, T.M.S. (1997a) *Red Blood Cell Substitutes: Principles, Methods, Products and Clinical Trials,* Vol. I (Monograph). Karger/ Landes Systems, Basel, Switzerland (available for free online viewing at www. artcell.mcgill.ca or www.artificialcell.info).

Chang, T.M.S. (1997b) Artificial cells. In: Renalo D., (ed.), *Encyclopedia of Human Biology,* 2nd edn. pp. 457–463. Academic Press, Inc., San Diego, California, USA.

Chang, T.M.S. (1997c) Artificial cells and bioencapsulation in bioartificial organs. *New York Acad Sci* **831**:249–259.

Chang, T.M.S. (1997a) Blood substitutes based on perflurocochemicals. *Blood Substitutes: Principles, Methods, Products and Clinical Trials,* Chap. VII. pp. 82–86. Karger, Basel, Switzerland.

Chang, T.M.S. (1997b) Future perspectives of blood substitutes. *Blood Substitutes: Principles, Methods, Products and Clinical Trials,* Chap. VIII. pp. 88–105. Karger, Basel, Switzerland.

Chang, T.M.S. (1997c) How safe are modified hemoglobin blood substitutes? *Blood substitutes: Principles, Methods, Products and Clinical Trials,* Chap. V. pp. 49–68. Karger, Basel, Switzerland.

Chang, T.M.S. (1997d) Introduction. *Blood Substitutes: Principles, Methods, Products and Clinical Trials*, Chap. I. pp. 1–2. Karger, Basel, Switzerland.

Chang, T.M.S. (1997e) Should there be a priority on blood substitutes for national blood supply policy? *Blood Substitutes: Principles, Methods, Products and Clinical Trials*, Chap. IX. pp. 111–114, Karger, Basel, Switzerland.

Chang, T.M.S. (1997f) What are modified hemoglobin blood substitutes? *Blood Substitutes: Principles, Methods, Products and Clinical Trials*, Chap. III. pp. 9–31, Karger, Basel, Switzerland.

Chang, T.M.S. (1997g) What are the functional and efficacy properties of modified hemoglobin? *Blood Substitutes: Principles, Methods, Products and Clinical Trials*, Chap. VI. pp. 32–47, Karger, Basel, Switzerland.

Chang, T.M.S. (1997h) What are the present modified hemoglobin blood substitutes being tested in clinical trial? *Blood Substitutes: Principles, Methods, Products and Clinical Trials*, Chap. VI. pp. 73–81, Karger, Basel, Switzerland.

Chang, T.M.S. (1997i) Why do we need blood substitutes? *Blood Substitutes: Principles, Methods, Products and Clinical Trials*, Chap. II. pp. 3–8, Karger, Basel, Switzerland.

Chang, T.M.S. (1997j) An urgent need to include blood substitute as a priority area in any national policies on blood supply. *Artif Cells Blood Substit Immobil Biotechnol* **25**:i–ii.

Chang, T.M.S. (1997k) Future perspectives of red blood cell substitutes. In: Rudolph, A., Feuerstein, G., and Rabinovici, R. (ed.), *Fundamental Principles and Clinical Application of Red Blood Cell Substitutes*. pp. 465–474. Marcel Dekker Publisher, New York.

Chang, T.M.S. (1997l) Live *E. coli* cells to treatment uremia: replies to letters to the editor. *Nat Med* **3**:2–3.

Chang, T.M.S. (1997m) Recent and future developments in modified hemoglobin and microencapsulated hemoglobin as red blood cell substitutes. *Artif. Cells Blood Substit Immobil Biotechnol* **25**:1–24.

Chang, T.M.S. (1997n) Red blood cell substitutes based on modified hemoglobin. In: Lanza, R.P., Langer, R., and Chick, W.L. (eds.), *Principles of Tissue Engineering*. pp. 517–526. Academic Press/Landes Co. Austin.

Chang, T.M.S., and Yu, W.P. (1997) Biodegradable polymer membrane containing hemoglobin for blood substitutes. USA Patent 5670173.

Chang, T.M.S., Greenberg, G., and Tsuchida, E. (eds.) (1997) Blood substitutes (special issue). *Artif Cells Blood Substit Immbol Biotechnol* **25**:1–241.

Chang, T.M.S. (1998a) Artificial cells with emphasis on cell encapsulation of genetically engineered cells. *Artif Organs* **22**:958–965.

Chang, T.M.S. (1998b) Method for microencapsulation of enzymes, cells and genetically engineered microorganisms in tissue engineering methods and protocols. In: Morgan, J.R., and Yarmush, M.L. (eds.), *Meth Mol Med* **18**:315–330.

Chang, T.M.S. (1998c) Modified hemoglobin-based blood substitutes: crosslinked, recombinant and encapsulated hemoglobin. *Vox Sanguinis* **74**(Suppl. 2):233–241.

Chang, T.M.S. (1998d) Introduction. *Blood Substitutes: Principles, Methods, Products and Clinical Trials,* Vol. 2. pp. xv–vvii. Karger, Basel, Switzerland.

Chang, T.M.S. (1998e) *Red Blood Cell Substitutes: Principles, Methods, Products and Clinical Trials,* Vol. 2. Karger/Landes Systems, Basel, Switzerland.

Chang, T.M.S. (1998f) Editorial: is there a role for first generation blood substitutes in the resuscitation of hemorrhagic shock? *Artif Cells Blood Substit Immobil Biotechnol* **26**:i–iii.

Chang, T.M.S. (1998g) Editorial: new concepts and theories. *Artif Cells Blood Substit Immobil Biotechnol* **26**:vii–viii.

Chang, T.M.S. (1998h) Modified hemoglobin blood substitutes: present status and future perspectives. *Biotechnol Ann Rev* **4**:75–112.

Chang, T.M.S. (1998i) Pharmaceutical and therapeutic applications of artificial cells including microencapsulation. Opening article, theme edition on microspheres and microcapsules *Eur Pharm Biopharm* **45**:3–8.

Chang, T.M.S. (1998j) Red blood cell substitutes in "Blood Preservation, Artificial Blood & Perioperative Fluid Management." Seminars in Anesthesia, Perioperative Medicine and Pain, Vol 17, Issue 67, No **3**:208–215.

Chang, T.M.S., D'Agnillo, F., and Razack, S. (1998) Crosslinked hemoglobin superoxide dismutase-catalase: a second generation hemoglobin based blood substitute with antioxidant activities. *Blood Substitutes: Principles, Methods, Products and Clinical Trials,* Vol. 2. pp. 178–196, Karger, Basel, Switzerland.

Chang, T.M.S., and Prakash, S. (1998a) Microencapsulated genetically engineered microorganisms for clinical application. International PTC application for Europe, Japan, USA and Canada. (Removal of urea and ammonia.)

Chang, T.M.S., and Prakash, S. (1998b) Microencapsulated genetically engineered cells: comparison with other strategies and recent progress. *Mol Med Today* **4**:221–227.

Chang, T.M.S., and Yu, W.P. (1998) Nanoencapsulation of hemoglobin and red blood cell enzymes based on nanotechnology and biodegradable polymer. *Blood Substitutes: Principles, Methods, Products and Clinical Trials,* Vol. 2. pp. 216–231. Karger, Basel, Switzerland.

Chang, T.M.S. (1999a) Foreword. In: Kuhtreiber, W.M., Lanza, R.P., and Chick, W.L. (eds.), *Handbook of Cell Encapsulation.* pp. 11–13. Birkhauser & Springer-Verlag.

Chang, T.M.S. (1999b) Artificial blood: a prospective. *Trends Biotechnol* **17**:61–67.

Chang, T.M.S. (1999c) Artificial cells, encapsulation and immobilization. *Ann NY Acad Sci* **875**:71–84.

Chang, T.M.S., and Prakash, S. (1999) Removal of urea in uremia and ammonia in liver failure with emphasis on the use of artificial cells containing genetically engineered cells. In: Kuhtreiber, W.M., Lanza, R.P., and Chick, W.L. (eds.), *Handbook of Cell Encapsulation.* pp. 379–416. Birkhauser & Springer-Verlag.

Chang, T.M.S., and Yu, W.P. (1999) Polymeric biodegradable hemoglobin nanocapsule as a new red blood cell substitute. In: Tsuchida, E. (ed.), *Present and Furture Perspectives of Blood Substitutes.* pp. 161–169.

Chang, T.M.S. (2000a) Red blood cell substitutes. Best practice and research. *Clin Haematol* **13**:651–668.

Chang, T.M.S. (2000) Artificial cell biotechnology in medical applications. *J Blood Purification* **18**:91–96.

Chang, T.M.S. (2000b) Artificially boosting the blood supply. *Chemistry and Industry*, 17 April, 281–284.

Chang, T.M.S. (2000c) Editorial: is there a need for blood substitutes in the new millennium and what can we expect in the way of safety and efficacy? *Artif Cells Blood Substit Immobil Biotechnol* **28**:i–vii.

Chang, T.M.S., D'Agnillo, F., Yu, W.P. and Razack, S. *et al.* (2000) New generations of blood subsitutes based on polyhemoglobin-SOD-CAT and nanoencapsulation. *Adv Drug Delivery Rev* **40**:213–218.

Chang, T.M.S., and Malave, N. (2000) The development and first clinical use of semipermeable microcapsules (artificial cells) as a compact artificial kidney. (special millenium republication of previous key papers). *Ther Apheresis* **4**:108–116.

Chang, T.M.S. (2001) Bioencapsulated hepatocytes for experimental liver support. *J Hepatol* **34**:148–149.

Chang, T.M.S., and Prakash, S. (2001a) Microencapsulated genetically engineered microorganisms for clinical application. USA Patent 6, 217, 859.

Chang, T.M.S., and Prakash, S. (2001b) Procedure for microencapsulationof enzymes, cells and genetically engineered microorganisms. *Mol Biotechnol* **17**:249–260.

Chang, T.M.S., and Yu, W.P. (2001) Biodegradable polymeric nanocapsules and uses thereof. US Provisional Patent Application. No 60/316, 001.

Chang, T.M.S., and Prakash, S. (2001c) Microencapsulated genetically engineered microorganisms for clinical application. Japanese Patent 3228941.

Chang, T.M.S. (2002a) Present status of modified hemoglobin as blood substitutes and oral therapy for end stage renal failure using artificial cells containing genetically engineered cells. *Ann N Y Acad Sci* **944**:362–372.

Chang, T.M.S. (2002b) Red blood cell substitutes. In: Atala, A., and Lanza, R. (eds.), *Methods of Tissue Engineering.* pp. 965–975. Academic Press.

Chang, T.M.S. (2002c) Oxygen carriers. *Curr Opin Invest Drugs* **3**:1187–1190.

Chang, T.M.S. (2002d) Red blood cell substitutes based on modified hemoglobin. In: Lanza, R.P., Langer, R., and Chick, W.L. (eds.), *Principles of Tissue Engineering.* pp. 601–610. Academic Press.

Chang, T.M.S., and Yu, B.L. (2002) Composition for inhibiting tumour growth and methods thereof. US Provisional Patent Application 60/364, 581.

.Chang, T.M.S. (2003a) Artificial cells for replacement of metablic organ functions. *Artif Cells Blood Substit Biotechnol* **31**:151–162.

Chang, T.M.S. (2003b) New generations of red blood cell substitutes. *J Intern Med* **253**:527–535.

Chang, T.M.S., and Yu, B.L. (2003) Composition for inhibiting tumour growth and methods thereof. US Patent Application.

Chang, T.M.S., and Lister, C. (2003) Screening test for modified hemoglobin blood substitute before use in human. Canadian Patent.

Chang, T.M.S., Powanda, D., and Yu, W.P. (2003) Ultrathin polyethylene-glycol-polylactide copolymer membrane nanocapsules containing polymerized Hb and enzymes as nano-dimension red blood cell substitutes. *Artif Cells Blood Substit Biotechnol* **3**:231–248.

Chang, T.M.S., Powanda, D., and Yu, W.P. (2003a) Biodegradable polymeric nanocapsules and uses thereof. International Patent Application No. PCT/CA02/01331—August 2002. WO 03/017989 A1.

Chang, T.M.S., Powanda, D., and Yu, W.P. (2003b) Analysis of polyethyleneglycolpolylactide nano-dimension artificial red blood cells in maintaining systemic hemoglobin levels and prevention of methemoglobin formation. *Artif Cells Blood Substit Biotechnol* **31**:231–248.

Chang, T.M.S. (2004a) A new red blood cell substitute. *Crit Care Med* **32**: 612–613.

Chang, T.M.S. (2004b) Artificial cells bioencapsulation in macro, micro, nano and molecular dimensions. *Artif Cells Blood Substit Biotechnol* **32**:1–23.

Chang, T.M.S. (guest ed.) (2004c) Special issue on blood substitutes: present and future. *Artif Organs*.

Chang, T.M.S. (2004d) Artificial cells for cell and organs replacement. *Artif Organs* **28**:265–270.

Chang, T.M.S. (2004e) Hemoglobin based red blood cell substitutes. *Artif Organs* **28**:289–294.

Chang, T.M.S., and Tsuchida, E. (2004) Guest editorial on blood substitutes: present and future. *Artif Organs* **28**:287–288.

Chang, T.M.S., Powanda, D., and Yu, W.P. (2004a) Biodegradable polymeric nanocapsules and uses thereof. US Patent Application, February 27.

Chang, T.M.S., Powanda, D., and Yu, W.P. (2004b) Biodegradable polymeric nanocapsules and uses thereof. Chinese Patent Application.

Chang, T.M.S. (2005a) Methods for microencapsulation of enzymes and cells. *Meth Biotechnol* **17**:289–306.

Chang, T.M.S. (2005b) PEG-PLA biodegradable hemoglobin nanocapsules as rbc substitutes. In: Winslow, R. (ed.), *Blood Substitutes*, pp. 523–531. Academic Press, San Diego.

Chang, T.M.S. (2005c) Polyhemoglobin-enzyme complexes. In: Winslow, R. (ed.), *Blood Substitutes*. pp. 451–459. Academic Press, San Diego.

Chang, T.M.S. (2005d) Artificial red blood cell substitutes. In: Ratner, B.D., Hoffman, A.S., Lemmons, J.E., et al. (eds.), *Biomaterials Science: An Introduction to Materials in Medicine*. Academic Press, San Diego, CA.

Chang, T.M.S. (2005e) Hemoglobin-based red blood cell substitutes. In: Ma, P.X., and Elisseeff, J. (eds.), *Scaffolding in Tissue Engineering*. pp. 473–480. Taylor & Francis, New York.

Chang, T.M.S. (2005f) Red blood cell substitutes: past, present and future. In: Kobayashi, K., Tsuchida, E., and Horinouchi, H. (eds.), *Artificial Oxygen Carriers*. pp. 22–33. Springer-Verlag Publisher.

Chang, T.M.S. (2005g) The role of artificial cells in cell and organ transplantation in regenerative medicine. *Panminerva Med* **47**:1–9.

Chang, T.M.S. (2005h) Artificial cells for blood substitutes, enzyme therapy, cell therapy and drug delivery. In: Nedovic, V., and Willaert, R. (eds.), *Cell Immobil Biotechnol*. pp. 249–255. Springer, Netherland.

Chang, T.M.S. (2005i) Therapeutic applications of polymeric artificial cells. *Nat Rev: Drug Discov* **4**:221–235.

Chang, T.M.S. (2005) Red blood cell substitutes: past, present and future. In: Kobayashi, K.E., Tsuchida, E., and Horinouch, H. (eds.), *Artificial Oxygen Carriers.* pp. 22–33. Spring-Verlag Publisher.

Chang, T.M.S. (2005) Methods for microencapsulation of enzymes and cells. *Meth Biotechnol* **17**:289–306.

Chang, T.M.S. (2005) Artificial cells for blood substitutes, enzyme therapy, cell therapy and drug delivery. In: Nedovic, V., and Willaert, R. (eds.), *Cell Immobilization Biotechnol.* pp. 249–255. Springer, Netherland.

Chang, T.M.S. (2005) Artificial red blood cell substitutes. In: Ratner, B.D., Hoffman A.S., Lemmons, J.E., *et al.* (eds.), *Biomaterials Science: An Introduction to Materials in Medicine.* Academic Press, San Diego, California.

Chang, T.M.S. (2006a) Blood substitutes based on bionanotechnology. *Trends Biotechnol* **24**:372–377.

Chang, T.M.S. (2006b) Polyhemoglobin-enzyme complexes. In: Winslow, R. (ed.), *Blood Substitutes.* pp. 451–459.

Chang, T.M.S. (2006c) PEG-PLA biodegradable hemoglobin nanocapsules as rbc substitutes. In: Winslow, R. (ed.), *Blood Substitutes,* pp. 523–531.

Chang, T.M.S. (2006) Hemoperfusion: beginning, present status and future perspectives. Opening paper in "Hemoperfusion" China (in press).

Chang, T.M.S. (2006) Artificial cells in medicine with emphasis on blood substitutes. In: Prakash, S. (ed.), *Artificial Cells and Cell Therapy* (in press).

Chapman, K.W., Snell, S.M., Jesse, R.G., *et al.* (1992) Pilot scale production of pyrogen-free modified human hemoglobin for research. *Biomater Artif Cells Immobil Biotechnol* **20**:415–422.

Chawla, A.S., and Chang, T.M.S. (1973) Nonthrombogenic polymeric membrane prepared by gamma-radiation grafting of heparin. *Am Chem Soc* **33**:379–385.

Chawla, A.S., and Chang, T.M.S. (1974) Nonthrombogenic surface by radiation grafting of heparin: preparation, *in vitro* and *in vivo* studies. *Int J Biomater Med Dev & Artif Organs* **2**:157–169.

Chawla, A.S., and Chang, T.M.S. (1975a) A new method for the preparation of nonthrombogenic surface by radiation grafting of heparin: preparation and *in vitro* studies. In: Gregor, H.P. (ed.), *Biomedical Applications of Polymers.* pp. 147–157. Plenum Press, New York.

Chawla, A.S., and Chang, T.M.S. (1975b) Use of solubility parameters for the preparation of hemodialysis membranes. *J Appl Polymer Sci* **19**:1723–1730.

Chawla, A.S., and Chang, T.M.S. (1985–1986) *In vivo* degradation of poly(lactic acid) of different molecular weights. *Biomater Med Dev Artif Organs* **13**:153–162.

Chen, H., Ouyang W, Jones M., *et al.* (2005) *In vitro* analysis of APA microcapsules for oral delivery of liver bacteria cells. *J Microencapsul* **22**:539–547.

Chi, O.Z., Lu, X., Wei, H.M., *et al.* (1996) Hydroxyethyl starch solution attenuates blood-brain disruption caused by intracarotid injection of hyperosmolar mannitol in rats. *Anesth Analg* **83**:336–341.

Chirito, E., Reiter, B., Lister, C., *et al.* (1978) Assessment of two rat models of acute hepatic failure for testing artificial liver devices. In: Chang, T.M.S. (ed.), *Artificial Kidney, Artificial Liver and Artificial Cells.* pp. 239–243. Plenum Press, New York.

Chirito, E., Lister, C., and Chang, T.M.S. (1979) Biochemical, hematological and histological changes in a fulminant hepatic failure rat model for artificial liver assessment. *Artif Organs* **3**:42–46.

Chow, K.M., Liu, Z.C., Prakash, S., *et al.* (2003a) Free and microencapsulated *lactobacillus* and effects of metabolic induction on urea removal. *Artif Cells Blood Substit Biotechnol* **4**:425–434.

Chow, K.M., Liu, Z.C., and Chang, T.M.S. (2003b) Animal remnant kidney model of chronic renal failure. *HK J Nephrology* **5**:57–64.

Cirone, P., Bourgeois, M., Austin, R.C. *et al.* (2002) A novel approach to tumor suppression with microencapsulated recombinant cells. *Hum Gene Ther* **13**:1157–1166.

Clark, L.C. Jr, and Gollan, F. (1996) Survival of mammals breathing organic liquids equilibrated with oxygen at atmospheric pressure. *Science* **152**:1755.

Cliff, R.O., Kwasiborski, V., and Rudolph, A.S. (1995) A comparative study of the accurate measurement of endotoxin in liposome encapsulated hemoglobin. **23**:331–336.

Cohan, S.L., Winchester, J.F., and Gelfand, M.C. (1982) Treatment of intoxications by charcoal hemadsorption. *Drug Metab Rev* **13**:681.

Cohan, S.L., Winchester, J.F., and Gelfand, M.C. (1978) Hemoperfusion. *Ann Rev Pharmacol Toxicol* **27**:169.

Cole, D.J., Drummond, J.C., Patel, P.M., *et al.* (1996) Effect of oncotic pressure of diaspirin cross-linked hemoglobin (DCLHb) on brain injury after temporary focal cerebral ischemia in rats. *Anesth Analg* **83**:342–347.

Cole, D.J., McKay, L., Jacobsen, W.K., *et al.* (1997) Effect of subarachnoid administration of a-a diaspirin crosslinked hemoglobin on cerebral blood flow in rats. *Artif Cells Blood Substit Immobil Biotechnol* **25**:95–104.

Cole, D.J., Nary, J.C., Drummond, J.C., *et al.* (1997) a-a diaspirin crosslinked hemoglobin, nitric oxide, and cerebral ischemic injury in rats. *Artif Cells Blood Substit Immobil Biotechnol* **25**:141–152.

Cole, D.J., Przybelski, R.J., Schell, R.M. and Martin, R.D. (1995) Diaspirin crosslinked hemoglobin (DCLHb^{TM}) does not affect the anesthetic potency of isoflurane in rats. *Artif Cells Blood Substit Immobil Biotechnol* **23**:89–99.

Conover, C.D., Lejeune, L., Linberg, R., *et al.* (1996) Transitional vacuole formation following a bolus infusion of PEG-hemoglobin in the rat. *Artif Cells Blood Substit Immobil Biotechnol* **24**:599–611.

Conover, C.D., Malatesta, P., Lejeune, L., *et al.* (1996) The effects of hemodilution with polyethylene glycol bovine hemoglobin (PEG-Hb) in a conscious porcine model. *J Investig Med* **44**:238–246.

Cordopatri, F., Boncinelli, S., Marsili, M., *et al.* (1982) Effects of charcoal haemoperfusion with prostacydin on the coagulation-fibrinolysis system and platelets. *Int J Artif Organs* **5**:243.

Coromili, V., and Chang T.M.S. (1993) Polydisperse dextran as a diffusing test solute to study the membrane permeability of alginate polylysine microcapsules. In: Oncelet, D., and Lievremont, M. (Guest eds.), *Biomater Artif Cells Immobil Biotechnol* **21**:427–444.

Cousineau, J., and Chang, T.M.S. (1977) Formation of amino acid from urea and ammonia by sequential enzyme reaction using a microencapsulated multi-enzyme system. *Biochem Biophys Res Commun* **79**:24–3l.

Cousineau, J., and Chang, T.M.S. (1978) Conversion of urea and ammonium to an amino acid using sequential enzymatic reaction with microencapsulated multienzyme system. In: Chang, T.M.S. (ed.), *Artificial Kidney, Artificial Liver and Artificial Cells*. pp. 249–254. Plenum Press, New York.

Cruise, G.M., et al. (1999) *In vitro* and *in vivo* performance of porcine islets encapsulated in interfacially photopolymerized poly(ethylene glycol) diacrylate membranes. *Cell Transplant* **8**:293–306.

D'Agnillo, F., and Chang, T.M.S. (1997a) Modified hemoglobin blood substitute from cross-linked hemoglobin-superoxide dismutase-catalase. US patent 5, 606, 025.

D'Agnillo, F., and Chang, T.M.S (1997b) Hemoglobin-enzyme complexes. Canadian Patent 2, 135, 739.

D'Agnillo, F., and Chang, T.M.S. (1997c) Production of hydroxyl radical generation in a rat hindlimb model of ischemia-reperfusion injury using crosslinked hemoglobinsuperoxide dismutase-catalase. *Artif Cells Blood Substit Immobil Biotechnol* **25**:163–180.

D'Agnillo, F., and Chang, T.M.S. (1998a) Polyhemoglobin-superoxide dismutase-catalase as a blood substitute with antioxidant properties. *Nat Biotechnol* **16**:667–671.

D'Agnillo, F., and Chang, T.M.S. (1998b) Absence of hemoprotein-associated free radical events following oxidant challenge of crosslinked hemoglobin-superoxide dismutasecatalase. *Free Radic Bio Med* **24**:906–912.

D'Agnillo, F., and Chang, T.M.S. (1993) Cross-linked hemoglobin-superoxide dismutasecatalase scavenges oxygen-derived free radicals and prevents methemoglobin formation and iron release. *Biomater Artif Cells Immobil Biotechnol* **21**:609–622.

Daka, J.N., and Chang, T.M.S. (1989) Bilirubin removal by the pseudoperoxidase activity of free and immobilized hemoglobin and hemoglobin co-immobilized with glucose oxidase. *Biomater Artif Cells Artif Organs* **17**:553–562.

Daka, J.N., Laidler, K.G., Sipehia, R., and Chang, T.M.S. (1988) Immobilized and kinetics of lactate dehydrogenase at a rotating nylon disk. *Biotechnol Bioeng* **32**:213–219.

Daka, N.J., Sipehia, R., and Chang, T.M.S. (1989) Enhanced oxidation of bilirubin by an immobilized tri-enzyme system of glucose oxidase, bilirubin oxidase and horseradish peroxidase. *Biochim Biophys Acta* **991**:487–489.

Dalle, B., et al. (1990) Improvement of the mouse ß–thalasemia upon erythropoietin delivery by encapsulated myoblasts. *Gene Ther* **6**:157–161.

Daull, P., Blouin, A., Cayer, J., et al. (2005) Profiling biochemical and hemodynamic markers using chronically instrumented, conscious and unrestrained rats undergoing severe, acute controlled hemorrhagic hypovolemic shock as an integrated *in vivo* model system to assess new blood substitutes. *Vasc Pharmacol* **43**:289–301.

Davey, M.G. (1989) Blood services in Canada: The Canadian Red Cross Society. In: Chang, T.M.S., and Geyer, R.P. (eds.), *Blood Substitutes*. pp. 51–53. Marcel Dekker Publisher, New York, USA.

Davies, A., and Magnum, A. (1995) Media release. January. Hemosol Inc.

de Vos P., *et al.* (2003) Association between macrophage activation and function of micro-encapsulated rat islets. *Diabetologia* **46**:666–673.

De Vos, P., and Marchetti, P. (2002) Encapsulation of pancreatic islets for transplantation in diabetes: the untouchable islets. *Trends Mol Med* **8**:363–366.

De Vos, P., Hamel, A.F., and Tatarkiewicz, K. (2002) Considerations for successful transplantation of encapsulated pancreatic islets. *Diabetologia* **45**:159–173.

Dellacherie, E., *et al.* (1992) Hemoglobin linked to polymeric effectors as red blood cell substitutes. *Biomater Artif Cells Immobil Biotechnol* **20**:309–318.

Deamer, D.W., and Bangham, A.D. (1976) Large-volume liposomes by an ether vaporization method. *Biochim Biophys Acta* **443**:629–634.

Deamer, D. (2005) A giant step towards artificial life? *Trends Biotechnol* **23**:336–338.

Detsch, O., Heesen, M., Muhling, J., *et al.* (1996) Isovolemic hemodilution with hydroxyethylstarch has no effect on somatosensory evoled potentials in healthy volunteers. *Act Anesthesiol Scand* **40**:665–670.

deVarennes, B., Chang, T.M.S., Symes, D.J.F., and Guerraty, A. (1988) Isolated heart preservation for 24 hours: is O_2 important? *J Biomater Artif Cells Artif Organs* **16**:357–358.

DeVenuto, F., and Zegna, A.I. (1982) Blood exchange with pyridoxalated-polymerized hemoglobin. *Surg Gynecol Obstet* **155**:342.

DeVenuto, F., Zuck, T.F., Zegna, A.I., and Moores, W.Y. (1997) Characteristics of stroma-free hemoglobin prepared by crystallization. *J Lab Clin Med* **89**:509.

Diaz-Buxo, J.A., Farmer, C.D., and Chandler, T.Y. (1978) Hemopertusion in the treatment of amitriptyline poisoning. *Trans Am Soc Artif Intern Organs* **24**:699.

Dietz, N.M., Joyner, M.J., and Warner, M.A. (1996) Blood substitutes: fluids, drugs, or miracle solutions? *Anesth Analg* **82**:390–405.

Dionne, K.E. *et al.* (1996) Transport characterization of membranes for immuno-isolation. *Biomaterials* **17**:257–266.

Discher, B.M., Won, Y., David, S.E., *et al.* (1999) Polymersomes: tough vesicles made from diblock copolymers. *Science* **284**:1143–1144.

Dittmer, J., Prusty, S., Ichikura, T., *et al.* (1992) Intravascular retention and distribution of DBBF crosslinked stroma-free hemoglobin in the mouse. *Biomater Artif Cells Immobil Biotechnol* **20**:751–756.

Dixit, V., and Chang, T.M.S. (1981) Soluble liver extracts as supplement to detoxification by hemoperfusion. *Int J Artif Organs* **4**:291–294.

Dixit, V., and Chang, T.M.S. (1982) Effects of prostaglandine on the survival time of fulminant hepatic failure rats. *Int J Artif Organs* **5**:388.

Dixit, V., and Chang, T.M.S. (1985) Preliminary report on effects of prostaglandine on brain edema in fulminant hepatic failure rats. *Int J Artif Organs* **8**:55–56.

Dixit, V., and Chang, T.M.S. (1987) Effects of prostaglandine on brain edema and liver histopathology in a galactosamine-induced fulminant hepatic failure rat model. *J Biomater Artif Cells Artif Organs* **15**:559–574.

Dixit, V., and Chang, T.M.S. (1990) Brain edema and blood brain barrier in galactosamine induced fulminant hepatic failure rats: animal models for evaluation of liver support systems. *J Am Soc Artif Int Organs* **36**:21–27.

Djordjevich, L., and Miller, I.F. (1980) Synthetic erythrocytes from lipid encapsulated hemoglobin. *Exp Hematol* **8**:584.

Domokos, G., Jopski, B., and Schmidt, K.H. (1992) Prepration, properties and biological function of liposome encapsulated hemoglobin. *Biomater Artif Cells Immobil Biotechnol* **20**:345–354.

Douglas, S., Hawkins, Julie, Park, R., *et al.* (2004) Asparaginase pharmacokinetics after intensive polyethylene glycol-conjugated L-asparaginase therapy for children with relapsed acute lymphoblastic leukemia. *Clin Cancer Res* **10**:5335–5341.

Dudziak, R., and Bonhard, K. (1980) The development of hemoglobin preparations for various indications. *Anesthesist* **29**:181.

Dueck, C.L., Neufeld, R.J., and Chang, T.M.S. (1986) Hydrodynamics of a fluidized bed reactor for urea hydrolysis by microencapsulated urease. *Can J Chem Eng* **64**:540–546.

Duncan, R. (2003) The dawning era of polymer therapeutics. *Nat Rev Drug Discov* **2**:347–360.

Dunea, G., and Kolff, W.J. (1965) Clinical experience with the Yatzidis charcoal artificial kidney. *Trans Am Soc Artif Intern Organs* **11**:178.

Dunham, E.T., and Glynn, I.M. (1961) Adenosine triphosphatase activity and the active movements of alkali metal ion. *J Physiol* **156**:274.

Dunlap, E., Farrell, L., Nigro, C., *et al.* (1995) Resuscitation with diaspirin crosslinked hemoglobin in a pig model of hemorrhagic shock. *Artif Cells Blood Substit Immobil Biotechnol* **23**:39–61.

Dupuis, N.P., Kusumoto, T., Robinson, M.F., *et al.* (1995) Restoration of tumor oxygenation after cytotoxic therapy by a perflubron emulsion/carbogen breathing. *Artif Cells Blood Substit Immobil Biotechnol* **23**:423–429.

Duvivier-Kali, V.F., Omer, A., Parent R.J., *et al.* (2001) Complete protection of islets against allorejection and autoimmunity by a simple barium-alginate membrane. *Diabetes* **50**:1698–1705.

Eisman, J.M., and Schnaare, R.L. (1996) The formation of a cross-linked carboxyhemoglobin membrane at an organic-aqueous interface. *Artif Cells Blood Substit Immobil Biotechnol* **24**:185–196.

Eldrige, J., Russell, R., Christenson, R., *et al.* (1996) Liver function and morphology after resuscitation from severe hemorrhagic shock with hemoglobin solutions or autologous blood. *Crit Care Med* **24**:663–671.

Emerich, D.F., *et al.* (1997) Protective effect of encapsulated cells producing neurotrophic factor CNTF in a monkey model of Huntington's disease. *Nature* **386**:395–399.

Ergan, F., Thomas, D., and Chang, T.M.S. (1984) Selection and microencapsulation of an "NADH-oxidizing" bacterium and its use for NAD regeneration. *Appl Biochem Biotechnol* **10**:61–71.

Ergan, F., Thomas, D., and Chang, T.M.S. (1987) Immobilization of the "NADH-oxidase" function of the bacterium *Leuconostoc Mesenteroides* in matrix supported microcapsules for continuous cofactor regeneration. *Ann N Y Acad Sci* **501**:372–376.

Espinosa-Melendez, E., Zelman, M., Barre, P., *et al.* (1982) Oxystarch modified by boiling: effects of oral administration in stable hemodialysis patients. *Proceedings of the International Symposium on Hemoperfusion and Artificial Organs.* pp. 83–85.

Espinosa-Melendez, E., and Chang, T.M.S. (1983) State-of-the-art in hemoperfusion. *Nephrologica Mexicana* **4**:51–55.

Espinosa-Melendez, E., Bourgouin, P.A., and Chang, T.M.S. (1986) Combined hemoperfusion/hemodialysis in the treatment of uremic patients with vascular access problems. *In J Artif Organs* **9**:309–312.

Espinosa-Melendez, E., and Chang, T.M.S. (1980) ACAC (albumin-collodion coated activated charcoal) hemoperfusion: *in vitro* removal and clearance of theophylline. *J Dial* **4**:21–29.

Espinosa-Melendez, E., Lal, S., Nair, N.P.V., and Chang, T.M.S., *et al.* (1979) Plasma beta endorphin levels in chronic schizophrenic patients and normal controls. *Artif Organs* **3**:375–376.

Estep, T.N., Gonder, J., Bornstein, I., *et al.* (1992) Immunogenicity of diaspirin crosslinked hemoglobin solutions. *Biomater Artif Cells Immobil Biotechnol* **20**:603–610.

Faassen, A., Sundby, S.R., Panter, S.S., *et al.* (1988) Hemoglobin: lifesaver and an oxidant. How to tip the balance. *Biomater Artif Cells Artif Organs* **16**: 93–104.

Faithfull, N.S. (1992) Oxygen delivery from fluorocarbon emulsions — aspects of convective and diffusive transport. *Biomater Artif Cells Artif Organs* **20**:797–804.

Faithfull, N.S. (1994) Mechanisms and efficacy of fluorochemical oxygen transport and delivery. In: Winslow, R.M. (Guest ed.), *Artif Cells Blood Substit Immobil Biotechnol* **22**:687–694.

Faivre, B., Labaeye, V., Menu, P., and Labrude, P. (1995) Oxygen carrier, using guinea pig isolated bowel model. *Artif Cells Blood Substit Immobil Biotechnol* **23**:495–504.

Farmer, M.C., Rudolph, A.S., Vandegriff, K.D., *et al.* (1988) Lipsome-encapsulated hemoglobin: oxygen binding properties and respiratory function. *Biomater Artif Cells Artif Organs* **16**:289–299.

Farmer, M.C., Ebeling, A., Marshall, T., *et al.* (1992) Validation of virus inactivation by heat treatment in the manufacture of diaspirin crosslinked hemoglobin. *J Biomater Artif Cells Immobil Biotechnol* **20**:429–434.

Feinstein, R.N., Howard, J.B., Braun, J.T., Seahalm, J.E. (1966) *Genetics* **53**:923–933.

Feinstein, R.N., Braun, J.T., and Howard, J.B. (1966) *J Lab Clin Merl* **68**:952–957.

Feinstein, R.N., Suter, H., and Jaroslow, B.N. (1968) *Science* **159**:638–640.

Feola, M., Gonzalez, H., Canizaro, P.C., and Bingham, D. (1983) Development of bovine stroma-free hemoglobin solution as a blood substitute. *Surg Gynecol Obstet* **157**:399.

Feola, M., Simoni, J., Ruc Tran, D.V.M., and Canizaro, P.C. (1988) Mechanisms of toxicity of hemoglobin solutions. *Biomater Artif Cells Artif Organs* **16**:217–226.

Feola, M., and Simoni, J. (1991) Biocompatibility of hemoglobin solutions: the effects of contaminants, hemoglobin and hemoglobin derivatives. *Biomater Artif Cells Immobil Biotechnol* **19**:382.

Flaim, S.F., Hazard, D.R., Hogan, J., and Peters, R.M. (1991) Characterization and mechanism of side-effects of Imagent BP (highly concentrated fluorocarbon emulsion) in swine. *Invest Radiol* **26**(Suppl.):S125–S128 (discussion).

Flothe, L., and Otting, F. (1984) Superoxide dismutase assays. *Meth Enzymol* **105**:93–104.

Frankel, H.L., Nguyen, H.B., Shea-Donohue, T., *et al.* (1996) Diaspirin cross-linked hemoglobin is efficacious in gut resuscitation as measured by a GI tract optode. *J Trauma* **40**:231–240 (discussion, 241).

Fratantoni, J.C. (1991) Points to consider in the safety evaluation of hemoglobin based oxygen carriers. *Transfusion* **31**:369–371.

Fratantoni, J.C. (1994) Points to consider on efficacy evaluation of hemoglobin and perfluorocarbon based oxygen carriers. *Transfusion* **34**:712–713.

Fronticelli, C., Bellelli, A., and Briniger, W.S. (2004) Approach to the engineering of Hb based oxygen carrier. *Transfu Med* **5**:516–520.

Frey, L., and Messmer, K. (1996) Transfusion Therapy. *Curr Opin Anaesthesiol* **9**:183–187.

Freyburger, G., Dubreuil, M., Boisseau, M.R., and Janvies, G. (1996) Rheological properties of commonly used plasma substitutes during preoperative normovolaemic acute haemodilution. *Br J Anaesth* **76**:519–525.

Fronticelli, C., Brinigar, W., Grycznski, Z., and Bucci, E. (1991) Engineering of low oxygen affinity hemoglobins. *Biomater Artif Cells Immobil Biotechnol* **19**:385.

Fu, Y.M., Yu, Z.X., Ferrans, V.J., *et al.* (1997) Tyrosine and phenylalanine restriction induces G0/G1 cell cycle arrest in murine melanoma *in vitro* and *in vivo*. *Nutr Cancer* **29**:104–113.

Gaber, B.P., and Farmer, M.C. (1984) Encapsulation of hemoglobin in phospholipid vesicles: preparation and properties of a red blood cell surrogate. *Prog Clin Biol Res* **165**:179–190.

Garofalo, F., and Chang, T.M.S. (1989) Immobilization of *P. pictorum* in open pore agar, alginate polylysine-alginate microcapsules for serum cholesterol depletion. *J Biomater Artif Cells Artif Organs* **17**:271–290.

Garofalo, F.A., and Chang, T.M.S. (1991) Effects of mass transfer and reaction kinetics on serum cholesterol depletion rates of free and immobilized *P. pictorum*. *Appl Biochem Biotechnol* **27**:75–91.

Gazzard, B.G., Portmann, B.A., Weston, M.J., *et al.* (1974) Charcoal haemoperfusion in the treatment of fulminant hepatic failure. *Lancet* **1**:1301–1307.

Gelfand, M.C., Winchester, J.F., Knepshield, J.H., *et al.* (1977) Treatment of severe drug overdose with charcoal hemoperfusion. *Trans Am Soc Artif Intern Organs* **23**:599.

Gennaro, M., Mohan, C., and Ascer, E. (1996) Perfluorocarbon emulsion prevents eicoasanoid release in skeletal muscle ischemia and reperfusion. *Cardiovasc Surg* **4**:399–404.

Geyer, R.P., Bolin, and Nemo, G.J. (eds.) (1983) *Advances in Blood Substitute Research*. pp. 1–468. Publisher Alan R. Liss Inc., New York.

Geyer, R.P. (1984) Blood-replacement preparations. In: Kirk-Othmer (ed.), *Encyclopedia of Chemical Technology*, Suppl. Vol. 3rd edn. John Wiley & Sons, Inc.

Geyer, R.P., Monroe, R.G., and Taylor, K. (1968) Survival of rats totally perfused with a fluorocarbon-detergent preparation. In: Norman, J.C., Folkman, J., Hardison, W.G., *et al.* (eds.), *Organ Perfusion and Preservation*. pp. 85–96. Appleton Century Crofts, New York.

Gibson, T.P., Lucas, S.V., Nelson, H.A., *et al.* (1978) Hemoperfusion removal of digoxin from dogs. *J Lab Cun Med* **91**:673.

Gill, I., and Ballesteros, A. (2000) Bioencapsulation within synthetic polymers (Part 2): non-sol-gel protein-polymer biocomposites. *Trends Biotechnol* **18**:469–479.

Gimson, A.E.S., Braud, S., Mellon, P.J., *et al.* (1982) Earlier charcoal haemoperfusion in fulminant hepatic failure. *Lancet* **2**:681.

Golubev, A.M., Magomedov, M.A., Leonteva, T.A., *et al.* (1996) The biomicroscopic evaluation of the blood microcirculatory bed in the mesentary of the small intestine during experimental perftoran infusion under normovolemia. *Morfologiia* **109**:36–40.

Goodin, T.H., Grossbard, E.B., Kaufman, R.J., *et al.* (1994) A perfluorochemical emulsion for prehospital resuscitation of experimental hemorrhagic shock: a prospective, randomized controlled study. *Crit Care Med* **22**:680–689.

Gotoh, K., Morioka, T., and Nishi, K. (1992) Effects of pyridoxalated hemoglobin polyoxyethylene conjugate (PHP) on pulmonary vascular responsiveness to various vasoactive substances in isolated perfused rat lungs. *Biomater Artif Cells Immobil Biotechnol* **20**:721–722.

Gould, S.A., Sehgal, L.R., Sehgal, H.L., and Moss, G.S. (1995) The development of hemoglobin solutions as red cell substitutes: hemoglobin solutions. *Transfu Sci* **16**:5–17.

Gould, S.A., and Moss, G.S. (1996) Clinical development of human polymerized hemoglobin as a blood substitute. *World J Surg* **20**:1200–1207.

Gould, S.A., Moore, E.E., Moore, F.A., *et al.* (1995) The clinical utility of human polymerized hemoglobin as a blood substitute following trauma and emergent surgery. *J Trauma* **39**:157.

Gould, S.A., *et al.* (1998) The clinical development of human polymerized hemoglobin, in blood substitutes. In: Chang, T.M.S. (ed.), *Principles, Methods, Products and Clinical Trials*, Vol. 2, pp. 12–28. Basel Karger.

Gould, S.A., *et al.* (2002) The life-sustaining capacity of human polymerized hemoglobin when red cells might be unavailable. *J Am Coll Surg* **195**:445–452.

Greenburg, A.G., Hayashi, R., and Krupenas, I. (1979) Intravascular persistence and oxygen delivery of pyridoxalated stroma-free hemoglobin during gradations of hypotension. *Surgery* **86**:13.

Greenburg, A.G. (1997) Use of donor blood in the surgical setting: potentials for application of blood substitutes. *Artif Cells Blood Substit Immobil Biotechnol* **25**: 25–30.

Greenburg, A.G., and Kim, H.W. (1992) Evaluating new red cell substitutes: a critical analysis of toxicity models. *Biomater Artif Cells Immobil Biotechnol* **20**:575–581.

Greenburg, A.G. (1995) Nitrosyl hemoglobin formation *in vivo* after intravenous administration of a hemoglobin-based oxygen carrier in endotoxemic rats. *Artif Cells Blood Substit Immobil Biotechnol* **23**:271–276.

Greenwald, R.B., Pendri, A., Martinez, A., *et al.* (1996) PEG thiazolidine-2-thoone, a novel reagent for facile protein modification: conjugation of bovine hemoglobin. *Bioconj Chem* **7**:638–641.

Gregoriadis, G. (eds.) (1976) *Drug Carriers in Biology and Medicine*. Academic Press, Inc., New York.

Griffiths, A.D., and Tawfik, D.S. (2000) Man-made enzymes — from design to *in vitro* compartmentalization. *Curr Opin Biotechnol* **11**:338–353.

Griffiths, E., Cortes, A., Gilbert, N., *et al.* (1995) Hemoglobin based blood substitutes and sepsis. *Lancet* **345**:158–160.

Growe, G.H., Chang, T.M.S., Slichter, S.J., *et al.* (1993) New concepts in blood product usage. *Ann R Coll Phys Surg Can* **26**:106–108.

Grunwald, J., and Chang, T.M.S. (1978) Nylon polyethyleneimine microcapsules for immobilizing multienzymes with soluble dextran-NAD+ for the continuous recycling of the microencapsulated dextran-NAD+. *Biochem Biophys Res Commun* **81**:565–570.

Grunwald, J., and Chang, T.M.S. (1979) Continuous recycling of NAD+ using an immobilized system of collodion microcapsules containing dextran-NAD+, alcohol dehydrogenase, and malic dehydrogenase. *J Appl Biochem* **1**:104–114.

Grunwald, J., and Chang, T.M.S. (1980) Microencapsulated multienzyme system and the recycling of NAD+ bound to soluble dextran. In: Sideman, S., and Chang, T.M.S. (eds.), *Hemoperfusion: Artificial Kidney and Liver Support and Detoxification*. pp. 389–395. Hemisphere Publishing Corp., Washington, DC.

Grunwald, J., and Chang, T.M.S. (1981) Immobilization of alcohol dehydrogenase, malic dehydrogenase and dextran-NAD+ within nylon-polyethyleneimine microcapsules: preparation and cofactor recycling. *J Mol Catal* **11**:83–90.

Gu, K.F., and Chang, T.M.S. (1987) Conversion of ammonia or urea to L-leucine, L-valine and L-isoleucine by artificial cells, immobilized multienzyme system. *Int J Biomater Artif Cells Artif Organs* **15**:297–304.

Gu, K.F., and Chang, T.M.S. (1988a) Conversion of ammonia or urea into L-leucine, L-valine and L-isoleucine using artificial cell immobilising multienzyme system and dextran-NADH+. I. Glucose dehydrogenase for cofactor recycling. *J Am Soc Artif Intern Organs* **11**:24–28.

Gu, K.F., and Chang, T.M.S. (1988b) Conversion of -ketoglutarate into L-glutamic acid with urea as ammonium source using multienzyme system and dextran-NAD+ immobilised by microencapsulation with artificial cells in a bioreactor. *J Bioeng Biotechnol* **32**:363–368.

Gu, K.F., and Chang, T.M.S. (1988c) Multienzyme system and Dextran-NAD immobilized by microencapsulation within artificial cells in a reactor for conversion of urea into L-glutamic acid. In: Moo-Young, M. (ed.), *Bioreactor Immobilized Enzymes and Cells*. pp. 59–62. Elsevier Press, New York.

Gu, K.F., and Chang, T.M.S. (1990a) Conversion of ammonia or urea into essential amino acids, L-leucine, L-valine and L-isoleucine using artificial cells containing an immobilized multienzyme system and dextran-NAD+. 2. Yeast alcohol dehydrogenase for coenzyme recycling. *J Biotechnol Appl Biochem* **12**:227–236.

Gu, K.F., and Chang, T.M.S. (1990b) Conversion of ammonia or urea into essential amino acids. L-leucine, L-valine and L-isoleucine using artificial cells containing an immobilized multienzyme system and dextran-NAD+. IV. Malate dehydrogenase for co-enzyme recycling. *J Catal* **62**:331–339.

Gu, K.F., and Chang, T.M.S. (1990c) Production of essential L-branched-chained amino acids, in bioreactors containing artificial cells immobilized multienzyme systems and dextran-NAD+. *Appl Biochem Biotechnol* **26**:263–269.

Gu, K.F., and Chang, T.M.S. (1991) Conversion of urea into essential amino acids, L-leucine, L-valine and L-isoleucine using artificial cells containing an immobilized multienzyme system and dextran-NAD+. III. L-lactic dehydrogenase for co-enzyme recycling. *J Appl Biochem Biotechnol* **27**:75–91.

Gudi, S.R.P., Clark, C.B., and Frangos, J.A. (1996) Fluid flow rapidly activates G proteins in human endothelial cells. Involvement of G proteins in mechanicochemical signal transduction. *Circ Res* **79**:834–839.

Gulati, A., Sharma, A.C., and Burhop, K.E. (1994) Effect of stroma-free hemoglobin and diasprin cross-linked hemoglobin on the regional circulation and systemic hemodynamics. *Life Sci* **55**:827–837.

Gulati, A., Sharma, A.C., and Singh, G. (1996) Role of endothelin in the cardiovascular effects of disapirin crosslinked and stroma reduced hemoglobin. *Crit Care Med* **24**:137–147.

Gunn, C.H. (1938) Hereditary scholuric jaundice in a new mutant strain of rats. *J Hereditary* 137–139.

Günzburg, W.H., and Salmons, B. (2001) Novel clinical strategies for the treatment of pancreatic carcinoma. *Trends Mol Med* 30–37.

Gupta, R.B., and Kompell, T.B. (eds.) (2006) Nanoparticle technology for drug delivery. pp. 1–403. Taylor & Francis, New York.

Hagihara, Y., *et al.* (1997) Transplantation of xenogeneic cells secreting beta-endorphin for pain treatment: analysis of the ability of components of complement to penetrate through polymer capsules. *Cell Transplant* **6**:527–530.

Hakim, R.M., Schulman, I.M., and Lazarus, J.M. (1985) *Abstracts Am Soc Nephrol* **18**:65A.

Hakim, R.M., Milford, E., Himmelfarb, J., *et al.* (1990) Extracorporeal removal of antiHLA antibodies in transplant candidates. *Am J Kidney Dis* **16**:423.

Halle, J.P., *et al.* (1994) Studies on small (< 300 microns) microcapsules: II — Parameters governing the production of alginate beads by high voltage electrostatic pulses. *Cell Transplant* **3**:365–372.

Hanczyc, M.M., Fujikawa, S.M., Szostak, J.W. (2003) Experimental models of primitive cellular compartments: encapsulation, growth and division. *Science* **302**:618–622.

Hanzawa, K., Ohzeki, H., Moro, H., *et al.* (1997) Effects of partial blood replacement with pyridoxylated hemoglobin polyoxyethylene conjugate solution on transient cerebral ischemia in gerbil. *Artif Cells Blood Substit Immobil Biotechnol* **25**:105–114.

Harris, J.M., and Chess, R.B. (2003) Effect of pegylation on pharmaceuticals. *Nat Rev Drug Discov* **2**:214–221.

Hartgerink, J.D. *et al.* (1998) Peptide nanotubes and beyond. *Chem Eur J* **4**:1367–1372.

Hasse, C., Klöck, G., Schlosser, A., *et al.* (1997) Parathyroid allotransplantation without immunosuppression. *Lancet* **350**:1296–1297.

Hertzman, C.M., Keipert, P.E., and Chang, T.M.S. (1986) Serum antibody titers in rats receiving repeated small subcutaneous injections of hemoglobin or polyhemoglobin: a preliminary report. *Int J Artif Organs* **9**:179–182.

Hess, J.R., MacDonald, V.W., and Winslow, R.M. (1992) Dehydration and shock: an animal model of hemorrhagic and resuscitation of battlefield injury. *Biomater Artif Cells Immobil Biotechnol* **20**:499–502.

Hess, J.R., Macdonald, V.W., and Brinkley, W.W. (1993) Systemic and pulmonary hypertension after resuscitation with cell free hemoglobin. *J Appl Physiol* **74**:1769–1778.

Hess, J.R. (1995) Review of modified hemoglobin research at letterman: attempts to delineate the toxicity of cell-free tetrameric hemoglobin. *Artif Cells Blood Substit Immobil Biotechnol* **23**:277–289.

Hobbhahn, J., Vogel, H., Kothe, N., *et al.* (1985) Hemodynamics and transport after partial and total blood exchange with pyridoxalated polyhemoglobin in dogs. *Acta Anesth Scand* **29**:537–543.

Hodakowski, G.T., Page, R.D., Harringer, W., *et al.* (1992) Ultra-pure polyemerized bovine hemoglobin blood substitute: effects on the coronary circulation. *Biomater Artif Cells Immobil Biotechnol* **20**:669–672.

Hodgins, D.S. (1971) *J Biol Chem* **9**:2977–2985.

Hoffman, S.J., Looker, D.L., Roehrich, J.M., *et al.* (1990) Expression of fully functional tetrameric human hemoglobin in *Escherichia coli*. *Proc Natl Acad Sci USA* **87**:8521–8525.

Hogan, M.C., Willford, D.C., Keipert, P.E., *et al.* (1992) Increased plasma O_2 solubility improves O_2 uptake of *in situ* dog muscle working maximally. *J Appl Physiol* **74**:2470–2475.

Hogan, M.C., Kurdak, S.S., Richardson, R.S., and Wagner, P.D. (1994) Partial substitution of red blood cells with free hemoglobin solution does not improve maximal O_2 uptake of working *in situ* dog muscle. *Adv Exp Med Biol* **361**:375–378.

Holland, P.V. (1996) Viral infections and the blood supply. *New Engl J Med* **334**:1734–1735.

Horiuchi, T., Ohta, Y., Hashimoto, K., *et al.* (1992) Machine perfusion of isolated kidney at $37°C$ using pyridoxalated hemoglobin-polyoxyethylene (PHP) solution, UW solution and its combination. *Biomater Artif Cells Immobil Biotechnol* **20**:549–557.

Hoskins, J.A., and Gray, J. (1982) *Res Commun Chem Pathol Pharmacol* **35**:275–282.

Hsia, J.C. (1989) Pasteurizable, freeze-driable hemoglobin-based blood substitute. US Patent 857, 636.

Hsia, J.C. (1991) o-Raffinose polymerized hemoglobin as red blood cell substitute. *Biomater Artif Cells Immobil Biotechnol* **19**:402.

Hsia, J.C., Song, D.L., Er, S.S., and Wong, L.T.L. (1992) Pharmacokinetic studies on a raffinose-polymerized human hemoglobin in the rat. *Biomater Artif Cells Immobil Biotechnol* **20**:587–596.

Hsia, N., and Everse, J. (1996) The cytotoxic activities of human hemoglobin and diaspirin cross-linked hemoglobin. *Artif Cells Blood Substit Immobil Biotechnol* **24**:533–551.

Hughes, G.S. Jr., Yancey, E.P., Albrecht, R., *et al.* (1995a) Hemoglobin-based oxygen carrier preserves submaximal exercise capacity in humans. *Clin Pharmacol Ther* **58**:434–443.

Hughes, G.S., Francom, S.F., Antal, E.J., *et al.* (1995b) Hematologic effects of a novel hemoglobin-based oxygen carrier in normal male and female subjects. *J Lab Clin* **126**:444–451.

Hughes, G.S. Jr., Francom S.F., Antal, E.J., *et al.* (1996a) Effects of a novel hemoglobin-based oxygen carrier on percent oxygen saturation as determined with arterial blood gas analysis and pulse oximetry. *Ann Emerg Med* **27**:164–169.

Hughes, G.S. Jr., Antal, E.J., Locker, P.K., *et al.* (1996b) Physiology and pharmacokinetics of a novel hemoglobin-based oxygen carrier in humans. *Crit Care Med* **24**:756–764.

Hunkeler, D. (2001) Allo transplants xeno: as bioartificial organs move to the clinic. In: Hunkeler, D., Cherrington, A., *et al.* (eds.), *Bioartificial Organs* **944**:1–6. *Ann N Y Acad Sci.*

Hunkeler, D., *et al.* (2003) Bioartificial organ grafts: a view at the beginning of the third millennium. *Artif Cells Blood Substit Immobil Biotechnol* **31**:365–382.

Hunt, A.C., and Burnette, R.R. (1983) Neohemocytes. In: Bolin, R.B., Geyer, R.P., and Nemo, G.J. (eds.), *Advances in Blood Substitute Research.* pp. 59–70. Alan R Liss Inc., New York.

Hunt, A.C., Burnette, R.R., MacGregor, R.D., *et al.* (1985) Neohemocytes, *Science* **230**:1165.

Ilan, E., and Chang, T.M.S. (1986) Modification of lipid-polyamide microcapsules for immobilization of free cofactors and multienzyme system for the conversion of ammonia to glutamate. *Appl Biochem Biotechnol* **13**:221–230.

Ilan, E., and Chang, T.M.S. (1987) Conversion of ammonia to glutamate by L-glutamate dehydrogenase, alcohol dehydrogenase and NAD+ immobilized within lipid-polyamide polyethyleneimine microcapsules. *Adv Exp Med Biol* **223**:189–192.

Ilan, E., Morton, P., and Chang, T.M.S. (1992) Bovine hemoglobin anaerobically reacted with divinyl sulfone: a potential source for hypothermic oxygen carriers. *Biomater Artif Cells Immobil Biotechnol* **20**:246–255.

Ilan, E., Morton, P., and Chang, T.M.S. (1993) The anaerobic reaction of bovine hemoglobin with divinyl sulfone: structural changes and functional consequences. *Biochimica Biophysica Acta* **1163**:257–265.

Ilan, E., and Chang, T.M.S. (1994a) Human hemoglobin anaerobically reacted with divinyl sulfone: a source for oxygen carriers. In: Winslow, R.M. (Guest ed.), *Artif Cells Blood Substit Immobil Biotechnol* **22**:687–694.

Ilan, E., and Chang, T.M.S. (1994b) Human hemoglobin anaerobically reacted with divinyl sulfone: a source for oxygen carriers. In: Winslow, R. (Guest ed.), *J Artif Cells Blood Substit Immobil Biotechnol* **22**:687–694.

Intaglietta, M. (1994) Hemodilution and blood substitutes. In: Winslow, R.M. (Guest ed.), *Artif Cells Blood Substit Immobil Biotechnol* **22**:137–144.

Intaglietta, M., Johnson, P.C., and Winslow, R.M. (1996) Microvascular and tissue oxygen distribution. *Cardiovasc Res* **32**:632–643.

Ito, Y., and Chang, T.M.S. (1992) *In vitro* study of multicellular hepatocytes spheroid formed in microcapsules. *J Artif Organs* 16:422–426.

Ivanistski, G.R., and Vorobev, S.I. (1996) Organization of mobile structures in blood stream: basis for the function of perfluorocarbon artificial blood. *Biofizika* **41**: 178–190.

Iwashita, Y., Yabuki, A., Yamaji, K., *et al.* (1988) A new resuscitation fluid "stabilized hemoglobin" preparation and characteristics. *Biomater Artif Cells Artif Organs* **16**:271–280.

Iwashita, Y. (1992) Relationship between chemical properties and biological properties of pyridoxalated hemoglobin-polyoxyethylene. *Biomater Artif Cells Immobil Biotechnol* **20**:299–308.

Izumi, Y., Sakai, H., Kose, T., *et al.* (1997) Evaluation of the capabilities of a hemoglobin vesicle as an artificial oxygen carrier in a rat exchange transfusion model. *ASAIO J* **43**:289–297.

Jamieson, G.A., and Greenwalt, T.J. (eds.), *Blood Substitutes and Plasma Expanders.* pp. 340. Alan R Liss Inc., New York.

Jaquel, P., *et al.* (1995) Metablism. *Clini Exp* **44**:18–26.

Jeong, S.T., and Byun, S.M. (1996) Decreased agglutinability of methoxy-polyethylene glycol attached red blood cells: significance as a blood substitute. *Artif Cells Blood Substit Immobil Biotechnol* **24**:503–511.

Jia, L., Bonaventura, C., Bonaventrua, J., and Stamler, J.S. (1996) S-nitrosohemoglobin: a dynamic activity of blood involved in vascular control. *Nature* **380**:221–226.

Jing, M., Ledvina, M.A., Bina, S., *et al.* (1995) Effects of halogenated and non-halogenated anesthetics on diaspirin crosslinked hemoglobin^{TM}-induced contractions of porcine pulmonary veins. *Artif Cells Blood Substit Immobil Biotechnol* **23**:487–494.

Jones M., Chen, H., Ouyang, *et al.* (2004) Microencapsulated genetically engineered *Lactobacillus plantarum* 80 (pcbh1) for bile acid deconjugation. *J Biomed Biotechnol* **1**:61–69.

Jing, M., Panico, F.G., Panico, J.L., *et al.* (1996) Diaspirin cross-linked hemoglobin does not alter isolated human umbilical artery or vein tone. *Artif Cells Blood Substit Immobil Biotechnol* **24**:621–628.

Joki, T., *et al.* (2001) Continuous release of endostatin from microencapsulated engineered cells for tumor therapy. *Nat Biotechnol* **19**:35–39.

Joyner, M.J., and Faust, R.J. (1996) Blood Substitutes: what is the target? In: Winslow, R.M., Vandegriff, K.D., and Intaglietta, M. (eds.), *Blood Susbtitutes: New Challenges*. pp. 15–33. Birkhäuser, Boston.

Kan, P., Chen, W.K., and Lee, C.J. (1996) Simulation of oxygen saturation of hemoglobin solution, RBC suspension and hemosome by a neural network system. *Artif Cells Blood Substit Immobil Biotechnol* 24:143–151.

Kashani, S.A., and Chang, T.M.S. (1988) Release of hepatic stimulatory substance from cultures of free and microencapsulated hepatocytes: preliminary report. *J Biomater Artif Cells Artif Organs* 16:741–746.

Kashani, S., and Chang, T.M.S. (1991a) Physical chemical characteristics of hepatic stimulatory factor prepared from cell free supernatant of hepatocyte cultures. *Biomater Artif Cells Immobil Biotechnol* 19:565–578.

Kashani, S., and Chang, T.M.S. (1991b) Effects of hepatic stimulatory factor released from free or microencapsulated hepatocytes on galactosamine induced fulminant hepatic failure animal model. *Biomater Artif Cells Immobil Biotechnol* 19:579–598.

Kasper, S.M., Walter, M., Grune, F., *et al.* (1996) Effects of hemoglobin-based oxygen carrier (HBOC-201) on hemodynamics and oxygen transport in patients undergoing preoperative hemodilution for effective abdominal aortic surgery. *Anesth Analg* 83:921–927.

Katsambas, A., and Nicolaidou, E. (1996) Cutaneous malignant melanoma and sun exposure. *Arch Dermatol* 132:444–450.

Kaufman, R.J. (1995) Clinical development of perfluorocarbon-based emulsions as red cell substitutes. In: Winslow, R.M., Vandergriff, K.D., and Intaglietta, M. (eds.), *Blood Substitutes: Physiological Efficacy*. Birkhäuser, Boston.

Kawasaki, C., Nishi, R., Uekihara, S., *et al.* (2000) Charcoal hemoperfusion in the treatment of phenytoin overdose. *Am J Kidney Dis* 35:323–326.

Kaziuka, E.N., and Chang, T.M.S. (1979) *In vitro* assessment of the removal of phenols by ACAC hemoperfusion. *Int J Artif Organs* 2:215–221.

Keipert, P., Minkowitz, J., and Chang, T.M.S. (1982) Cross-linked stroma-free polyhemoglobin as a potential blood substitute. *Int J Artif Organs* 5:383–385.

Keipert, P., and Chang, T.M.S. (1983) *In vivo* assessment of pyridoxalated cross-linked polyhemoglobin as an artificial red cell substitute in rats. *Trans Am Soc Artif Intern Organs* 29:329–333.

Keipert, P., and Chang, T.M.S. (1984) Preparation and *in vitro* characteristics of a blood substitute based on pyridoxylated polyhemoglobin. *Appl Biochem Biotechnol* 10:133–141.

Keipert, P.E., and Chang, T.M.S. (1985) Pyridoxylated polyhemoglobin as a blood substitute for resuscitation of lethal hemorrhagic shock in conscious rats. *Biomater Med Dev Artif Organs* 13:1–15.

Keipert, P.E. (1995) Use of oxygent, a perfluorochemical-based oxygen carrier, as an alternative to intraoperative blood transfusion. *Artif Cells Blood Substit Immobil Biotechnol* 23:381–394.

Keipert, P.E., Verosky, M., and Triner, L. (1988) Metabolism, distribution, and excretion of HbXL: a non-dissociating interdimerically crosslinked hemoglobin with exceptional oxygen offloading capability. *Biomater Artif Cells Artif Organs* **16**:643–645.

Keipert, P.E., and Chang, T.M.S. (1987) *In vivo* effects of total and partial isovolemic exchange transfusion in fully conscious rats using pyridoxylated polyhemoglobin solution as a colloidal oxygen-delivery blood substitute. *Vox Sang* **53**:7–14.

Keipert, P.E., and Chang, T.M.S. (1988) Pyridoxylated-polyhemoglobin solution: a low viscosity oxygen-delivering blood replacement fluid with normal oncotic pressure and long-term storage feasibility. *J Biomater Artif Cells Artif Organs* **16**:185–196.

Keipert, P.E., Faithfull, N.S., Roth, D.J., *et al.* (1996) Supporting tissue oxygenation during acute surgical bleeding using a perfluorochemical-based oxygen carrier. *Adv Exp Med Biol* **388**:603–609.

Khanna, R., and Chang, T.M.S. (1990) Characterization of L-histidine ammonia-lyase immobilized by microencapsulation in artificial cells: preparation kinetics, stability, and *in vitro* depletion of histidine. *Int J Artif Org* **13**:189–195.

Kim, H.W., Chen, F., and Greenburg, A.G. (1992) Hemoglobin-based double (exchange transfusion — carbon clearance) model for testing post-resuscitation reticuloendothelial function. *Biomater Artif Cells Immobil Biotechnol* **20**:777–780.

Kim, H.W., and Greenburg, A.G. (1995) Hemoglobin mediated vasoactivity in isolated vascular rings. *Artif Cells Blood Substit Immobil Biotechnol* **23**:303–309.

Kim, H.W., and Greenburg, A.G. (1997) Ferrous hemoglobin scavenging of endothelium derived nitric oxide is a principal mechanism for hemoglobin mediated vasoactivities in isolated rat thoracic aorta transfusion. *Artif Cells Blood Substit Immobil Biotechnol* **25**:121–134.

Kim, H.W., Breiding, P., and Greenburg, A.G. (1997a) Enhanced modulation of hypotension in endotoxemia by concomitant nitric oxide synthesis inhibition and nitric oxide scavenging. *Artif Cells Blood Substit Immobil Biotechnol* **25**:153–162.

Kjellstrand, C., Borges, H., Pru, C., *et al.* (1981) On the clinical use of micro-encapsulated zirconium phosphate-urease for the treatment of chronic uremia. *Trans Am Soc Artif Intern Organs* **27**:24–30.

Klein, H.G. (1994) Oxygen carriers and transfusion medicine. *Artif Cells Blood Substit Immobil Biotechnol* **22**:123–135.

Klinkmann, H., Falkenhagen, D., and Chang, T.M.S. (eds.) (1990) Hemoperfusion, sorbent and immobilized bioreactants (Special issue). *Int J Biomater Artif Cells Artif Organs* **18**:455–568.

Kluger, R., Jones, R.T., and Shih, D.T. (1994) Crosslinking hemoglobin by design: lessons from using molecular clamps. *Artif Cells Blood Substit Immobil Biotechnol* **22**:415–428.

Kobayashi, K., Izumi, Y., Yoshizu, A., *et al.* (1997) The oxygen carrying capability of hemoglobin vesicles evaluated in rat exchange transfusion models. *Artif Cells Blood Substit Immobil Biotechnol* **25**:357–366.

Kobayashi, K., *et al.* (2003) Essential *Bacillus subtilis* genes. *Proc Natl Acad Sci* **100**:4678–4683.

Kochetygov, N.I., Gorkoun, A.V., Gerbut, K.A., *et al.* (1996) Study of therapeutic efficiency of new blood substitutes in experimental hemorrhagic shock. *Artif Cells Blood Substit Immobil Biotechnol* **24**:685–691.

Kolff, W.J., and Berk, H.T.J. (1944) The artificial kidney: a dialyser with a great area. *Acta Med Scand* **117**:121.

Koo, J., and Chang, T.M.S. (1993) Secretion of erythropoietin from microencapsulated rat kidney cells: preliminary results. *Int J Artif Organs* **16**:557–560.

Kreimeier, U., Messamer, K., Krishnamurti, C., and Alving, B. (1996) Biological consequences of cross-linked hemoglobin in animal models of surgery and endotoxemia. In: Winslow, R.M., Vandegriff, K.D., and Intaglietta, M. (eds.), *Blood Sunstitutes: New Challenges*. pp. 99–111. Birkhäuser, Boston.

Kulitreibez, W.M., Lauza, P.P., and Cuicks, W.L. (eds.) (1999) *Cell Encapsul Technol Ther* **1**:450. Burkhauser, Boston.

Kumar, A., Sen, A.P., Saxena, P.R., and Gulati, A. (1997) Resuscitation with diaspirin crosslinked hemoglobin increases cerebral and renal blood perfusion in hemorrhaged rats. *Artif Cells Blood Substit Immobil Biotechnol* **25**:85–94.

Kumar, R. (1995) Recombinant hemoglobins as blood substitutes: a biotechnology perspective. *Proc Soc Exp Biol Med* **208**:150–158.

Lacy, P.E., Hegre, O. D, Gerasimidi-Vazeou, A., et al. (1991) Maintenance of normoglycemia in diabetic mice by subcutaneous xenografts of encapsulated islets. *Science* **254**:1782–1784.

Lagasse, E., Connors, H., Al-Dhalimy, M., et al. (2000) Purified hematopoietic stem cells can differentiate into hepatocytes *in vivo*. *Nat Med* **6**:1229–1234.

Lalla,.R., Ning, J., and Chang, T.M.S. (1989) Effects of pyridoxalated polyhemoglobin and stroma-free hemoglobin on ADP-induced platelet aggregation. *J Biomater Artif Cells Artif Organs* **17**:363–369.

Langer, R., Blackshear, P.J., Chang, T.M.S., et al. (1986) Enzyme and drug delivery systems. *Trans Am Soc Artif Intern Organs* **32**:639–645.

Langermans, J.A., van Vuren-van der Huslt, M., and Bleeker, W.K. (1996) Safety evaluation of a polymerized hemoglobin solution in a murine infection model. *J Lab Clin Med* **127**:428–434.

Larsson, P.O., and Mosbach, K. (1974) The preparation and characterization of a water-soluble conenzymically active dextran-NAD$^+$. *FEBS Lett* **46**:119–122.

LaVan, D.A., Lynn, D.M., and Langer, R. (2002) Moving smaller in drug discovery and delivery. *Nat Rev Drug Discov* **1**:77–84.

Lear, J.D. et al. (1997) Electrostatic effects on ion selectivity and rectification in designed ion channel peptides. *J Am Chem Soc* **119**:3212–3217.

Lee, R., Neya, K., Svizzero, T.A., and Vlahares, G.J. (1995) Limitations of the efficacy of hemoglobin-based oxygen-carrying solutions. *J Appl Physiol* **79**:236–242.

Legallais, C., David, B., and Doré, E. (2001) Bioartificial livers (BAL): current technological aspects and future developments. *J Memb Sci* **181**:81–95.

Leppaniemi, A., Soltero, R., Burris, D., et al. (1996) Early resuscitation with low-volume PolyDCLHB is effective in the treatment of shocked induced by penetrating vascular injury. *J Trauma* **40**:242–248.

Leslie, S.B., Puvvada, S., Ratna, B.R., and Rudolph, A.S. (1996) Encapsulation of hemoglobin in a bicontinuous cubic phase lipid. *Biochim Biophys Acata* **1285**:246–254.

Lesney, M.S. (2001) Going cellular. *Mod Drug Disc* **4**:45–46.

Levin, J., Noth, R.I., Kaca, W., *et al.* (1996) Hemoglobin-endotoxin interactions. In: Winslow, R.M., Vandegriff, K.D., and Intaglietta, M. (eds.), *Blood Sunstitutes: New Challenges.* pp. 185–202. Birkhäuser, Boston.

Li, T., Yu, R., Zhang, H.H., Liang, W.G., *et al.* (2006) A method for purification and viral inactivation of human placenta Hb. *Artif Cells Blood Substit Biotechnol* **34**:175– 188.

Li, X.Z., Zhang, X.W., and Liu, Q. (2005) Determination of the molecular weight distribution of the PEGylated bovine Hb (PEG-bHb). *Artif Cells Blood Substit Biotechnol* **33**:13–28.

Lim, F., and Sun, A.M. (1980) Microencapsulated islets as bioartificial endocrine pancreas. *Science* **210**:908–909.

Lin, C.C., Chan, T.Y., and Deng, J.F. (2004) Clinical features and management of herb-induced aconitine poisoning. *Ann Emerg Med* **43**:574–579.

Lin, C.C., Chou, H.L., and Lin, J.L. (2002) Acute aconitine poisoned patients with ventricular arrhythmias successfully reversed by charcoal hemoperfusion. *Am J Emerg Med* **20**:66–67.

Liu, D., Mori, A., and Huang, L. (1992) Role of liposome size and RES blockade in controlling biodistrbution and tumour uptake of GM1-containing liposomes. *Biochim Biophus Acta* **1104**:95–101.

Liu, J., *et al.* (2002) Study on a novel strategy to treatment of phenylketonuria. *Artif Cells Blood Substit Immobil Biotechnol* **30**:243–258.

Liu, J.P., Gluud, L.L., Als-Nielsen, B., and Gluud, C. (2004) Artificial and bioartificial support systems for liver failure. *Cochrane Database Syst Rev* **1**:CD003628.

Liu, Z.C., and Chang, T.M.S. (2000) Effects of bone marrow cells on hepatocytes: when cocultured or co-encapsulated together. *Artif Cells Blood Substit Immobil Biotechnol* **28**:365–374.

Liu, Z.C., and Chang, T.M.S. (2002) Increased viability of transplantation hepatocytes when coencapsulated with bone marrow stem cells using a novel method. *Artif Cells Blood Substit Immobil Biotechnol* **30**:99–11.

Liu, Z.C., and Chang, T.M.S. (2003) Coencapsulation of stem sells and hepatocytes: *in vitro* conversion of ammonia and *in vivo* studies on the lowering of bilirubin in Gunn rats after transplantation. *Int J Artif Organs* **26**:491–497.

Liu, Z.C., Chow, K.M., and Chang, T.M.S. (2003) Evaluation of two protocols of uremic rat model: partial nephrectomy and infarction. *Renal Failure* **25**:935–943.

Liu, Z.C., and Chang, T.M.S. (2005) Transplantation of bioencapsulated bone marrow stem cells improves hepatic regeneration and survival of 90% hepatectomized rats: a preliminary report. *Artif Cells Blood Substit Biotechnol* **33**:405–410.

Liu, Z.C., and Chang, T.M.S. (2006a) Transdifferentiation of bioencapsulated bone marrow cells into hepatocyte-like cells in the 90% hepatectomized rat model. *J Liver Transplant* **12**:566–572.

Liu, Z.C., and Chang, T.M.S. (2006b) Polymeric artificial cells for coencapsulation of hepatocytes and bone marrow stem cells: *in vitro* and *in vivo* analysis. *Biotechnol Ann Rev* (in press).

Liu, Z.C., and Chang, T.M.S. (2006c) Artificial cell microencapsulated stem cells in regenerative medicine, tissue engineering and cell therapy. In: Pedraz (ed.), *Cell Encapsulation* (in press).

Lloyd, K.G., Homykiewic, O., Davidson, L., *et al.* (1981) Biochemical evidence of dysfunction of brain neurottansmitter ub the Lesch-Nyhan syndrome. *New Engl J Mea* **305**:1106–1111.

Lloyd-George, I., and Chang, T.M.S. (1993) Free and microencapsulated *Erwinia herbicola* for the production of tyrosine. In: Oncelet, D., and Lievremont, M. (Guest eds.), *Biomater Artif Cells Immobil Biotechnol* **21**:323–334.

Lloyd-George, I., and Chang, T.M.S. (1995) Characterization of free and alginate-polylysine-alginate microencapsulated *Erwinia herbicola* for the conversion of ammonia, pyruvate and phenol into L-tyrosine and L-DOPA. *J Bioeng Biotechnol* **48**:706–714.

Lok, C. (2001) Blood product from cattle wins approval for use in humans. *Nature* **410**:855.

Looker, D., Abbott-Brown, D., Cozart, P., *et al.* (1992) A human recombinant hemoglobin designed for use as a blood substitute. *Nature* **356**:258–260.

Looker, D., Durfee, S., Shoemaker, S., *et al.* (1991) Production of recombinant hemoglobin specifically engineered to enhance delivery and circulating half-life: a recombinant cell-free blood substitute. *Biomater Artif Cells Immobil Biotechnol* **19**:418.

Lopez Lago, A.M., *et al.* (2002) Paraquat poisoning and hemoperfusion with activated charcoal. *Ann Intern Med* **19**:310–312.

Lorch, J.A., and Garella, S. (1979) Hemoperfusion to treat intoxications. *Ann Intern Med* **91**:301.

Lörh, M., Hoffmeyer, A., Kröger, J.C., *et al.* (2001) Microencapsulated cell-mediated treatment of inoperable pancreatic carcinoma. *Lancet* **357**:1591–1592.

Lörh, M., *et al.* (2001) Microencapsulated cell-mediated treatment of inoperable pancreatic carcinoma. *Lancet* **357**:1591–1592.

Lowe, K.C., Anthony, P., and Davey, M.R. (1995a) Enhanced protoplast growth at the interface between oxygenated fluorocarbon liquid and aqueous culture medium supplemented with pluronic F-68. *Artif Cells Blood Substit Immobil Biotechnol* **23**:417–422.

Lowe, K.C., Furmidge, B.A., and Thomas, S. (1995b) Haemolytic properties of pluronic surfactants and effects of purification. *Artif Cells Blood Substit Immobil Biotechnol* **23**:135–139.

Lutz, J., Kettemann, M., Racz, I., and Noth, V. (1995) Several methods utilized for the assessment of biocompatibility of perfluorochemicals. *Artif Cells Blood Substit Immobil Biotechnol* **23**:407–415.

Mackay, Z., Shugufta, Q., Din, M., and Guru, A.A. (1996) Hemodilution in complicated high velocity vascular injuries of limbs. *J Cardiovasc Surg* (Torino), **37**:217–221.

Malcolm, D., Kissinger, D., and Garrioch, M. (1992) Diaspirin cross-linked hemoglobin solution as a resuscitative fluid following severe hemorrhage in the rat. *Biomater Artif Cells Immobil Biotechnol* **20**:495–498.

Mallick, A., and Bodenham, A.R. (1996) Modified haemoglobins as oxygen transporting blood substitutes. *Br J Hosp Med* **55**:443–448.

Malsch, N.H. (ed.) (2005) *Biomedical nanotechnology.* pp. 1–209. Taylor & Francis, New York.

Manjula, B.N., Smith, P.K., Malavalli, A., and Acharya, A.S. (1995) Intramolecular crosslinking of oxy hemoglobin by bis sulfosuccinimidyl suberate and sebacate: generation of crosslinked hemoglobin with reduced oxygen affinity. *Artif Cells Blood Substit Immobil Biotechnol* **23**:311–318.

Manning, J.M. (1996) Examples of chemical modification and recombinant DNA approaches with hemoglobin. *Transfus Clin Biol* **3**:109–111.

Manning, J.M. (1994) Random chemical modification of hemoglobin to identify chloride binding sites in the central dyad axis: their role in control of oxygen affinity. In: Winslow, R.M. (ed.), *Artif Cells Blood Substit Immobil Biotechnol* **22**:199–206.

Marchand, G., Dunlap, E., Farrell, L., *et al.* (1996) Resuscitation with increasing doses of diaspirin crosslinked hemoglobin in swine. *Artif Cells Blood Substit Immobil Biotechnol* **24**:469–487.

Marks, D.H., Patressi, J., and Chaudry, I.T. (1985) Effects of pyridoxalated stabilized stroma-free hemoglobin solution on the clearance of intravascular lipid by the reticuloendothelial system. *Circ Shock* **16**:165–172.

Marshall, T., Weltzer, J., Hai, T., *et al.* (1992) Trace element analysis in diaspirin cross-linked hemoglobin solutions. *Biomater Artif Cells Immobil Biotechnol* **20**:453–456.

Martin, A.M., Gibbins, J.K., Oduro, A., and Herbert, R. (1978) Clinical experience with cellulose-coated carbon hemoperfusion. In: Chang, T.M.S. (ed.), *Artificial Kidney Artificial Liver and Artificial Cells.* pp. 143–152. Plenum, New York.

Matheson, B., Kwansa, H.E., Rebel, A., and Bucci, E. (2002) Vascular responses to infusions of a non-extravasating Hb polymer. *J Appl Physio* **93**:1479–1486.

Mathews, A.J., Durfee, S.L., Looker, D.L., *et al.* (1991) Functional properties of potential blood substitutes: protein engineering of human hemoglobin. *Biomater Artif Cells Immobil Biotechnol* **19**:431.

Matsumura, H., Araki, H., Morioka, T., and Nishi, K. (1992) Pyridoxalated hemoglobin polyoxyethylene conjugate (PHP) on the endothelium-dependent relaxation in rat mesemteric arterioles. *Biomater Artif Cells Immobil Biotechnol* **20**:679–682.

Matsushita, M., Yabuki, A., Malchesky, P.S., *et al.* (1988) *In vivo* evaluation of a pyridoxalated-hemoglobin-poloxyethylene conjugate. *Biomater Artif Cells Artif Organs* **16**:247–260.

Matsushita, S., Sakakibara, Y., Jikuya, T., *et al.* (1992) Effect of stabilized hemoglobin as a component of cardioplegia on warm ischemic heart. *J Biomater Artif Cells Immobil Biotechnol* **20**:703–708.

McCall, W.G. (1996) Physiological and practical considerations of fluid management. *CRNA* **7**:62–70.

McCohaney, P.J., and Dixon, F.J. (1966) *Int Arch Allergy* **29**:185–189.

McDonagh, P.F., and Wilson, D.S. (1995) The initial response of blood leukocytes to incubation with perfluorocarbon blood substitute emulsions. *Artif Cells Blood Substit Immobil Biotechnol* **23**:439–447.

McNeish, J. (2004) Embryonic stem cells in drug discovery. *Nat Rev Drug Discov* **3**:70–80.

Meadows, C.L., Gher, M.E., Quintero, G., and Lafferty, T.A. (1993) *J Periodontol* **64**:103–109.

Meadows, G.G., Pierson, H.F., Abdallah, R.M., and Desai, R. (1982) Dietary influence of tyrosine and phenylalanine on the response of B16 melanoma to carbidopa-levodopa methyl ester chemotherapy. *Cancer Res* **42**:3056–3063.

Hagan B. (1999) Designed membrane channels and pores. *Curr Opin Biotechnol* **10**:94–103.

Melkikh, A.V., and Seleznev, V.D. (2005) Models of active transport of ions in biomembranes of various types of cells. *J Theor Biol* **234**:403–412.

Melkikh, A.V., and Seleznev, V.D. (2006) Requirements on models and models of active transport of ions in biomembranes. *Bull Math Biol* **68**:385–399.

Menu, P., Faivre, B., Labrude, P., *et al.* (1992) Possible importance of chromatographic purification position in a blood substitute elaboration process. *Biomater Artif Cells Immobil Biotechnol* **20**:443–446.

Menu, P., Donner, M., Faivre, B., *et al.* (1995) *In vitro* effect of dextran-benzene-tetra-carboxylate hemoglobin on human blood rheological properties. *Artif Cells Blood Substit Immobil Biotechnol* **23**:319–330.

Messmer, K. (1988) Characteristics, effects and side effects of plasma substitutes, In: Lowe, K.C. (ed.), *Blood Substitutes: Preparation, Physiology and Medical Applications.* pp. 51–70, Ellis Horword Ltd, Chichester, England and distributed by VCH Publisher Ltd, Cambridge, UK.

Michalopoulos, G., Houck, K.D., Dolan, M.L., and Luetteke, N.C. (1984) Control of hepatocyte replication by two serum factors. *Cancer Res* **44**:4414–4419.

Miller, I.F., Mayoral, J., Djordjevich, L., and Kashani, A. (1988) Hemodynamic effects of exchange transfusions with liposome-encapsulated hemoglobin. *Biomater Artif Cells Artif Organs* **16**:281–288.

Minato, N., Sasaki, T., Sakuma, I., *et al.* (1992) Potential clinical applications of the oxygen carrying solutions. *Biomater Artif Cells Immobil Biotechnol* **20**:221–232.

Mitsuno, T., and Naito, R. (eds.) (1979) Perfluorochemical blood substitutes. *Excerpta Med.* Amsterdam.

Mitsuno, T., and Ohyanagi, I. (1985) Present status of clinical studies of fluosol-DA (20%) in Japan. In: Tremper, K.K. (ed.), *Perfluorochemical Oxygen Transport.* pp. 169–184. Little Brown & Co, Boston.

Mobed, M., and Chang, T.M.S. (1991) Preparation and surface characterization of carboxymethylchitus incorporated submicron bilayer lipid membrane artificial cells (liposomes) encapsulating hemoglobin. *Biomater Artif Cells Immobil Biotechnol* **19**:731–744.

Mobed, M., and Chang, T.M.S. (1992) Purification and quantitative determination of carboxymethylchitin-incorporated into submicron bilayer-lipid membrane artificial cells (liposomes) encapsulating hemoglobin. *Biomater Artif Cells Immobil Biotechnol* **20**:329–336.

Mobed, M., Nishiya, T., and Chang, T.M.S. (1992) Preparation of carbo xymethylchitinin corporated submicron bilayer-lipid membrane artificial cells (liposomes) encapsulating hemoglobin. *Biomater Artif Cells Immobil Biotechnol* **20**:325–328.

Mobed, M., and Chang, T.M.S. (1996a) The importance of standardization of carboxymethylchitin concentration by the dye-binding capacity of alcian blue before adsorption on liposomes. *Artif Cells Blood Substit Immobil Biotechnol* **24**:107–120.

Mobed, M., and Chang, T.M.S. (1996b) Use of chintinase for surface characterization of carbozymethy and carboxtmethyl/clycolchitin coated liposomes for potential medical applications. *Chitin Enzymol* **2**:313–315.

Mobed, M., and Chang, T.M.S. (1998a) Colloidal stability of sterically stabilized grafted stealth liposomes and physically adsorbed carboxymethylchitin/carboxymethylglycolchitin liposomes for biological applications. *J Microencapsul* **15**:595–607.

Mobed, M., and Chang, T.M.S. (1998b) Adsorption of chitin derivatives onto liposomes: optimization of adsorption conditions. *J Microencapsul* **15**:595–607.

Mobed, M., and Chang, T.M.S. (1998c) Comparison of polymerically stabilized PEG grafted liposomes and physically adsorbed carboxymethyl/chitin and carboxytmethyl/clygochitin liposomes for biological applications. *Biomaterials* **19**:1167–1177.

Mohsini, K., Lister, C., and Chang, T.M.S. (1980) The effects of homologous cross-circulation and *in situ* liver perfusion on fulminant hepatic failure rats. *Artif Organs* **4**:171–175.

Mok, W., Chen, D.E., and Mazur, A. (1975) Covalent linkage of subunits of hemoglobin. *Fed Proc* **34**:1458.

Monnard, P.A., and Deamer, D.W. (2001) Nutrient uptake by protocells: a liposome model system. *Orig Life Evol Blosph* **31**:147–155.

Morley, D., and Chang, T.M.S. (1987) Effect of membrane thickness of collodion coated activated charcoal on adsorption of small and middle molecules. *J Biomater Artif Cells Artif Organs* **15**:617–632.

Moss, G.S., DeWoskin, R., Rosen, A.L., et al. (1976) Transport of oxygen and carbon dioxide by hemoglobin-saline solution in the red cell-free primate. *Surg Gynecol Obstet* **142**:357.

Moss, G.S., Gould, S.A., Sehgal, L.R., et al. (1988) Hemoglobin solution — from tetramer to polymer. *Biomater Artif Cells Artif Organs* **16**:57–69.

Motterlini, R., Foresti, R., Vandegriff, K., and Winslow, R.M. (1995) The autoxidation of alpha-alpha cross-linked hemoglobin: a possible role in the oxidative stress to endothelium. *Artif Cells Blood Substit Immobil Biotechnol* **23**:291–301.

Mouelle, P., Labrude, P., Grandgeorge, M., and Vigneson, C. (1992) A temporary blood substitute based on dextran hemoglobin conjugates. III. Effects on guinea-pig heart perfusion and myocardial ischemia reperfusion. *Biomater Artif Cells Immobil Biotechnol* **20**:697–702.

Mullett, C.J., and Polak, M.J. (1995) Vasoreactivity of Fluosol^{TM} perfluorocarbon emulsion vs Earl's balanced salt/albumin perfusates in the isolated perfused rat lung. *Artif Cells Blood Substit Immobil Biotechnol* **23**:449–457.

Murray, J.A., Ledlow, A., Launspach, J., *et al.* (1995) The effects of recombinant human hemoglobin on esophageal motor functions in humans. *Gastroenterology* **109**:1241–1248.

Myhre, B.A. (1990) The first recorded blood transfusions: 1656 to (1668) *Transfusion* **30**:358.

Naito, R., and Yokoyama, K. (1978) An improved perfluorodecalin emulsion. In: Jamieson, G.A., and Greenwalt, T.J. (eds.), *Blood Substitutes and Plasma Expanders.* pp. 81. Alan R. Liss Inc., New York.

Nakai, K., Abe, A., Matsuda, N., *et al.* (1992) Development of analytical methods to evaluate SFH biomaterials. *Artif Cells Immobil Biotechnol* **20**:446–452.

Nakoa, T., Nagano, K., Adachi, K., and Nakoa, M. (1963) Separation of two adenosine triphosphatases from erythrocyte membrane. *Biochem Biophys Res Commun* **13**:444.

Nasielski, P., and Chang, T.M.S. (1978) Coated charcoal hemoperfusion for the removal of galactose in rats. *Int J Artif Organs* **1**:143–147.

Nasielski, P., and Chang, T.M.S. (1978) Microencapsulated charcoal hemoperfusion forgalactosemia. In: Chang, T.M.S. (ed.), *Artificial Kidney Artificial Liver and Artificial Cells.* pp. 255–257. Plenum Press, New York.

Neufeld, R., Arbeloa, M., and Chang, T.M.S. (1984) Design of a fluidized bed reactor for microencapsulated urease. *Appl Biochem Biotechnol* **10**:109–119.

Nho, K., Glower, D., Bredehoeft, S., Shankar, H., *et al.* (1992) PEG-bovine hemoglobin: safety in a canine dehydrated hypovolemic-hemorrhagic shock model. *Biomater Artif Cells Immobil Biotechnol* **20**:511–524.

Nho, K., Linberg, R., Johnson, M., *et al.* (1996) PEG-hemoglobin: an efficient oxygen-delivery system in the rat exchange transfusion and hypovolemic shock models. *Artif Cells Blood Substit Immobil Biotechnol* **22**:795–803.

Ni, Y., Klein, D.H., and Song, D. (1996) Recent development in pharmacokinetic modeling of perfluorocarbon emulsion. *Artif Cells Blood Substit Immobil Biotechnol* **24**:81–90.

Ning, J., and Chang, T.M.S. (1988) Effects of stroma-free Hb and polyhemoglobin on complement activation blood cell counts and coagulation factors in rats. *Biomater Artif Cells Artif Organs* **16**:651–652.

Ning, J., and Chang, T.M.S. (1990a) Effects of homologous an heterologous stroma-free hemoglobin and polyhemoglobin on complement activation, leukocytes and platelets. *Biomater Artif Cells Artif Organs* **18**:219–233.

Ning, J., and Chang, T.M.S. (1990b) *In vivo* effects of stroma-free hemoglobin and polyhemoglobin on coagulation factors in rats. *Int J Artif Organs* **13**:509–516.

Ning, J., and Chang, T.M.S. (1990c) Measurement of complement activation by CH50 in rats. *Biomater Artif Cells Artif Organs* **18**:203–218.

Ning, J., Anderson, P.J., and Biro, G.P. (1992) Resuscitation of bled dogs with pyridoxalated-polymerized hemoglobin solution. *Biomater Artif Cells Immobil Biotechnol* **20**:525–530.

Nishiya, T., and Chang, T.M.S. (1989) Effects of cholesterol and melittin on spontaneous lipid transfer between liposomes. *Chem Phys Lipids* **51**:83–90.

Nishiya, T., Dasgupta, M., Okumura, Y., and Chang, T.M.S. (1988) Circular dichroism study of membrane dynamics focused on effect of monosialoganglioside. *Biochem J* **104**:62–65.

Nishiya, T., Okumura, Y., and Chang, T.M.S. (1988) Synthesis and properties of circular dichroism active phospholipids. *Chem Phys Lipids* **49**:69–79.

Nishiya, T., and Chang, T.M.S. (1994) Toxicity of liposomes containing low mol% of dienoyl phosphocholine to blood: use of carboxymethyl chitin to reduce toxicity. *J Artif Cells Blood Substit Immobil Biotechnol* **22**:883–888.

Nishiya, T., Lam, R.T.T., Eng, F., *et al.* (1995) Mechanistic study on toxicity of positively charged liposomes containing stearylamine to blood. *Artif Cells Blood Substit Immobil Biotechnol* **23**:505–512.

Nishiya, T., and Jain, B. (1996) Study on *in vitro* stability of polymerized liposomes. *Artif Cells Blood Substit Immobil Biotechnol* **24**:43–50.

Niu, Z., Jia, S.R., Zhang, D.Y. *et al.* (1980) The effects of hemoperfusion in galactosamine induced FHF rats. *Chongqing Med Coll Bull* (Nov):1–6.

Nylan, W.L. (1974) Heritable disorders of amino acids metabolism. In: Nylan, W.L. (eds.), *Patterns of Clinical Expression and Genetic Variations.* pp. 109–159. John Wiley and Sons Ltd., Sussex, England.

Oberholzer, T. *et al.* (1995) Polymerase chain reaction in liposomes. *Chem Biol* **2**:677–682.

Oberholzer, T. *et al.* (1999) Protein expression inliposomes. *Biochem Biophys Res Commun* **261**:238–241.

Odaka, M., Tabata, Y., Kobayashi, H., *et al.* (1978) Cinical experience of bead-shaped charcoal hemoperfusion in chronic renal failure and fulminant hepatic failure. In: Chang, T.M.S. (ed.), *Artificial Kidney, Artificial Liver and Artificial Cells.* pp. 79–88. Plenum Press, New York.

Odaka, M., Hirasawa, H., Kobayashi, H., *et al.* (1980) Clinical and fundamental studies of cellulose coated bead-shaped charcoal hemoperfusion in chronic renal failure. In: Sideman, S. and Chang, T.M.S. (eds.), *Hemoperfusion: Kidney and Liver Support and Detoxification,* Part I. pp. 45–55. Hemisphere, Washington, DC.

O'Donnell, J.K., Swanson, M., Pilder, S., *et al.* (1992) Production of human hemoglobin in transgenic swine. *Biomater Artif Cells Immobil Biotechnol* **20**:149.

Ogden, J.E., Woodrow, J., Perks, K., *et al.* (1992) Expression and assembly of functional human hemoglobin in SACCHAROMYCES CEREVISIAE. *Biomater Artif Cells Immobil Biotechnol* **20**:473–477.

O'Grady, J.G., Gimson, A.E., O'Brien, C.J., *et al.* (1988) Controlled trials of charcoal hemoperfusion and prognostic factors in fulminant hepatic failure. *Gastroenterology* **94**:1186.

Ohyanag Ogden, J.E., and Parry, E.S. (1995) The development of hemoglobin solutions as red blood cell substitutes. *Int Anesthesiol Clin* **33**:115–129.

Okada, N., *et al.* (1997) Cytomedical therapy for IgG1 plasmacytosis in human interleukin-6 transgenic mice using hybridoma cells microencapsulated in alginate-poly(L)lysine-alginate membrane. *Biochim Biophys Acta* **1360**:53–63.

Olson, J.S. (1994) Genetic engineering of myoglobin as a simple prototype for hemoglobin-based blood substitutes. *Artif Cells Blood Substit Immobil Biotechnol* **22**:429–442.

Olson, J.S., Eich, R.F., Smith, L.P., *et al.* (1997) Protein engineering strategies for designing more stable hemoglobin-based blood substitutes. *Artif Cells Blood Substit Immobil Biotechnol* **25**:227–241.

Olsen, J.S., Foley, E.W., and Rogge, C. (2004) Nitric oxide scavenging and the hypertensive effect of Hb-based blood substitutes. *Free Radic Biol Med* **36**:685–697.

Omer, A. *et al.* (2003) Survival and maturation of microencapsulated porcine neonatal pancreatic cell clusters transplanted into immunocompetent diabetic mice. *Diabetes* **52**:69–75.

Orive, G., Hernandez, R.M., Gascon, A.R., *et al.* (2003) Cell encapsulation: promise and progress. *Nat Med* **9**:104–107.

Orive, G., Hernández, R.M., Gascón, A.R., *et al.* (2004) History, challenge and perspectives of cell microencapsulation. *Trends Biotechnol* **22**:87–92.

O'Shaughnessy, J.A. (2003) Pegylated liposomal doxorubicin in the treatment of breast cancer. *Clin Breast Cancer* **4**:318–328.

Ortmanis, A., Neufeld, R.J., and Chang, T.M.S. (1984) Study of microencapsulated urease in a continuous feed, stirred tank reactor. *Enzyme Microb Technol* **6**:135–139.

Ouchterlony, O. (1967) In: Weir, D.M. (ed.), *Handbook of Experimental Immunology*. p. 655. Blackwell Scientific Publications, Oxford.

Pagano, R., and Thompson, T.E. (1968) Spherical lipid bilayer membranes, electrical and isotopic studies of ion permeability. *J Mol Biol* **38**:41.

Palmour, R.M., Goodyer, P., Reade, T., and Chang, T.M.S. (1989) Microencapsulated xanthine oxidase as experimental therapy in Lesch-Nyhan disease. *Lancet* **2**:687–688.

Panter, S.S., Vandegriff, K.D., Van, P.O., and Egan, R.F. (1994) Assessment of hemoglobin-dependent neurotoxicity: alpha-alpha cross-lined hemoglobin. *Artif Cells Blood Substit Immobil Biotechnol* **22**:687–694.

Park, Y.K., Abuchowski, A., Davis, S., and Davis, F. (1981) Pharmacology of *Escherichia coli*-L-asparaginase polyethylene glycol adduct. *Anticancer Res* **1**:373–376.

Pardridge, W.M. (2002) Drug and gene targeting to the brain with molecular Trojan horses. *Nat Rev. Drug Discov* **1**:131–139.

Pavlik, P.A., Boyd, M.K., and Olsen, K.W. (1996) Molecular dynamics of a hemoglobin crosslinking reaction. *Biopolymers* **39**:615–618.

Payne, J.W. (1973) Glutaraldehyde crosslinked protein to form soluble molecular weight markers. *Biochem J* **135**:866–873.

Pearce, L.B., and Gawryl, M.S. (1998) Overview of preclinical and clinical efficacy of Biopure's HBOCs. In: Chang, T.M.S. (ed.), *Blood Substitutes: Principles, Methods, Products and Clinical Trials*, Vol. 2, pp. 82–98. Karger, Basel.

Pearce, L.B., Gawryl, M.S., Rentko, V.T., *et al.* (2006) HBOC-201 [Hb Glutamer-250 (Bovine), Hemopure]: clinical studies. In: Winslow, R. (ed.), *Blood Substitutes*. pp. 437–450. Academic Press San Diego.

Peerless, S.J., Nakamura, R., Rodriguez-Salazar, A., and Hunter, I.G. (1985) Modification of cerebral ischemia with fluosol. *Stroke* **16**:38.

Peng, A., Meng, F.Q., Sun, L.F., *et al.* (2004) Therapeutic efficacy of charcoal hemoperfusion in patients with acute severe dichlorvos poisoning. *Acta Pharmacol Sin* **25**:15–21.

Perutz, M.F. (1980) Stereochemical mechanism of oxygen transport by hemoglobin. *Proc R Soc Lond B* **208**:135.

Perutz, M.F. (1989) Myoglobin and hemoglobin: role of distal residues in reactions with haem ligands. *Trends Biochem Sci* **14**:42–44.

Philips, W.T., *et al.* (1999) Polyethylene glyco-modified liposome-encapsulated hemoglobin: a long circulating red cell substitute. *J Pharm Exp Ther* **288**:665–670.

Phillips, R., Mawhinney, T., Harmata, M., and Smith, D. (1995) Characterization of *Gallus Domesticus*-N-acetyl-galactosaminidase blood group A_2 activity. *Artif Cells Blood Substit Immobil Biotechnol* **23**:63–80.

Phillips, W.T., Rudolph, A.S., and Klipper, R. (1992) Biodistribution studies of liposome encapsulated hemoglobin (LEH) studied with a newly developed 99m-technetium liposome label. *Biomater Artif Cells Immobil Biotechnol* **20**:757–760.

Phiri, J.B., Senthil, V., and Grossweiner, L.I. (1995) Diffuse optics determination of hemoglobin derivatives in red blood cells and liposome encapsulated hemoglobin. *Artif Cells Blood Substit Immobil Biotechnol* **23**:23–38.

Photos, P.J., Bacakova, L., Discher, B., Bates, F.S., *et al.* (2003) Polymer vesicles *in vivo*: correlations with PEG molecular weight. *J Control Release* **90**:323–334.

Piskin, E., Evren, V., Piskin, K., *et al.* (1982) Kinetics of ammonia removal by gas desorption. *Proceedings, International Symposium on Hemoperfusion and Artificial Organs.* pp. 89–91.

Piskin, E., and Chang, T.M.S. (1979) Urea and ammonium removal based on alkalinization and removal of free ammonia. *Int J Artif Organs* **2**:211–214.

Piskin, E., and Chang, T.M.S. (1980) A new process for removal of urea. *Proc Eur Soc Artif Organs* **7**:127–131.

Piskin, E., and Chang, T.M.S. (1980) Effects of pH and urea concentration on urea removal from dialysate by enzymatic conversion and gas absorption using a capillary hemodialyzer. *Int J Artif Organs* **3**:36–40.

Piskin, E., and Chang, T.M.S. (1981) Effects of temperature on the removal of urea as ammonia by enzymatic conversion and gas absorption using an oxygenator. *Int J Artif Organs* **4**:45–48.

Piskin, E., and Chang, T.M.S. (1981) Urea and ammonia removal from dialysate in an oxygenator: effects of dialysate and air flow rates. *J Memb Sci* **9**:343–350.

Piskin, E., and Chang, T.M.S. (1980) Effects of air and fluid flow rates on removing different concentrations of urea and ammonia from fluid by gas absorption in a capillary hemodialyzer. *J Dial* **4**:31–42.

Piskin, E., Evren, V., Ozdural, A.R., and Chang, T.M.S. (1981) Design of a packed-bed gas desorption column for the removal of urea as ammonia. *Artif Organs* **5**(Supp.):624–627.

Piskin, E., and Chang, T.M.S. (eds.) (1982) Hemoperfusion and artificial organs, Artificial Organs Society.

Piskin, E., and Chang, T.M.S. (eds.) (1983) Past, present and future of artificial organs, Meteksan Publisher Co., Ankara, Turkey.

Pitard, B. *et al.* (1996) ATP synthesis by the F_0F_1 ATP synthase from thermophilic *Bacillus* PS3 reconstituted into liposomes with bacteriorhodopsin. 1. Factors defining the optimal reconstitution of ATP synthases with bacteriorhodopsin. *Eur J Biochem* **235**:769–778.

Pliura, D.H., Wiffen, D.E., Ashraf, S.S., *et al.* (1996) Purification of hemoglobin by displacement chromatography. US Patent 5,545,328.

Poli de Figueiredo, L.F., Elgjo, G.I., Mathru, M., *et al.* (1997) Hypertonic acetate-aahemoglobin for small volume resuscitation of hemorrhagic shock, *Artif Cells Blood Substit Immobil Biotechnol* **25**:61–74.

Powanda, D., and Chang, T.M.S. (2002) Cross-linked polyhemoglobin-superoxide dismutase-catalase supplies oxygen without causing blood brain barrier disruption or brain edema in a rat model of transient global brain ischemia-reperfusion. *Artif Cells Blood Substit Immobil Biotechnol* **30**:25–42.

Powell, C.C., Schultz, S.C., and Malcolm, D.S. (1996) Diaspirin crosslinked hemoglobin ($DCLHB^{TM}$): more effective than lactated Ringer's solution in restoring central venous oxygen saturation after hemorrhagic shock in rats. *Artif Cells Blood Substit Immobil Biotechnol* **24**:197–200.

Poznansky, M.J., and Chang, T.M.S. (1974) Comparison of the enzyme kinetics and immunological properties of catalase immobilized by microencapsulation and catalase in free solution for enzyme replacement. *Biochim Biophys Acta* **334**:103–115.

Prakash, S., and Chang, T.M.S. (1993) Genetically engineered *E. coli* cells containing *K. aerogenes* gene, microencapsulated in artificial cells for urea and ammonia removal. *Biomater Artif Cells Immobil Biotechnol* **21**:629–636.

Prakash, S., and Chang, T.M.S. (1995) Kinetic studies of microecnapsulated genetically engineered *E. coli* cells containing *K. aerogenes* gene for urea and ammonia removal. *Biotechnol Bioeng* **46**:621–626.

Prakash, S., and Chang, T.M.S. (1996a) Microencapsulated genetically engineered live *E. coli* DH5 cells administered orally to maintain normal plasma urea level in uremic rats. *Nat Med* **2**:883–887.

Prakash, S., and Chang, T.M.S. (1996b) Microencapsulated genetically engineered *E. coli* DH5 cells for plasma urea and aommonia removal based on: 1. column bioreactor and 2. oral administration in uremic rats. *Artif Cells Blood Substit Immobil Biotechnol* **24**:201–218.

Prakash, S., and Chang, T.M.S. (1996c) Microencapsulated genetically engineered microorganisms for clinical application. Br Priority Patent No 9601333–9.

Prakash, S., and Chang, T.M.S. (1998) Growth and survival of renal failure rats that received oral microencapsulated genetically engineered *E. coil* DH5 cells for urea removal. *Artif Cells Blood Substit Immobil Biotechnol* **26**:35–51.

Prakash, S., and Chang, T.M.S. (1999a) Artificial cell microcapsules containing genetically engineered *E. coli* DH5 cells for *in vitro* lowering of plasma potassium, phosphate, magnesium, sodium, chloride, uric acid, cholesterol, and creatinine: a preliminary report. *Artif Cells Blood Substit Immobil Biotechnol* **28**:475–482.

Prakash, S., and Chang, T.M.S. (1999b) Microencapsulated genetically engineered *E. coli* DH5 cells for the removal of undesired electrolytes and/or metabolites. USA Provisional Application. Serial Number 60/131, 468.

Prakash, S., and Chang, T.M.S. (1999c) Growth kinetics of genetically engineered *E. coli* DH5 cells in artificial cell apa membrane microcapsules: preliminary report. *Artif Cells Blood Substit Immobil Biotechnol* **27**:291–301.

Prakash, S., and Chang, T.M.S. (2000a) Artificial cells microencapsulated geneticaly engineered *E. coli* DH5 cells for the removal of undesired electrolyes and/or metabolites. International Application No: PCT/CA00/00482.

Prakash, S., and Chang, T.M.S. (2000b) Artificial cells microencapsulated genetically engineered *E. coil* DH5 cells for the lowering of plasma creatinine *in vitro* and *In vivo*. *Artif Cells Blood Substit Immobil Biotechnol* **28**:397–408.

Prakash, S., and Chang, T.M.S. (2000c) Lowering of uric acid *in vitro* and *in vivo* by artificial cell microcapsulated genetically engineered *E. coli* DH5 cells. *Int J Artif Organs* **23**:429–435.

Prichard, S., Chirito, E., Chang, T.M.S., and Sinderman, A.D. (1977) Microencapsulated charcoal hemoperfusion: a possible therapeutic adjunct in digoxin toxicity. *J Dial* **1**:367–377.

Prichard, S., Chirito, E., Sniderman, A.D., and Chang, T.M.S. (1978) Use of albumin cellulose nitrate microencapsulated charcoal hemoperfusion: in acute digoxin toxicity in dogs. In: Chang, T.M.S. (ed.), *Artificial Kidney, Artificial Liver and Artificial Cells* pp. 245–247. Plenum Press, New York.

Pristoupil, T.I., Sterbikova, J., Vrana, M., *et al.* (1992) Heart protection by cardioplegic solutions containing oxyhemoglobin pretreated by carbontetrahloride and freeze-drying with sucrose. *Biomater Artif Cells Immobil Biotechnol* **20**:709–720.

Quebec, E.A., and Chang, T.M.S. (1995) Superoxide dismutase and catalase crosslinked to polyhemoglobin reduces methemoglobin formation *in vitro*. *Artif Cells Blood Substit Immobil Biotechnol* **23**:693–706.

Rabiner, S.F., Helbert, J.R., Lopas, H., and Friedman, L.H. (1967) Evaluation of stroma free hemoglobin solution as a plasma expander. *J Exp Med* **126**:1127.

Rabinovici, R., Neville, L.F., Rudolph, A.S., and Feurstein, G. (1995) Hemoglobin-based oxygen-carrying resuscitation fluids (editorial comments). *Crit Care Med* **23**:801–804.

Rasmussen, S. *et al.* (2003). Bridging nonliving and living matter. *Artif Life* **9**:269–316.

Raymond, M.C., Neufeld, R.J., and Poncelet, D. (2004) Encapsulation of brewers yeast in chitosan coated carrageenan microspheres by emulsification/thermal gelation. *Artif Cells Blood Substit Immobil Biotechnol* **32**:275–291.

Razack, S., D'Agnillo, F., and Chang, T.M.S. (1997) Crosslinked hemoglobinsuperoxide dismutase-catalase scavenges free radicals in a rat model of intestinal ischemia-reperfusion injury. *Artif Cells Blood Substit Immobil Biotechnol* **25**:181–192.

Read, T.A., *et al.* (2001) Local endostatin treatment of gliomas administered by microencapsulated producer cells. *Nat Biotechnol* **19**:29–34.

Riess, J. (Guest ed.) (1994) Blood substitutes and related products: the fluorocabon approach. *Artif Cells Blood Substit Immobil Biotechnol* **22**:945–1511.

Riess, J.G. (1991) Fluorocabron-based *in vivo* oxygen transport and delivery systems. *Vox Sang* **61**:225–239.

Riess, J.G., and Krafft, M.P. (1997) Advanced fluorocarbon-based systems for oxygen and drug delivery, and diagnosis. *Artif Cells Blood Substit Immobil Biotechnol* **25**:43–52.

Rioux, F., Drapeau, C., and Marceau, F. (1995) Recombinant human hemoglobin (rHb1.1) selectively inhibits vasorelaxation elicied by nitric oxide donors in rabbit isolated aortic rings. *J Cardiovasc Pharmacol* **25**:587–594.

Robinson, M.F., Dupuis, N.P., Kusumoto, T., *et al.* (1995) Increased tumor oxygenation and radiation sensitivity in two rat tumors by a hemoglobin-based oxygen carrying preparation. *Artif Cells Blood Substit Immobil Biotechnol* **23**:431–438.

Rockwell, S., Kelley, M., and Mattrey, R. (1992) Preclinical evaluation of oxygent J as an adjunct to radiotherapy. *Biomater Artif Cells Immobil Biotechnol* **20**:883–893.

Rohlfs, R.J., and Vandergriff, K.D. (1996) Non-phospholipid liposomes: a novel method for the preparation of hemoglobin-containing lipid vesicles. In: Winslow, R.M., Vandegriff, K.D., and Intaglietta, M. (eds.), *Blood Substitutes: New Challenges.* pp. 163–184. Birkhäuser, Boston.

Rokstad, A.M., Holtan, S., Strand, B., *et al.* (2002) Microencapsulation of cells producing therapeutic proteins: optimizing cell growth and secretion. *Cell Transplant* **11**:313–324.

Rollwagen, F.M., Gafney, W.C., Pacheo, N.D., *et al.* (1996) Multiple responses to admistration of liposome-encapsulaed hemoglobin (LEH): effects on hematopoiesis and serum IL-6 levels. *Exp Hematol* **24**:429–436.

Rosenthal, A.M., and Chang, T.M.S. (1971) The effect of valinomycin on the movement of rubidium across lipid coated semipermeble microcapsules. *Proc Can Fed Biol Soc* **14**:44.

Rosenthal, A.M., and Chang, T.M.S. (1980) The incorporation of lipid and Na+-K+-ATPase into the membranes of semipermeable microcapsules. *J Memb Sci* **6**:329–338.

Roth, R.I., Levin, J., Chapman, K.W., *et al.* (1933) Production of modified crosslinked cell-free hemoglobin for human use: the role of quantitative determination of endotoxin contamination. *Transfusikon* **33**:919–924.

Roth, R.I., and Kaca, W. (1994) Toxicity of hemoglobin solutions: hemoglobin is a lipopolysacchride (LPS) binding protein which enhances LPS biological activity. *Artif Cells Blood Substit Immobil Biotechnol* **22**:387–398.

Roth, R.I., and Levin, J. (1994) Hemoglobin, a newly recognized lipopolysaccharide (LPS)-binding protein that enhances LPS biological activity. *J Biol Chem* **269**:25078–25084.

Rozga, J., Hozman, M.D., Ro, M.S., *et al.* (1993) Development of a hybrid bioartificialliver. *Ann Surg* **217**:502.

Rudloph, A.S. (1995) Encapsulation of hemoglobin in liposomes. In: Winslow, R.M., Vandergriff, K.D., and Intaglietta, M. (eds.), *Blood Substitutes: Physiological Basis of Efficacy.* Bikhäuser, Boston.

Rudolph, A.S. (1994) Encapsulated hemolgobin: current issues and future goals. *Artif Cells Blood Substit Immobil Biotechnol* **22**:347–360.

Rudolph, A.S., Rabinovici, R., and Feuerstein, G.Z. (eds.) (1997) *Red Blood Cell Substitutes*, Vol. 1, p. 48. Marcel Dekker, Inc., New York.

Runge, T.M., McGinity, J.W., Frisbee, S.E., *et al.* (1997) Enhancement of brain pO_2 during cardiopulmonary bypass using a hyperosmolar oxygen carrying solution. *Artif Cells Blood Substit Immobil Biotechnol* **25**:297–308.

Russo, P.A., Mitchell, G.A., and Tanguay, R.M. (2001) Tyrosinemia: a review. *Pediatric Dev Pathol* **4**:212–221.

Ruth, D., (2003) The dawning era of polymer therapeutics. *Nat Rev Drug Discov* **2**:347–360.

Safos, S., and Chang, T.M.S. (1995) Enzyme replacement therapy in ENU2 phenylketonuric mice using oral microencapsulated phenylalanine ammonia-lyase: a preliminary report. *Artif Cells Blood Substit Immobil Biotechnol* **23**:681–692.

Saitoh, Y., Taki, T., Arita, N., *et al.* (1995) Cell therapy with encapsulated xenogeneic tumor cells secreting beta-endorphin for treatment of peripheral pain. *Cell Transplant* **S1**:S13–S17.

Sakai, H., Hamada, K., Takeoka, S., *et al.* (1996) Physical properties of hemoglobin vesicles as red cell substitutes. *Biotechnol Prog* **12**:119–125.

Sakai, H., Takeoka, S., Park, S.I., *et al.* (1997) Surface modification of hemoglobin vesicles with poly(ethylene glycol) and effects on aggregation, viscosity, and blood flow during 90% exchange transfusion in anesthetized rats. *Bioconj Chem* **8**:23–30.

Sakai, H., Horinouchi, H., Tomiyama, K., *et al.* (2001) Hemoglobin-vesicles as oxygen carriers: influence on phagocytic activity and histopathological changes in metabolism. *Am J Pathol* **159**:1079–1088.

Sakai, H., Masada, Y., Takeoka, S., and Tsuchida, S. (2002) Characteristics of bovine Hb as a potential source of Hb vesicles for an artificial oxygen carrier. *J Biochem* **131**:611–617.

Sakai, H., Horinouchi, H., Masada, Y., *et al.* (2004a) Metabolism of hemoglobin-vesicles (artificial oxygen carriers) and their influence on organ functions in a rat model. *Biomaterials* **25**:4371–4325.

Sakai, H., Masada, Y., Horinouchi, H., *et al.* (2004b) Physiologic capacity of reticuloendothelial system for degradation of hemoglobin-vesicles (artificial oxygen carriers) after massive intravenous doses by daily repeated infusions for 14 days. *J Pharmacol Exp Ther* **311**:874–884.

Sakai, H., Masada, Y., Horinouchi, H., *et al.* (2004c) Hemoglobin-vesicles suspended in recombinant human serum albumin for resuscitation from hemorrhagic shock in anesthetized rats. *Crit Care Med* **32**:539–545.

Sakai, H., Masada, Y., Horinouchi, H., *et al.* (2004d) Physiologic capacity of reticuloendothelial system for degradation of hemoglobin-vesicles (artificial oxygen carriers) after massive intravenous doses by daily repeated infusions for 14 days. *J Pharmacol Exp Ther* **311**:874–884.

Sakai, S., Ono, T., Ijima, H., and Kawakami, K. (2001) Synthesis and transport characterization of alginate/aminopropyl-silicate/alginate micro-capsule: application to bioartificial pancreas. *Biomaterials* **22**:2827–2834.

Sanders, K.E., Ackers, G., and Sligar, S. (1996) Engineering and design of blood substitutes. *Curr Opin Struct Biol* **6**:534–540.

Sarkissian, C.N., Shao, Z., Blain, F., *et al.* (1999) A different approach to treatment of phenylketonuria: phenylalanine degradation with recombinant phenylalanine ammonia lyase. *Proc Natl Acad Sci* **96**:2339–2344.

Savitsky, J.P., Doozi, J., Black, J., and Arnold, J.D. (1978) A clinical safety trial of stroma free hemoglobin. *Clin Pharm Ther* **23**:73.

Scheller, F.W. *et al.* (2001) Research and development in biosensors. *Curr Opin Biotechnol* **12**:35–40.

Schrieber, G.B., Busch, M.P., Kleinman, S.H., and Korelitz, J.J. (1996) The risk of transfusion transmitted viral infections. *New Eng J Med* **334**:1685–1690.

Schuldt, U., and Hunkeler, D. (2000) Characterization methods for microcapsules. *Minerva Biotecn* **12**:249–264.

Schultz, S.C., Powell, C.C., Bernard, E., and Malcom, D.S. (1995) Diaspirin crosslinked hemoglobin (DCLHb) attenuates bacterial translocation in rats. *Artif Cells Blood Substit Immobil Biotechnol* **23**:647–664.

Schultz, S.C., Powell, C.C., Burris, D.G., *et al.* (1994) The efficacy of diaspirin cross-linked hemoglobin solution resuscitation in a model uncontrolled haemorrhage. *J Trauma* **37**:408–412.

Schwinger, C., *et al.* (2002) High throughput encapsulation of murine fibroblasts in alginate using the JetCutter technology. *J Microencapsul* **19**:273–280.

Scriver, C. (1989) The hyperphenylalaninemias. In: Scriver, C.R., Beaudet, A.L., Sly, W.S., and Valle, D. (eds.), *The Metabolic Basis of Inherited Disease*, 6th edn., Vol. 1, pp. 495–546. McGraw-Hill Co. New York.

Sehgal, L.R., Rosen, A.L., Gould, S.A., *et al.* (1980) *In vitro* and *in vivo* characteristics of polymerized pyridoxalated hemoglobin solution. *Fed Proc* **39**:2383.

Sehgal, L.R., Gould, S.A., Rosen, A.L., *et al.* (1984) Polymerized pyridoxalated hemoglobin: a red cell substitute with normal oxygen capacity. *Surgery* **95**:433.

Sehgal, L.R., Rosen, A.L., Gould, S.A., *et al.* (1983) Preparation and *in vitro* characteristics of polymerized pyridoxalated hemoglobin. *Transfusion* **23**:158.

Sehgal, L.R., Sehgal, H.L., Rosen, A.L., *et al.* (1988) Characteristics of polymerized pyridoxalated hemoglobin. *Biomater Artif Cells Artif Organs* **16**:173–183.

Sekiguchi, S., Ito, K., Kobayash, M., *et al.* (1988) Preparation of virus-free pyridoxalated hemoglobin from the blood of HBV or HTLVI healthy carriers. *Biomater Artif Cells Artif Organs* **16**:113–121.

Sekiguchi, S. (1992) Studies on the quality control of stroma free hemoglobin. *Biomater Artif Cells Immobil Biotechnol* **20**:407–414.

Sekiguchi, S. (1997) The impact of red cell substitutes on the blood service in Japan. *Artif Cells Blood Substit Immobil Biotechnol* **25**:53–60.

Setola, V., and Roth, B.L. (2003) Why mice are neither miniature humans nor small rats: a cautionary tale involving 5-hydroxytryptamine-6 serotonin receptor species variants. *Mol Pharmacol* **64**:1277–1278.

Shah, N., and Mehra, A. (1996) Modeling of oxygen uptake in perfluorocarbon emulsions. Some comparisons with uptake by blood. *ASAIO J* **42**:181–189.

Shedlovsky, A., McDonald, J.D., Symula, D., and Dove, W.F. (1993) Mouse models of human phenylketonuria. *Genetics* **134**:1205–1210.

Sherwood, R.L., McCormick, D.L., Zheng, S., and Beissinger, R.L. (1995) Influence of steric stabilization of liposome-encapsulated hemoglobin on *Listeria monocytogenes* host defence. *Artif Cells Blood Substit Immobil Biotechnol* **23**:665–679.

Shew, R., and Deamer, D. (1983) A novel method for encapsulating macromolecules in liposomes. *Biochim Biophys Acta* **816**:1–8.

Shi, Z.Q., and Chang, T.M.S. (1982) The effects of hemoperfusion using coated charcoal or tyrosinase artificial cells on middle molecules and tyrosine in brain and serum of hepatic coma rats. *Trans Am Soc Artif Intern Organs* **28**:205–209.

Shi, Z.Q., and Chang, T.M.S. (1983) *In vitro* adsorption spectrum of plasma amino acids by coated charcoal hemoperfusion. *Int J Artif Organs* **6**:267–270.

Shi, Z.Q., and Chang, T.M.S. (1984a) Amino acid disturbances in experimental hepatic coma rats. *Int J Artif Organs* **7**:197–202.

Shi, Z.Q., and Chang, T.M.S. (1984b) Effect of charcoal hemoperfusion on amino acid disturbance in experimental hepatic coma. *Int J Artif Organs* **7**:203–208.

Shi, Z.Q., and Chang, T.M.S. (1985) Amino acid disturbance in experimental hepatic coma: the effects of charcoal hemoperfusion. In: Chang, T.M.S., and Ho, B.L. (eds.), *Hemoperfusion and Artificial Organs*. pp. 301–310. China Academic Publisher, China.

Shih, D.T.B., Noboru, N, Dickey, B., *et al.* (1991) Function and stability of low-affinity variant and engineered hemoglobins. *Biomater Artif Cells Immobil Biotechnol* **19**:483.

Shoemaker, S., Stetler, G., Looker, D., *et al.* (1991) Recombinant human hemoglobin has been engineered to decrease toxicity and improve efficacy. *Biomater Artif Cells Immobil Biotechnol* **19**:484–489.

Shoemaker, S., Gerber, M., Evans, G., *et al.* (1994) Initial clinical experience with a rationally designed genetically engineered recombinant human hemoglobin. *Artif Cells Blood Substit Immobil Biotechnol* **22**:457–465.

Shorr, R.G., Viau, A.T., and Abuchowski, A. (1996) Phase 1B safety evaluation of PEG hemoglobin as an adjuvant to radiation therapy in human cancer patients. *Artif Cells Blood Substit Immobil Biotechnol* **24**:407.

Shu, C.D., and Chang, T.M.S. (1980) Tyrosinase immobilized within artificial cells for detoxification in liver failure. I. Preparation and *in vitro* studies. *Int J Artif Organs* **3**:287–291.

Shu, C.D., and Chang, T.M.S. (1981) Tyrosinase immobilized within artificial cells for detoxification in liver failure: II. *In vivo* studies in fulminant hepatic failure rats. *Int J Artif Organs* **4**:82–84.

Shu, C.D., and Chang, T.M.S. (1982a) Artificial cells immobilized tyrosinase for detoxifications in liver failure. *Chongqing Med Coll Bull* **1**:1–4.

Shu, C.D., and Chang, T.M.S. (1982b) Effects of hemoperfusion on plasma cortisone levels in galactosamine induced fulminant hepatic failure rats. *Proc Symp Artif Liver*.

Shu, C.D., and Chang, T.M.S. (1983a) Middle molecules in the serum, brain, liver and urine of galactosamine-induced fulminant hepatic failure rats. *Int J Artif Organs* **6**:273–277.

Shu, C.D., and Chang, T.M.S. (1983b) Galactosamine induced fulminant hepatic failure rats: changes in middle molecules. *Chinese J Gastroenterol* **3**:102–104.

Shum, K.L., Leon, A., Viau, A.T., *et al.* (1996) The physiological and histopathological response of dogs to exchange transfusion with polyethylene glycol-modified bovine hemoglobin (PEG-Hb). *Artif Cells Blood Substit Immobil Biotechnol* **24**:655–683.

Sideman, S., and Chang, T.M.S. (1979) Hemoperfusion: kidney and liver supports and detoxification. *Artif Organs* **3**:393–402.

Sideman, S., and Chang, T.M.S. (1980a) Hemoperfusion: state of the art and future requirements. *Artif Organs* **4**:70–75.

Sideman, S., and Chang, T.M.S. (eds.) (1980b) Hemoperfusion: I. Artificial kidney and liver support and detoxification. Hemisphere Publishing Corp., Washington, DC., USA.

Sideman, S., and Chang, T.M.S. (eds.) (1981) Hemoperfusion: II. Devices and clinical applications, Samuel Neaman Institute for Advanced Studies in Science and Technology, Technion, Haifa, Israel.

Silverstein, M.D., Ford, C.A., Lysaght, M.J., and Henderson, C.W. (1974) Treatment of severe fluid overload by ultrafiltration. *New Engl J Med* **291**:747–751.

Simoni, J., Simoni, G., Garcia, E.L., *et al.* (1995) Protective effect of selenium on hemoglobin mediated lipid peroxidation *in vivo*. *Artif Cells Blood Substit Immobil Biotechnol* **23**:469–486.

Simoni, J., Simoni, G., Lox, C.D., *et al.* (1997) Expression of adhesion molecules and Von Willebrand factor in human coronary artery endothelial cells incubated with differently modified hemoglobin solutions. *Artif Cells Blood Substit Immobil Biotechnol* **25**:211–226.

Simoni, J., Simoni, G., Lox, C.D., *et al.* (1997) Modified hemoglobin solution with desired pharmacological properties does not activate nuclear transcription factor NF-kappa B in human vascular endothelial cells. *Artif Cells Blood Substit Immobil Biotechnol* **25**:193–210.

Singh, S.M., McCormick, B.B., Mustata, S., *et al.* (2004) Extracorporeal management of valproic acid overdose: a large regional experience. *J Nephrol* **17** 43–49.

Sipehia, R., Bannard, R.A.B., and Chang, T.M.S. (1986a) Adsorption of large lipophilic molecules with exclusion of small hydrophilic molecules by microencapsulated activated charcoal formed by coating with polyethylene membrane. *J Memb Sci* **29**:277–286.

Sipehia, R., Chawla, A.S., and Chang, T.M.S. (1986b) Enhanced albumin binding to polypropylene beads via anhydrous ammonia gaseous phase. *Biomaterials* **1**:471–473.

Sipehia, R., Chawla, A.S., Daka, J.N., *et al.* (1988) Immobilization of enzymes on polypropylene beads surfaces by anhydrous ammonia gaseous plasma technique. *J Biomed Mater Res* **22**:417–422.

Sipehia, R., Bannard, R.A.B., and Chang, T.M.S. (1989a) Poly(vinylidene fluoride) orpoly(vinylidene chloride/vinyl chloride) — coated activated charcoal for the adsorption of large lipophilic molecules with exclusion of small hydrophilic molecules. *J Memb Sci* **47**:293–301.

Sipehia, R., Daka, J.N., Chawla, A.S., and Chang, T.M.S. (1989b) Immobilisation of protein and enzymes onto functionalised polypropylene surfaces by a gaseous plasma modification technique. In: Burrington, J.D., and Clark, D.S. (eds.), *Biocatalysis and Biomimetics*, Chapter 12. Washington, DC.

Sipehia, R., Garfinkle, A., Jackson, W.B., and Chang, T.M.S. (1990) Towards an artificial cornea: surface modification of optically clear, oxygen permeable soft contact lens materials by ammonia plasma modification technique for the enhanced attachment and growth of corneal epithelial cells. *Int J Biomater Artif Cells Artif Organs* **18**:643–656.

Siu-Chong, E., and Chang, T.M.S. (1974) *In vivo* effects of intraperitoneally injected L-asparaginase solution and L-asparaginase immobilized within semipermeable nylon microcapsules with emphasis on blood L-asparaginase, 'body' L-asparaginase, and plasma L-asparagine levels. *Enzyme* **18**:218–239.

Siu-Chong, E., and Chang, T.M.S. (1977) L-Asparaginase as a model for enzyme therapy of substrate-dependent tumors. In: Chang, T.M.S. (ed.), *Biomedical Applications of Immobilized Enzymes and Proteins,* Vol. 1, pp. 105–120. Plenum Press, New York.

Sloviter, H., and Kamimoto, T. (1967) Erythrocyte substitute for perfusion of brain. *Nature* **216**:458.

Soon-Shiong, P., *et al.* (1994) Insulin independence in a type 1 diabetic patient after encapsulated islet transplantation. *Lancet* **343**:950–951.

Sou, K., Naito, Y., Endo, T., *et al.* (2003). Effective encapsulation of proteins into size-controlled phospholipids vesicles using freeze-thawing and extrusion. *Biotechnol Progr* **19**:1547–1552.

Sou, K., Klipper, R., Goins, B., *et al.* (2005) Circulation kinetics and organ distribution of Hb-vesicles developed as a red blood cell substitute. *J Pharmacol Exp Ther* **312**:702–709.

Spangrude, G.J., Heimfeld, S., and Weissman, I.L. (1988) Purification and characterization of mouse hematopoietic stem cells. *Science* **241**:58–62.

Spence, R.K. (1995) Perfluorocarbons in the twenty-first century: clinical applications as transfusion alternatives. *Artif Cells Blood Substit Immobil Biotechnol* **23**:367–380.

Sprung, J., Kindscher, J.D., Wahr, J.A., *et al.* (2002) The use of bovine Hb glutamer-250 (Hemopure) in surgical patients: results of a multicenter, randomized, single-blinded trial. *Anesth Analg* **94**:799–808.

Stabilini, R., Palazzini, G., Pietta, G.P., *et al.* (1983) A pyridoxalated polymerized hemoglobin solution as oxygen carrying substitute. *Int J Artif Organs* **6**:319.

Stein, W.D. (1967) *The Movement of Molecules Across Cell Membranes.* Academic Press, New York.

Steinberg-Yfrach, G., *et al.* (1998) Light-driven synthesis of ATP catalysed by F_0F_1-ATP synthase in an artificial photosynthetic membrane. *Nature* **392**:479–482.

Sun, Y.L., Ma, X.J., Zhou, D.B., *et al.* (1996) Normalization of diabetes in spontaneously diabetic cynomologus monkeys by xenografts of microencapsulated porcine islets without immunosuppression. *J Clin Invest* **98**:1417–1422.

Sundarum, P.V., Pye, E.K., Chang, T.M.S., *et al.* (1972) Recommendations for standardization of nomenclature in enzyme technology. *Biotechnol Bioeng,* Vol. 3. pp. 15–18.

Sussman, N.L., Gislason, G.T., and Kelly, J.H. (1994) Extracorporeal liver support. Application to fulminant hepatic failure. *J Clin Gastroenterol* **18**:320.

Szebeni, J. (1998) The interaction of liposomes with the complement system. *Crit Rev Ter Drug Carrier Syst* **15**:57–88.

Szebeni, J., Wassef, N.M., Rudolph, A.S., and Alving, C.R. (1995) Complement activation by liposome-encapsulated hemoglobin *in vitro*: the role of endotoxin contamination. *Artif Cells Blood Substit Immobil Biotechnol* **23**:355–363.

Szebeni, J., Hauser, H., Eskelson, C.D., and Winterhalter, K.H. (1988) Factors influencing the *in vitro* stability of artificial red blood cells based on hemoglobin-containing liposomes. *Biomater Artif Cells Artif Organs* **16**:301–312.

Szebeni, J., Wassef, N.M., Rudolph, A.S., and Alving, C.R. (1996) Complement activation in human serum by liposome-encapsulated hemoglobin: the role of natural anti-phospholipid antibodies. *Biochim Biophys Acta* **1285**:127–130.

Szostak, J.W. *et al.* (2001) Synthesizing life. *Nature* **409**:387–390.

Tabata, Y., and Chang, T.M.S. (1980) Comparisons of six artificial liver support regimes in fulminant hepatic coma rats. *Trans Am Soc Artif Intern Organs* **26**:394–399.

Tabata, Y., and Chang, T.M.S. (1982b) Intermittent vascular access for extracorporeal circulation in conscious rats: a new technique. *Artif. Organs* **6**:213–214.

Tabata, Y., Odaka, M., Hirasawa, H., *et al.* (1983) The effect of charcoal hemoperfusion on DNA in the liver of hepatic failure rats. In: Piskin, E., and Chang, T.M.S. (eds.), *Past, Present and Future of Artificial Organs.* pp. 255–259. Meteksan Publ. Co., Ankara.

Tabata, Y., Odaka, M., Hirasawa, H., *et al.* (1984) The effects of charcoal hemo-perfusion on the liver metabolism of fulminant hepatic failure rats. In: Atsumi, K., Maekawa, M., and Ota, K. (eds.), *Progress in Artificial Organs.* pp. 747–749. ISAO Press, Cleveland.

Tabata, T., Odaka, M., Hirasawa, H., *et al.* (1985) The effect of charcoal hemoperfusion on the adenine nucleotide of the liver in fulminant hepatic failure rats. In: Chang, T.M.S., and Ho, B.L. (eds.), *Hemoperfusion and Artificial Organs.* pp. 296–300. China Academic Publisher.

Tai, J., Kim, H.W., and Greenburg, A.G. (1997) Endothelin-1 is not involved in hemoglobin associated vaso-activities. *Artif Cells Blood Substit Immobil Biotechnol* **25**:135–140.

Takahashi, A. (1995) Characterization of neo red cells (NRCs), their function and safety *in vivo* tests. *Artif Cells Blood Substit Immobil Biotechnol* **23**:347–354.

Takaori, M., and Fukui, A. (1996) Treatment of massive hemorrhage with liposome encapsulated human hemoglobin (NRC) and hydroxyethyl starch (HES) in beagles. *Artif Cells Blood Substit Immobil Biotechnol* **24**:643–653.

Takeoka, S., Ohgushi, T., Sakai, H., *et al.* (1997) Construction of artificial methemoglobin reduction systems in hb vesicles. *Artif Cells Blood Substit Immobil Biotechnol* **25**:31–42.

Tam, S.C., Blumenstein, J., and Wong, J.T. (1976) Dextran hemoglobin. *Proc Natl Acad USA* **73**:2128.

Tan, S.A., *et al.* (1996) Rescue of motoneurons from axotomy-induced cell death by polymer encapsulated cells genetically engineered to release CNTF. *Cell Transplant* **5**:577–587.

Tanaka, J.I., Takino, H., and Malchesky, P.S. (1992) Does oxygen supply improve graft viability in liver preservation? *Biomater Artif Cells Immobil Biotechnol* **20**:545–548.

Tani, T., Chang, T.M.S., Kodama, M., and Tsuchiya, M. (1992) Endotoxin removed from hemoglobin solution using polymyxin-B immobilized fibre (PMX-F) followed by a new turbidometric endotoxin assay. *Biomater Artif Cells Immobil Biotechnol* **20**:394- 399.

Teicher, B.A. (1995) An overview on oxygen-carriers in cancer therapy. *Artif Cells Blood Substit Immobil Biotechnol* **23**:395–405.

Terman, D.S., Tavel, T., Petty, D., *et al.* (1977) Specific removal of antibody by extracorporeal circulation over antigen immobilized in colodion charcoal. *Clin Exp Immunol* **28**:180.

Terman, D.S., Buffaloe, G., Mattioli, C., *et al.* (1979) Extracorporeal immuno-absorption: initial experience in human systemic lupus erythematosus. *Lancet* **2**:824.

Terman, D.S., Garcia-Rinaldi, R., McCalma, R., *et al.* (1979). Modification of hyperacute renal xenograft rejection after extracorporeal immunoadsorption of heterospecific antibody. *Int J Artif Organs* **2**:35.

Terman, D.S. (1980) Extracorporeal immunoadsorbents for extraction of circulating immune reactants. In: Giordano, E. (ed.), *Sorbents and Their Clinical Applications.* pp. 470. Academic Press Inc., New York.

Theise, N.D., Nimmakayalu, M., Gardner, R., *et al.* (2000). Liver from bone marrow in humans. *Hepatology* **32**:11–16.

Thomas, M.J. (1996) Royal College of Physicians, Edinburgh: final consensus statement. Consensus conference on autologous transfusion, *Vox Sang* **70**:183–184.

Tominaga, M. *et al.* Pharmacological evaluation of portal venous isolation and charcoal haemoperfusion for high-dose intra-arterial chemotherapy of the pancreas. *Br J Surg* **84**:1072–1076.

Torchilin, V.P. (2005) Recent advances with liposomes as pharmaceutical carriers. *Nat Rev Drug Discov* **4**:145–160.

Traylor, R.J., and Pearl, R.G. (1996) Crystalloid versus colloid versus colloid: all colloids are not created equal. *Anesth Analg* **83**:209–212.

Tremper, K.K. (ed.) (1985) Perfluorochemical oxygen transport. *Int Anesthesiology Clinics*, Little, Brown & Co., Boston.

Tsai, A.G., Kerger, H., and Intaglietta, M. (1996) Microvasular oxygen distribution: effects due to free hemoglobin in plasma. In: Winslow, R.M., Vandergriff, K.D., and Intaglietta, M. (eds.), *Blood Substitutes: New Challenges.* pp. 124–131. Birkhäuser, Boston.

Tsuchida, E., Komatsu, T., Yanagimoto, T., and Sakai, H. (2002) Preservation stability and *in vivo* administration of albumin-heme hydrid solution as an entirely synthetic oxygen-carrier. *Polymer Adv Technol* **13**:845–850.

Tsuchida, E., Nishide, H., and Ohno, H. (1988) Liposome/heme as a totally synthetic oxygen carrier. *Biomater Artif Cells Artif Organs* **16**:313–319.

Tsuchida, E. (1994) Stabilized hemoglobin vesicles. *Artif Cells Blood Substit Immobil Biotechnol* **22**:467–479.

Tsuchida, E. (ed.) (1998) Present and future perspectives. *Blood Substitutes* Vol. 1, p. 267. Elsevier, Amsterdam.

Tsuchida, E., and Nishide, H. (1992) Synthesis and characterization of artificial red cell (ARC). *Biomater Artif Cells Immobil Biotechnol* **20**:337–354. US Pharmacopeia and National Formulary (2002) **1046**:2762–2790.

Uchida, N., and Weissman, I.L. (1992) Searching for hematopoietic stem cells: evidence that Thy-1.1lo Lin- Sca-1+ cells are the only stem cells in C57BL/Ka-Thy-1.1 bone marrow. *J Exp Med* **175**:175–184.

Uhlenkott, C.E., Huijzer, J.C., Cardeiro, D.J., *et al.* (1996) Attachment, invasion, chemotaxis, and proteinase expression of B16–BL6 melanoma cells exhibiting a low metastatic phenotype after exposure to dietary restriction of tyrosine and phenylalanine. *Clin Exp Metastasis* **14**:125–137.

Ulatowski, J.A., Asano, Y., Koehler, R.C., *et al.* (1997) Sustained endothelial dependent dilation in pial arterioles after crosslinked hemoglobin transfusion. *Artif Cells Blood Substit Immobil Biotechnol* **25**:115–120.

Ulatowski, J.A., Koehler, R.C., Nishikawa, T., *et al.* (1995) Role of nitric oxide scavenging in peripheral vasoconstrictor response to crosslinked hemoglobin. *Artif Cells Blood Substit Immobil Biotechnol* **23**:263–269.

Ulatowski, J.A., Nishikawa, T., Matheson-Urbaitis, N., *et al.* (1996). Regional blood flow alterations after bovine fumaryl-β-crosslinked hemoglobin transfusion and nitric oxide synthase inhibition. *Crit Care Med* **24**:558–565.

Uludag, H., De Vos, P., and Tresco, P.A. (2000) Technology of mammalian cell encapsulation. *Adv Drug Delivery Rev* **42**:29–64.

Usuba, A., Motoki, R., Suzuki, K., *et al.* (1992) Study of effect of the newly developed artificial blood "neo red cells (NRC)" on hemodynamics and blood gas transport in canine hemorrhagic shock. *Biomater Artif Cells Immobil Biotechnol* **20**:531–538.

Usuba, A., Motoki, R., Ogata, Y., *et al.* (1995) Effect and safety of liposome encapsulated hemoglobin "neo red cells (NRC)" as a perfusate for total cardiopulmonary bypass. *Artif Cells Blood Substit Immobil Biotechnol* **23**:337–346.

Vandegriff, K.D., and LeTellier, Y.C. (1994) A comparison of rates of heme exchanges; site-specifically cross-linked versus polymerized human hemoglobin. *Artif Cells Blood Substit Immobil Biotechnol* **22**:443–456.

Vandegriff, K.D., Rohlfs, R.J., and Winslow, R.M. (1988) Kinetics of ligand binding to crosslinked hemoglobin. *Biomater Artif Cells Artif Organs* **16**:647–649.

Vasilakakis, D.M., D'Haese, P.C., Lamberts, L.V., *et al.* (1992) Removal of alumino-xamine and ferrioxamine by charcoal hemoperfusion and hemodialysis. *Kidney Int* **41**:1400.

Vercellottii, G.M., Balla, G., Balla, J., *et al.* (1994) Heme and the vasculature: an oxidative hazard that induces antioxidant defence in the endothelium. *Artif Cells Blood Substit Immobil Biotechnol* **22**:687–694.

Vercoutere, W.S. *et al.* (2001) Rapid discrimination among individual DNA molecules at single nucleotide resolution using a nanopore instrument. *Nat Biotechnol* **19**:248–250.

Verpooten, G.A., and De Brae, M.E. (1984) Combined hemoperfusion hemodialysis in severe poisoning: kinetics of drug extraction. *Resuscitation* **11**:275.

Vogel, W.M., Leiberthal, C.S., Apstein, C.S., *et al.* (1988) Effects of stroma-free hemoglobin solutions on isolated perfused rabbit hearts and isolated perfused rat kidneys. *Biomater Artif Cells Artif Organs* **16**:227–236.

Vogel, W.M., Cassidy, G., and Valeri, C.R. (1992) Effects of o-raffinose-polymerized human hemoglobin on coronary tone and cardiac function in isolated hearts. *Biomater Artif Cells Immobil Biotechnol* **20**:673–678.

Vogel, W.M., Hsia, J.C., Briggs, L.L., *et al.* (1987) Reduced coronary vasoconstrictor activity of hemoglobin solutions purified by ATP-agarose affinity chromatography. *Life Sci* **41**:89–93.

Wagner, M.L., and Tamm, L.K. (2000) Tethered polymer-supported planar lipid bilayers for reconstitution of integral membrane proteins: silane-polyethyleneglycol-lipid as a cushion and covalent linker. *Biophys J* **79**:1400–1414.

Wahl, H.P., and Chang, T.M.S. (1986) Recycling of NAD+ cross-linked to albumin or hemoglobin immobilized with multienzyme systems in artificial cells. *J Mol Catal* **39**:147–154.

Wahr, J.A., Trouwborst, R.K., Spence, R.K., *et al.* (1994) A pilot study of the efficay of an oxygen carrying emulsion oxtgent^{TM}, in patients undergoing surgical blood loss. *Anesthesiology* **80**:A397.

Walder, J.A., Zaugg, R.H., Walder, R.Y., *et al.* (1979) Diaspirins that cross-link alpha chains of hemoglobin: Bis(3,5-dibromosalicyl) succinate and bis(3,5-dibormosalicyl)fumarate. *Biochemistry* **18**:4265–4270.

Wallace, E.L., Surgenor, D.M., Hao, H.S., *et al.* (1993) Collection and transfusion of blood and blood components in the United States (1989). *Transfusion* **33**:139–144.

Walter, S., and Chang, T.M.S. (1988) Chronotropic effects of stroma-free hemoglobin and polyhemoglobin on cultured myocardiocytes derived from newborn rats. *Biomater Artif Cells Artif Organs* **16**:701–703.

Walter, S.V., and Chang, T.M.S. (1990) Chronotropic effects of *in vivo* perfusion with albumin, stroma-free hemoglobin and polyhemoglobin solutions. *Biomater Artif Cells Artif Organs* **18**:283–299.

Wang, M.Y., Yu, Y.T., and Chang, T.M.S. (2005) New method for preparing more stable microcapsules for the entrapment of genetically engineered cells. *Artif Cells Blood Substit Biotechnol* **33**:257–269.

Wang, Y.C., Lee, C.J., Chen, W.K., *et al.* (1996) Alteration of cerebral microcirculation by hemodilution with hemosome in awake rats. *Artif Cells Blood Substit Immobil Biotechnol* **24**:35–42.

Williams, R. (1983) Fulminant hepatic failure, *Post Graduate* **59** (Suppl.):33.

Wilson, J.M., Young, A.B., and Kelly, W.N. (1983) Hypoxanthine-guanine phospho ribosylttansferase deficiency: the molecular basis of the clinical syndromes. *New Engl J Med* **309**:900–910.

Winchester, J.F., Apiliga, M.T., MacKay, J.M., and Kennedy, A.C. (1976) Combined hemodialysis-charcoal hemoperfusion in the dialysis patient. *Kidney Int* **10** (Suppl. 7):S315–S319.

Winchester, J.F. (1986) Management of iron overload. *Semin Nephrol* **4**(Suppl. 1):22.

Winchester, J.F. (ed.) (1996) *Replacement of Renal Function by Dialysis*, 4th edn. Kluwer Academic Publishers, Boston.

Winchester, J.F. (1990) Active methods for detoxification: oral sorbents, forced diuresis, hemoperfusion and hemodialysis. In: Haddad, L.M., and Winchester, J.F. (eds.),

Clinical Management of Poisoning and Drug Overdose, 2nd edn. p. 148. WB Saunders Co., Philadelphia, Pennysylvania.

Wingard, R.L., Lee, W.O., and Hakim, R.O. (1991) Extracorporeal treatment of familial hypercholesterolemia with monoclonal antibodies to low-density lipoproteins. *Am J Kidney Dis* **18**:559.

Winn, S.R. *et al.* (1994) Polymer-encapsulated cells genetically modified to secrete human nerve growth factor promote the survival of axotomized septal cholinergic neurons. *Proc Natl Acad Sci USA* **91**:2324–2328.

Winslow, R., and Chang, T.M.S. (eds.) (1990) Red blood cell substitutes (special issue). *Int J Biomater Artif Cells Artif Organs* **18**:133–342.

Winslow, R.M. (1992) Potential clinical applications for blood substitutes. *Biomater Artif Cells Immobil Biotechnol* **20**:205–220.

Winslow, R.M. (Guest ed.) (1994) Blood substitutes: modified hemoglobin. *Artif Cells Blood Substit Immobil Biotechnol* **22**:360–944(special issue).

Winslow, R.M. (1995) Blood substitutes: a moving target. *Nat Med* **1**:1212–1215.

Winslow, R.M. (1996a) Blood substitutes in development, pp. 27–37. Ashley Publications Ltd.

Winslow, R.M. (1996b) Blood substitute oxygen carriers designed for clinical applications. In: Winslow, R.M., Vandegriff, K.D., and Intaglietta, M. (eds.), *Blood Substitutes: New Challenges.* pp. 60–73. Birkhäuser, Boston.

Winslow, R.M. (1996c) Blood substitutes in development. *Exp Opin Invest* **5**:1443–1452.

Winslow, R.M. (ed.), (2006) *Blood Substitutes.* Academic Press, Amsterdam.

Wolfe, E.A., and Chang, T.M.S. (1987) Orally ingested microencapsulated urease and an adsorbent, zirconium phosphate, to remove urea in kidney failure. *Int J Artif Organs* **10**:269–275.

Wong, H., and Chang, T.M.S. (1986) Bioartificial liver: implanted artificial cells microencapsulated living hepatocytes increases survival of liver failure rats. *Int J Artif Organs* **9**:335–336.

Wong, H., and Chang, T.M.S. (1988) The viability and regeneration of artificial cell microencapsulated rat hepatocyte xenograft transplants in mice. *J Biomater Artif Cells Artif Organs* **16**:731–740.

Wong, H., and Chang, T.M.S. (1991a) A novel two-step procedure for immobilizing living cells in microcapsule for improving xenograft survival. *Biomater Artif Cells Immobil Biotechnol* **19**:687–698.

Wong, H., and Chang, T.M.S. (1991b) Microencapsulation of cells within alginate poly-L-lysine microcapsules prepared with standard single step drop technique: histologically identified membrane imperfections and the associated graft rejection. *Biomater Artif Cells Immobil Biotechnol* **19**:675–686.

Wong, J.T. (1988) Rightshifted dextran-hemoglobin as blood substitute. *Biomater Artif Cells Artif Organs* **16**:237–245.

Woodle, M.C., Newman, M.S., and Working, P.K. (1995) Biological properties of sterically stabilized liposomes. In: Lasic, D.D., and Maartine, F. (eds.), *Stealth Liposomes.* pp. 103–117. DRC Press, Boca Raton.

Xu, W., Liu, L., and Charles, I.G. (2002) Microencapsulated iNOS-expressing cells cause tumor suppression in mice. *FASEB J* **16**:213–215.

Yang, L., Cheng, Y., Yan, W.R. and Yu, Y.Y. (2004) Extracorporeal whole blood immunoadsorption of autoimmune myasthenia gravis by cellulose tryptopan adsorbent. *Artif Cells, Blood Substit Biotechnol* **32**:519–518.

Yatzidis, H. (1964) A convenient hemoperfusion micro-apparatus over charcoal for the treatment of endogenous and exogenous intoxications. *Proc Eur Dial Transplant Assoc* **1**:83.

Yoshizu, A., Izumi, Y., Park, S., *et al.* (2004) Hemorrhagic shock resuscitation with an artificial oxygen carrier hemoglobin vesicle (HbV) maintains intestinal perfusion and suppresses the increase in plasma TNFα. *ASAIO J* **50**:458–463.

Yu, B.L., and Chang, T.M.S. (2002) *In vitro* kinetics of encapsulated tyrosinase. *Artif Cells Blood Substit Immobil Biotechnol* **30**:533–546.

Yu, B.L., and Chang, T.M.S. (2004a) *In vitro* and *in vivo* effects of poly hemoglobintyrosinase on murine B16F10 melanoma. *Melanoma Res J* **14**:197–202.

Yu, B.L., and Chang, T.M.S. (2004b) Effects of long term oral administration of microencapsulated tyrosinase on maintaining decreased systemic tyrosine levels in rats. *J Pharm Sci* **93**:831–837.

Yu, B.L., and Chang, T.M.S. (2004c) Polyhemoglobin-tyrosinase, an oxygen carrier with murine B16F10 melanoma suppression properties: a preliminary report. *Artif Cells Blood Substit Biotechnol* **32**:293–302.

Yu, B.L., and Chang, T.M.S. (2004d) Effects of combined oral administration and intravenous injection on maintaining decreased systemic tyrosine levels in rats. *Artif Cells Blood Substit Biotechnol* **32**:129–148.

Yu, B.L., and Chang, T.M.S. (2004e) *In vitro* and *in vivo* enzyme studies of polyhemoglobintyrosinase. *Biotechnol Bioeng* **86**:835–841.

Yu, B.L., Liu, Z.C., and Chang, T.M.S. (2006) Polyhemoglobin with different percentage of tetrameric hemoglobin and effects on vasoactivity and electrocardiogram. *Artif Cells Blood Substit Biotechnol* **34**:159–175.

Yu, B.L., and Chang, T.M.S. (2006) Enzyme artificial cells: with emphasis on polyhemoglobin tyrosinase. In: Prakash, S. (ed.), *Artif Cells Cell Ther.*

Yu, W.P., and Chang, T.M.S. (1994) Submicron biodegradable polymer membrane hemoglobin nanocapsules as potential blood substitutes: a preliminary report. *J Artif Cells Blood Substit Immobil Biotechnol* **22**:889–894.

Yu, W.P., and Chang, T.M.S. (1996) Submicron polymer membrane hemoglobin nanocapsules as potential blood substitutes: preparation and characterization. *Artif Cells Blood Substit Immobil Biotechnol* **24**:169–184.

Yu, W.P., Wong, J., and Chang, T.M.S. (1998) Preparation and characterization of polylactic acid microcapsules containing ciprofloxacin for controlled release. *J Microencapsul* **15**:515–523.

Yu, W.P., Wong, J.P., and Chang, T.M.S. (1999) Biodegradable polylactic acid nanocapsules containning ciprofloxacin: preparation and characterization. *Artif Cells Blood Substit Immobil Biotechnol* **27**:263–278.

Yu, W.P., Wong, J.P., and Chang, T.M.S. (2000) Sustained drug release characteristics of biodegradable composite poly(d, l)lactic acid poly(l)lactic acid microcapsules containing ciprofloxacin. *Artif Cells Blood Substit Immobil Biotechnol* **28**:39–56.

Yu, Y.T., and Chang, T.M.S. (1980) The effects of polymer-solvent compositions on the formation of collodion membrane artificial cells. *Int J Biomater Med Devices Artif Organs* **8**:273–281.

Yu, Y.T., and Chang, T.M.S. (1981a) Ultrathin lipid-polymer membrane microcapsules containing multienzymes, cofactors and substrates for multistep enzyme reactions. *FEBS Letters (Fed of Euro Biochem Soc)* **125**:94–96.

Yu, Y.T., and Chang, T.M.S. (1981b) Lipid-polymer membrane artificial cells containing multienzyme systems, cofactors and substrates for the removal of ammonia and urea. *Trans Am Soc Artif Intern Organs* **27**:535–538.

Yu, Y.T., and Chang, T.M.S. (1982a) Immobilization of multienzymes and cofactors within lipid- polyamide membrane microcapsules for the multistep conversion of lipophilic and lipophobic substrates. *Enzyme Microb Technol* **4**:327–331.

Yu, Y.T., and Chang, T.M.S. (1982b) Multienzymes and cofactors immobilized within lipid polyamide membrane microcapsules for sequential substrate conversion. *Enzyme Eng* **6**:163–164.

Yuan, Z.Y., and Chang, T.M.S. (1986) Rat microsomes and cytosol immobilized by microencapsulation in artificial cells. *Int J Artif Organs* **9**:63–68.

Zambrowicz, B.P., and Sands, A.T. (2003) Knockouts model the 100 best-selling drugs — will they model the next 100? *Nat Rev Drug Discov* **2**:38–51.

Zhao, L., Smith, J.R., and Eyer, C.L. (1995) Effects of a 100% perfluorooctylbromide emulsion on ischemia/reperfusion injury following cardioplegia. *Artif Cells Blood Substit Immobil Biotechnol* **23**:513–531.

Zheng, Y., and Olsen, K.W. (1996) Tris(3, 5-dibromosalicyl) tricarballylate crosslinked hemoglobin: functional evaluation. *Artif Cells Blood Substit Immobil Biotechnol* **24**:587–598.

Zhou, M.X., and Chang, T.M.S. (1987) Effects of polylactic acid microcapsules containing prostaglandin E2 on the survival rates of grade II coma galactosamine-induced fulminant hepatic failure rats. *Biomater Artif Cells Artif Organs* **15**:549–558.

Zhou, M.X., and Chang, T.M.S. (1988) Control release of prostaglandin E2 from polylactic acid microcapsules, microparticles and modified microparticles. *J Microencapsul* **5**:27–36.

Zolotareva, E., and Chang, T.M.S. Enzyme stability of polyhemoglobin-superoxide dismutase-catalase and effects of temperature (in preparation).

Zolotareva, E., and Chang, T.M.S. Preparation of polyhemglobin containing low concentrations of tetrameric hemoglobin (in preparation).

Zuck, T.F. (1994) Difficulties in demonstrating efficacy of blood substitutes. *Artif Cells Blood Substit Immobil Biotechnol* **22**:945–954.

Zuck, T.F., and Riess, G. (1994) Current status of injectable oxygen carriers. *Crit Rev Clin Lab Sci* **31**:295–324.

Index

1957 report on artificial cells 335–354
50th anniversary of artificial cells 1–11

a

activated charcoal in artificial cells
25–27, 261–266
activation of prodrug 186–188
adsorbent 25–27, 261–266
albumin coated membrane 294, 295,
313
albumin coated artificial cells in
immunoadsorption 294, 295
alginate-polylysine-alginate method
217–219, 366–371
alginate-polylysine-alginate 2 step
method 219–221, 372, 373
alpha hemolysin in artificial cell
membrane 330
amino acids enterorecirculation
148–153
application to oral enzyme
therapy 153
classical theory 148
result shows extensive
recirculation 149
theory of extensive
enterorecirculation of amino
acids 150, 151
antibody coated membrane 313, 314

antigen coated membrane 313, 314
antioxidant 16, 17, 62–92
see polyhemoglobin with
antioxidant enzymes
antioxidant enzymes 16, 17, 62–92
see polyhemoglobin with
antioxidant enzymes
antioxidant properties 78, 79
absorbent spectra 81–83
degradation of hb 81
hydrogen peroxide scavenging 80
iron release 81, 83, 84
lipid peroxidation 84, 85
salicylate hydroxylation 84, 85
superoxide scavenging 79, 80
artificial cell basic features 12–14
artificial cells and molecular biology
see molecular biology
artificial cell hemoperfusion
see hemoperfusion
artificial cells for cell therapy
see cell therapy: basic principles
cell therapy: challenges
cell therapy: examples
cell therapy: genetically
engineered cells
cell therapy: hepatocytes
cell therapy: hepatocytes and stem
cells

cell therapy: microorganisms
artificial
cell therapy: stem cells
artificial cells for computer and
nanoscale robotics
see computer and nanoscale
robotics
artificial cell for drug delivery
see drug delivery
artificial cells for regenerative medicine
see Regenerative medicine: basic
Regenerative medicine:
hepatocytes
Regenerative medicine:
hepatocytes and stem cells
Regenerative medicine: stem cells
artificial cell membrane variations 4, 5,
8–11, 16, 17, 300–312
polymeric membrane artificial
cells 4, 5, 8–11, 300–304
lipid membrane artificial cells 8,
301, 304, 305
liposomes, lipid vesicles, 8, 304,
305
lipid-polymer membrane 8, 9,
6–11, 305–312
lipid-polymer membrane with
channels 9, 306–309
lipid-polymer membrane with
macrocylic carriers 9, 306–309
lipid-polymer membrane with
valinomycin 9, 306–309
NA-K-ATPase in membrane of
artificial cells 9, 310–312
Artificial cells (macro dimension)
preparation: drop method
1957 method 4, 343, 344
alginate-polylysine-alginate
366–371
alginate-polylysine-alginate: novel
2 step method 372, 373
lipid-polymer membrane 364
lipid-polymer membrane
macrocyclic carrier 365
macroporous membrane 374, 375

NA-K-ATPase in membrane 365,
366
polymer membrane 343, 344,
362, 363
Artificial cell (micro dimension)
preparation: emulsion method
1957 method 4, 344–348
cellulose nitrate membrane
355–357
double emulsion methods 361,
362
emulsion methods 4, 344–348,
355–375
lipid-polymer membrane that
retains ATP and NAD(P)H 360
polyamide membrane 357–359
Artificial cell (nano dimension) method:
see nano artificial red blood cells
Artificial cell (nanobiotechnology
assembling) method
see modified hemoglobin
polyhemoglobin
polyhemoglobin with antioxidant
enzymes
polyhemoglobin-asparaginase
polyhemoglobin-catalase-
superoxide dismutase
polyhemoglobin-tyrosinase
artificial cell surface properties
312–314
see surface properties
artificial cell terminologies 299
artificial cells with internal
compartments 8, 21, 22, 330
artificial cells with multienzymes and
cofactor recyling 8–10, 20, 320–327
ATP recycling 320, 321
dextran-nadh recyled 322–325
multienzymes in artificial cells
320–327
NAD(P)H recyling 320, 321
NADH recycling 324, 326, 327
urea into essential amino acids
324, 325

urease, glutamate dehydrogenase
and glucose-6-phosphate
dehydrogenase 321, 322
waste into useful products 324,
325
artificial red blood cells 2–5, 15–18,
31–61
see also
micro dimension artificial red
blood cells
modified hemoglobin
nano artificial red blood cells:
biodegradable
nano artificial red blood cells:
lipid vesicles
polyhemoglobin
polyhemoglobin with antioxidant
enzymes
artificial kidney machine 252
artificial peptide nucleic acid (pna) 332
ATP recyle 19, 20
availability of cells 203, 204

b

background 1, 2
basic features of artificial cells 12–30
bilayer lipid membrane 301, 304, 305
bioadsorbent: see artificial cell
hemoperfusion
biocompatibility 204, 205
bioencapsulation of cells
basic principles 8, 11–25,
195–198
methods of cell encapsulation
206–208, 217–244
see also
cell therapy: challenges 203–209
cell therapy: examples 198–203,
228–243
cell therapy: genetically
engineered cells 199–212
cell therapy: hepatocytes 228–236
cell therapy: hepatocytes and stem
cells 237–242

cell therapy: microorganisms
artificial 213–224
cell therapy: stem cells 245–251
biodegradable nanocapsules 17, 18,
316, 317
biodegradable nanoparticles 17, 18,
316, 317
biodegradable polymeric artificial cells
8–10, 12–14, 17, 18, 29, 110, 129,
316, 317
biotechnology, molecular biology, and
regenerative medicine 7
blood brain barrier 72, 73
blood brain barrier analysis, 90–92
blood substitutes 2–5, 15–18, 31–61
see also
micro dimension artificial red
blood cells
modified hemoglobin
nano artificial red blood cells:
biodegradable
nano artificial red blood cells:
lipid vesicles
polyhemoglobin
polyhemoglobin with antioxidant
enzymes
Bohr effect 117, 118
bone marrow stem cell 225–251
bone marrow stem cells preparation
222
brain edema 73, 74
brain edema analysis 90–92

c

capillary fibre cell encapulation 208
carbonic anhydrase 95–97
catalase
see enzyme therapy for
acatalasemia
polyhemoglobin with antioxidant
enzymes
catalase analysis 87
cell encapsulation
basic principles 8, 11–25, 195–198

see also cell therapy
cell encapsulation methods 206–208,
    217–244
    alginate-polylysine-alginate
        217–219
    alginate-polylysine-alginate 2 step
        month 219–221
    hepatocytes 221, 222
    bone marrow stem cells 222
    genetic engineered E. coli DH5
        cells 222, 223
    artificial cells with
        macromolecular cutoffs 223,
        224
    microorganism 223, 224
cell homogenate 18, 19
cell in membrane 205–207
cell physiology 28
cell protrusion 205–207
cell therapy: basic principles 8, 11–25,
    195–198
cell therapy: challenges
    availability of cells 203, 204
    biocompatibility 204, 205
    capillary fibre cell encapulation
        208
    cell encapsulation methods
        206–208
    cell in membrane 205–207
    cell protrusion 205–207
    mass transfer 204, 205
    oral administration 208, 209
cell therapy: examples
    erythropietin 203
    hepatocytes 199, 202, 203,
        228–243
    islets 198–202
    parathyroid 203
    renal cells 203
cell therapy: genetically engineered
    cells
    general 199, 200, 209
    growth hormone 210, 211
    monoclonal antibodies 210, 211
    factor XI, erythropoietin 210, 211

erythropoietin 210, 211
    neurological disorders 210, 211
    prodrug for tumour 211, 212
    pancreatic CA 211, 212
    tumour 211, 212
cell therapy: hepatocytes
    see regenerative medicine
cell therapy: hepatocytes and stem cells
    see regenerative medicine
cell therapy: microorganisms artificial
    bacterium to recyle NADH 213,
        214
    bactobacillus plantarum to
        remove bile 216
    E. coli DH5 to remove urea 215,
        222, 223
    erwinia herbicola bacterium to
        produce L-DOPA 213, 214
    method of encapsulation 217–224
    lactobacillus delbrueckii to
        remove urea 215, 216
    pseudomonas-pictorum to remove
        cholesterol 214, 215, 223, 224
cell therapy: stem cells
    see regenerative medicine
cellulose nitrate membrane artificial
    cells 4, 8–11, 344–348, 355–357
challenges of cells therapy 203–209
characterisation of nano artificial rbc
    113–115
chemical information processing
    systems 332
circulation time measruement 88, 89
circulation time of nano artificial rbc
    121–127
clearance of artificial cell
    hemoperfusion 263, 264
clinical treatment of poisoning
    252–256, 266–277
clinical trials in kidney failure
    252–257, 276–288
clinical trials of immmunosorbents
    252–254, 258, 294, 295
clinical use 5, 6
cofactor recycling 19, 20

comparision of modified hb 38–42
compartmentalization in artificial cell
8, 21, 22, 330
complement activation as screening test
48–52
    basic principle 49, 50
    use in research 50, 51
    use in industrial production 50, 51
    correlation to clinical symptoms
    51, 52
    clinical trials in human 51, 52
    C3a in-vitro screening test —
    blood from finger pricks 59
    C3a in-vitro screening test —
    human plasma 58, 59
    human plasma 48, 49
    method 58, 59
computer and nanoscale robotics 331,
332
    artificial peptide nucleic acid
    (PNA) 332
    chemical information processing
    systems 332
    computer, self-repairing 331, 332
    nanoscale robotics 331, 332
    new generation computer system
    331, 332
    programmable artificial cell
    evolution 331, 332
computer, self-reparing 331, 332
conjoint hemoperfusion-hemodialysis
283–286
conjugated hemoglobin 8–11, 16, 18,
34, 36, 38
    see modified hemoglobin
conjugation of polymer with proteins
8–11, 16, 18, 34, 36, 38, 170, 171
crosslinking asparaginase to
hemoglobin 168, 169
crosslinking of asparaginase to PEG
170, 171
cytosol in artificial cells 328

d

dextran-hemoglobin 9 see also
    conjugated hemoglobin
dextran-nadh recyled 322–325
dimensions 13, 14
DNA plasmid in artificial cell 329, 331
DNA templates in artificial cell 329
DNA-mimics in artificial cell 330
donor blood 3, 44
double emulsion methods 361, 362
drop method for large artificial cells 4,
5, 8, 9, 343, 344, 362–375
drug delivery 17–19, 28–30, 314–320
    basic principles 314, 315
    biodegradable nanocapsules 17,
    18, 316, 317
    biodegradable nanoparticles 17,
    18, 316, 317
    biodegradable polymeric artificial
    cells 8, 17, 18, 316, 317
    lipid vesicles 8–10, 18, 19,
    316–318
    liposomes 8–10, 18, 19, 316–318
    nanocapsules 18, 316, 317
    nanoparticles 18, 316, 317
    PEG-lipid vesicles 318, 319
    polymeric semipermeable
    microcapsules 316
    polymersomes 319, 320

e

early artificial cells 3–5, 7–11, 12–14,
335–354
ECG and tetrameric hemoglobin 40–42
embolism prevention 262
emulsion methods for artificial cells
334–348, 355–375
emulsion phase separation method
344–348
enterorecirculation of amino acids
148–153
    application to oral enzyme
    therapy 153

classical theory 148
result shows extensive
    recirculation 149
theory of extensive
    enterorecirculation of amino
    acids 150, 151
enzyme artificial cells for activating
    prodrugs 186–188
    clinical trial results 187, 188
enzyme artificial cells for tumours
    160–192
enzyme therapy basis 19, 20
enzyme therapy for acatalasemia
    133–135
    acatalasemic mice 135
    antioxidant catalase artificial cells
        133–141
    antioxidant effect of catalase
        artificial cells 135–137
    artificial cell immunoisolate
        catalase 137–140
    catalase antibody excluded by
        artificial cells 138–140
    catalase artificial cells for
        perfusion 134, 135
    catalase artificial cells not
        antigenic 137, 138
    catalase solution antigenic 137,
        138
    immunological studies 137–139
    implantation of catalase artificial
        cells 133–135
    perborate removal 135–137
enzyme therapy for Lesch-Nyhan
    disease 142–147
    cerebro-spinal fluid 144, 145
    enzyme kinetics 144, 145
    hypoxanthine 144, 145
    inosine 144, 145
    oral administration 142–147
    uric acid 144, 145
    xanthin oxidase artificial cells
        142–147
    xanthine 144, 145

enzyme therapy lymphosarcoma
    160–171
    6C3HED lymphosarcoma in mice
        162–168
    asparaginase activities 165, 166
    asparaginase retained by artificial
        cells 166, 167
    crosslinking asparaginase to
        hemoglobin 168, 169
    crosslinking of asparaginase to
        PEG 170, 171
    plasma asparagine level 164, 165
enzyme therapy for melanoma
    172–179
    B16F10 melanoma bearing mice
        model 191
    B16F10 melanoma cells culture
        191
    enzyme kinetics 184, 185
    method of preparation 189, 192
    oral administration 183–191
    polyhemoglobin-tyrosinase
        172–178
    tyrosinase artificial cells 179–186
enzyme therapy for phenylketonuria
    147–159
    application to oral enzyme
        therapy 153
    chemical induced
        phenylketonuria rat model 154
    congenital phenylketonuria mice
        model 157–159
    effects on intestinal phenylalanine
        153–155
    effects on cerebrospinal fluid
        153–155
    effects on growth 155–157
    effects on plasma PHE 153–155
    enterorecirculation of amino acids
        148, 150, 151
    enzyme replacement therapy 147,
        148
    genetic defect in phenylalanine
        hydroxylase 147–159

oral enzyme artificial cells to
deplete other amino acids 158,
159
oral pal artificial cell in PKU rats
152, 153
result shows extensive
recirculation 149
enzymes 19, 20
enzymes and multienzymes 8–10, 20,
118, 119, 121, 320–327
equivalent pore radius 302
erythropietin 203, 210, 211
ether evaporation method 4, 344–348,
355–357

f

factor XI 210, 211
fate of polylactide membrane 115–117
ferrylhemoglobin measurement 87, 88
first artificial red blood cells 2–5
first report on artificial cells 335–354
first routine clinical use of artificial cells
5, 6, 255, 256, 266–276
future perspectives 299–334

g

galactosamine hepatic failure rats 228,
229, 291–294
genetically engineered cells 24, 25,
222, 223
glucose-6-phosphate dehydrogenase,
glutamate dehydrogenase and urease
321, 322
glutamate dehydrogenase, urease and
glucose-6-phosphate dehydrogenase
321, 322
growth hormone 210, 211
Gunn rat model 240, 241
Gunn rats with hyperbilirubinemia
229–231

h

hemodialysis-hemoperfusion 283–286
hemoglobin toxicity 32
hemoglobin based red blood cell
substitutes
see
micro dimension artificial red
blood cells
modified hemoglobin
nano artificial red blood cells:
biodegradable
nano artificial red blood cells:
lipid vesicles
polyhemoglobin
polyhemoglobin with antioxidant
enzymes
hemoperfusion 6, 8, 9, 252–298
activated charcoal 261
clearance 263, 264
development and clinical trials
252–258
embolism prevented 262
hemoperfusion-hemodialysis
283–286
hemoperfusion-ultrafiltration
286–288
hydrodynamic 265, 266
middle molecules 283, 284, 287,
288, 290
permeability and transport
characteristics 258–261
platelets removal prevented
262–264
principle 252–254, 258–261
procedure for hemoperfusion 297,
298
procedures for hemoperfusion
device 295–298
hemoperfusion in hepatic coma
288–294
cross-circulation, liver perfusion,
blood exchange 291–294
galactosamine liver failure rats
291–294

hemoperfusion in liver failure 288–294
hepatic coma related chemicals 290
hepatic coma 257, 258
middle molecules 290
recovery in hepatic coma 288–290
results around the world in hepatic coma 290, 291
hemoperfusion in immunology 252–254, 258, 294, 295
immunoadsorption 294, 295
immunoadsorption 258
hemoperfusion in kidney failure 252–257, 276–288
hemoperfusion-hemodialysis 283–286
hemoperfusion-ultrafiltration 286–288
kidney failure patients 276–283
kidney failure 256, 257
middle molecules 283, 284, 287, 288
hemoperfusion in poisoning 252–256, 266, 277
adult patients 255–277
pediatric patient theophylline overdose 273–275
treating poisoning around the world 275, 276
hemoperfusion-hemodialysis 283–286
hemoperfusion-ultrafiltration 286–288
hemorrhagic shock and cerebral ischemia 90
hemorrhagic shock and intestine 67, 68
hemorrhagic shock rat model 52–54, 56–59
experimental designs 53, 54
method 59, 60
result 52, 53
hemorrhagic shock stroke rat model 67, 69–75
hepatectomized rats 241–243
hepatic coma recovery 291–294

hepatic coma related chemicals 290
hepatocyte viability 233–236
hepatocytes 199, 202, 203, 228–251
hepatocytes plus stem cells 237–241
hepatocytes preparation 221, 222
Hill coefficient 117, 118
historical milestone 7–11
hydrodynamic in hemoperfusion 265, 266
hydrogen peroxide scavenging 87
hyperbilirubinemia 229–231

i

ideas first proposed 7–11
immunoisolation of encapsulated cells 8, 11–25, 195–198, 231–236
immunoisolation of enzyme in artificial cells 19, 20, 97, 98, 137–140
immunorejection 231–233
immunosorbent 252–254, 258, 294, 295
intestine ischemia-reperfusion 67–69
hemorrhagic shock and intestine 67, 68
reperfusion on ischemic intestine 67–69
intracellular compartments in artificial cell 8, 21, 22, 330
intramolecularly crosslinked hemoglobin 35, 36, 38
see also modified hemoglobin
iron release measurement 88
ischemia reperfusion 62–92
see polyhemoglobin with antioxidant enzymes
islets in artificial cells 8, 11–25, 198–202

k

kidney failure 252–257, 276–288

l

lipid membrane artificial cells 8–10,
    18, 19, 27–29, 104–110, 301,
    304–309, 360, 365
lipid vesicles 8–10, 18, 19, 27–29,
    104–110, 301, 304–309, 360, 365
lipid-polymer membrane 27, 28,
    305–312, 364
lipid-polymer membrane with channels
    28, 306–309, 365
lipid-polymer membrane with
    macrocylic carrier 28, 306–309, 365
lipid-polymer membrane with
    valinomycin 306–309, 365
lipid-polymer retains and NAD(P)H
    360, 361
liposomes 8–10, 18, 19, 27–29,
    104–110, 301, 304–309, 360, 365
liposomes evolved into lipid membrane
    artificial cells 316–318
liver support 288–294
living artificial cell 329–331
lymphosarcoma 6C3HED in mice
    162–168

m

macro dimension 13, 14
macrocylic carrier 306–309
macromolecular cut off membrane
    223, 224, 374, 375
macroporous membrane 223, 224,
    374, 375
magnetic material 22
mass transfer 204, 205
melanoma B16-F10 bearing mice
    model 191
melanoma B16-F10 cells culture 191
melanoma enzyme therapy 172–179
membrane model systems 4, 5, 8–11,
    16–18, 300–312
membrane variations 4, 5, 8–11,
    16–18, 223, 224, 300–312

methemoglobin reductase system 19,
    119, 120
micro dimension 13, 14
micron dimension artificial red blood
    cells 93–103
    oxygen dissociation curive 95, 96
    carbonic anhydrase 95–97
    catalase 97
    immunological studies 97, 98
    polymeric membrane artificial rbc
        98, 99
    neuraminic acid and circulation
        time 99–101
    surface charge and circulation
        time 101, 102
    sulfonated nylon membrane 101,
        102
    polysaccharide incorporation 102,
        103
microorganisms 200, 213–216,
    222–224
microsomes in artificial cells 328
middle molecules 287, 288, 290
middle molecules and hemoperfusion
    283, 284
modified hemoglobin 2–5, 15–18,
    31–61
    comparision of modified
        hemoglobin 38–42
    conjugated hemoglobin 8–10, 16,
        18, 34, 36, 38
    ECG and tetrameric hemoglobin
        40–42
    hemorrhagic shock rat model
        52–54, 56–59
    intramolecularly crosslinked hb
        35, 36, 38
    polyhemoglobin in clinical trials
        36–38
    polyhemoglobin: see
        polyhemoglobin
    present status 35, 36
    recombinant human hb 35, 36, 38
    regional differences 53–56
    regulatory issues 53–56

safety, efficacy 53–56
tetrameric hemoglobin on
    vasoactivity and ECG 40–42
tetrameric hemoglobin <2% 43
theories of vasopressor effects 39
types 33–42
vasoactivity and tetrameric
    hemoglobin 40–42
vasopressor effects 39–43
modified hemoglobin vasoactivity
    39–43
    theories 39
    tetrameric hemoglobin and
        vasoactivity 40–42
    tetrameric hemoglobin and ECG
        40–42
    vasoactivity and tetrameric
        hemoglobin 40–42
    ECG and tetrameric hemoglobin
        40–42
    polyhemoglobin with <2%
        tetramerichemoglobin 43
molecular biology
    "living" artificial cell 329–331
    alpha hemolysin 330
    compartmentalization 8, 21, 22,
        330
    cytosol in artificial cells 328
    DNA plasmid 329, 331
    DNA templates 329
    DNA-mimics 330
    intracellular compartments 8, 21,
        22, 330
    microsomes in artificial cells 328
    plasmid 331
    polymerases in artificial cells
        328–331
    reproduce and divide 330
    ribosomal translation 329
    ribosomes 329
    ribosomes in artificial cells
        328–331
    self-replicating ribozyme 330
    simple RNA enzyme 329

T7 RNA polymerase and templates
    329
transcription/translocation in
    artificial cells 328–331
molecular dimension 13, 14
monoclonal antibodies 210, 211
multienzyme systems with cofactor
    recycling 8–10, 19, 20, 320–327
multienzymes in artificial cells 8–10,
    19, 20, 320–327

n

NA-K-ATPase in membrane of artificial
    cells 9, 28, 310–312
NA-K-ATPase membrane method 365,
    366
nano artificial red blood cells
    biodegradable 9, 10, 110–129
        amount of PLA 116
        Bohr effect 117, 118
        electromicroscopic appearance
            113
        fate of PLA membrane 115–117
        hemoglobin, polymer, lipid,
            specific gravity, 114, 115
        Hill coefficient 117, 118
        nano artificial rbc characterisation
            of 113–115
        nanoencapsulating efficiency
            hemoglobin 117
        oxygen affinity 117, 118
        PEG-PLA artificial rbc circulation
            time 121–127
        PEG-PLA copolymers 121–127
        polylesters as biodegradable
            polymer membrane 112, 113
        preparation of PEG-PLA nano
            artificial rbc 129
        preparation of PLA nano artificial
            rbc 128, 129
        relevance in human 127
        required properties 111, 112
        safty and efficacy 115–121
        size distribution 113, 114

steady shear viscosity 115
enzymes and multienzymes 118,
119, 121
methb reductase system 119, 120
reducing agents for methb 120
circulation time of nano artificial
rbc 121–127
nano artificial red blood cells: lipid
vesicles 8–10, 104–107
circulation time after infusion 107,
108
composition and preparation 105,
106
exchange transfusion 109
exchange transfusion 109
hemodilution 109
historical 104–105
nanobiotechnology and artificial cells
15–17
see nano artificial red blood cells
nanocapsules
nanoparticles
nanobiotechnology and assembling of
hemoglobin 16, 17
see modified hemoglobin
polyhemoglobin
nanobiotechnology and assembling of
hemoglobin with other enzymes 17,
18
see polyhemoglobin-asparaginase
see polyhemoglobin-tyrosinase
nanobiotechnology and nano artificial
red blood cells 15–17
see nano artificial red blood cells
nanobiotechnology and oxygen carrier
2–5, 15–18, 31–61
see
micro dimension artificial red
blood cells
modified hemoglobin
nano artificial red blood cells:
biodegradable
nano artificial red blood cells:
lipid vesicles
polyhemoglobin

polyhemoglobin with antioxidant
enzymes
nanobiotechnology and
polyhemoglobin
see polyhemoglobin
nanobiotechnology and
polyhemoglobin with antioxidant
enzymes
see polyhemoglobin with
anitoxidant enzymes
nano dimension 13, 14
nanobiosensors 312
nanocapsules 316, 317
nanoencapsulation efficiency for
hemoglobin 117
nanoparticles 29, 316, 317
nanoscale robotics based on artificial
cells 331, 332
nanotechnology 15–18
negative charge membrane 312
neuraminic acid and circulation time
99–101
neurological disorders 210, 211
new generation computer system based
on artificial cells 331, 332

O

oral administration 208, 209
oxygen affinity 117, 118
oxidants 16, 17, 62–92
see polyhemoglobin with
antioxidant enzymes
oxygen carrier with antioxidant
enzymes 16, 17, 62–92
see polyhemoglobin with
antioxidant enzymes
oxygen carriers 2–5, 15–18, 31–61
see
micro dimension artificial red
blood cells
modified hemoglobin
nano artificial red blood cells:
biodegradable

nano artificial red blood cells:
    lipid vesicles
polyhemoglobin
polyhemoglobin with antioxidant
    enzymes
oxygen dissociation curive 95, 96
oxygen radicals 16, 17, 62–92
    see polyhemoglobin with
    antioxidant enzymes
oxygen radical measurement 89, 90

p

pancreatic CA and activation of
    prodrug 211, 212
parathyroid in artificial cells 203
pediatric theophylline overdose
    hemoperfusion 273–275
PEG and membrane 313, 314
PEG-asparaginase 162, 170, 171
PEG-hemoglobin
    see modified hemoglobin
PEG-lipid vesicles 318, 319
PEG-PLA artificial red blood cell
    circulation time 121–127
PEG-PLA copolymers for nano artificial
    red blood cell 121–127
permeability and transport
    characteristics 258–260
permeability range 303
plasmid in artificial cell 331
platelets protection 262–264
poisoning 252–256, 266–277
polyamide membrane artificial cells of
    micro dimensions 357–359
polyhemoglobin 8–10, 15–18, 33, 34,
    36, 38–48
    see also modified hemoglobin
    antibody titers 45, 46
    antigenicity 45, 46
    comparision with other modified
        hemoglobin 38–42
    donor blood vs polyhemoglobin
        44

ECG and tetrameric hemoglobin
    40–42
hemorrhagic shock model 52–54,
    56–59
immunogenicity 45–48
infusion into immunized rats
    46–48
polyhb in clinical trials 36–38
polyhemoglobin cf donor blood
    44
tetrameric hemoglobin <2% 43
tetrameric hemoglobin on
    vasoactivity and ECG 40–42
theories of vasopressor effects 39
vasoactivity and tetrameric
    hemoglobin 40–42
vasopressor effects 39–43
polyhemoglobin methods 56–61
    C3a in-vitro screening test —
        blood from finger pricks 59
    C3a in-vitro screening test —
        human plasma 58, 59
    exchange transfusion method 61
    hemorrhagic shock method 59, 60
    low tetrameric hemoglobin
        method 56–58
    methb prevention 58
    molecular weight distribution 58
    preparative chromatography 57
polyhemoglobin-asparaginase 160–171
polyhemoglobin-catalase-superoxide
    dismutase
    see polyhemoglobin with
        antioxidant enzymes
polyhemoglobin-tyrosinase for
    melanoma 172–179
polyhemoglobin with antioxidant
    enzymes: in-vitro 16, 17, 86–92
    characterization 75–92
    in vitro characterization 75, 76
    oxygen dissociation curve 77, 78
    methemoglobin 77, 78
    antioxidant properties 78, 79
    absorbent spectra 81–83
    degradation of hemoglobin 81

hydrogen peroxide scavenging 80
iron release 81, 83, 84
lipid peroxidation 84, 85
polyhemoglobin-catalase-
    superoxide dismutase 16, 17,
    86–92
salicylate hydroxylation 84, 85
superoxide scavenging 79, 80
polyhemolgobin with antioxidant
    enzymes: in vivo 16, 17, 63–92
    blood brain barrier 72, 73
    brain edema 73, 74
    cat activity in plasma 65, 66
    discussions 66
    duration of hemorrhagic shock
        69–71
    duration of stroke and reperfusion
        injuries 71, 72
    hemorrhagic shock and intestine
        67, 68
    hemorrhagic shock and stroke rat
        model 67, 69–75
    intestine ischemia-reperfusion
        67–69
    intravenous injections 64–66
    plasma hb concentration 64, 65
    reperfusion on ischemic intestine
        67–69
    superoxide dismutase activity in
        plasma 65
polyhemoglobin with antioxidant
    enzymes: procedures 86–92
    blood brain barrier analysis 90–92
    brain edema analysis 90–92
    catalase analysis 87
    circulation time measurement 88,
        89
    ferrylhb measurement 87, 88
    hemorrhagic shock and cerebral
        ischemia model method 90
    hydrogen peroxide scavenging 87
    intestinal ischemia-reperfusion rat
        model method 89–92
    iron release measurement 88

oxygen radical measurementl 89,
    90
preparation 75, 85, 86
superoxide dismutase/catalase
    ratio 87
superoxide dismutase analysis 86,
    87
transient global cerebral
    ischemia-reperfusion rat model
    method 90
PEG-PLA membrane nano artificial red
    blood cells 9, 10, 12–14, 18,
    121–129
polylactide membrane nano artificial
    cells 9, 10, 12–14, 18, 110–129
polylesters as biodegradable polymer
    membrane 112, 113
polymer membrane method 343, 344,
    362, 363
polymerases in artificial cells 328–331
polymeric membrane artificial cells
    300–304
polymeric semipermeable
    microcapsules 28
polymersomes: polymeric membrane
    artificial cells 319, 320
polysaccharide incorporation 102, 103
polysacharide and membrane 313, 314
porosity of polymeric membrane
    artificial cells 300–304
preclinical studies of hemoperfusion
    266
predictions 333, 334
preparation of PLA nano artificial rbc
    128, 129
principle of hemoperfusion 252–254,
    258–261
procedure for hemoperfusion 297, 298
procedures for hemoperfusion device
    295–297
prodrug for tumour 211, 212
prodrugs activation 186–188
programmable artificial cell evolution
    331, 332

r

recombinant human hemoglobin 35, 36, 38
recovery of consciousness in hepatic coma 291–294
recycling NADH in lipid-polymer artificial cells 324, 326, 327
recycling of ATP in artificial cell 320, 321
recycling of NAD(P)H in artificial cell 320, 321
red blood cell problems 2–5
reducing agents for methemoglobin 120
Regenerative medicine: basic
    design of study 227
    examples 226
    principle 225, 226
Regenerative medicine: hepatocytes
    aggregation after implantation 233, 234
    galactosamine fulminant hepatic failure 228, 229
    Gunn rats with hyperbilirubinemia 229–231
    hepatocyte viability after implantation 234–236
    immunoisolation 231–236
    protect immunorejection 231–233
    recovery of hepatocyte viability 233
Regenerative medicine: hepatocytes and stem cells
    factors to be considered 242, 243
    Gunn rat model 240, 241
    hepatectomized rats 241–243
    in vitro viability of encapsulated cells 237–239
    in vivo viability of encapsulated cells 237–239
    survival after encapsulated hepatocytes 241, 242
    viability of in culture 237
Regenerative medicine: stem cells

blood chemistry 246
immunocytochemistry 248, 249
laparotomy and histology 247–249
PAS glycogen stain 248, 249
plasma hepatic growth factor (hgf) levels 246, 247
possible mechanisms 248–251
remnant liver weight 245, 246
survival of hepatectomized rats 244, 245
viability of stem cells in culture 244
regional differences 53–56
regulatory issues 53–56
relevance in human 127
renal cells in artificial cells 203
replicating biological cells 332, 333
replicating nature 332, 333
reproduce and divide 330
requirements for artificial red blood cell 111, 112
ribosomal translation in artificial cell 329
ribosomes in artificial cell 328–331
routine uses of hemoperfusion 275, 276

s

safety, efficacy 53–56
self-replicating ribozyme in artificial cell 330
simple RNA enzyme in artificial cell 329
spray dry method 340–343
stem cell 23, 24, 200, 212, 213, 225–251
stem cells in regeneration medicine 225–251
stroke-hemorrhagic shock rat model 67, 69–75
    blood brain barrier 72, 73
    brain edema 73, 74
    duration of hemorrhagic shock 69–71

duration of stroke and reperfusion injuries 71, 72
rat model 71
solutions on reperfusion injuries 72
substrate-dependent tumours 160–185, 189–194
sulfonated nylon membrane 101, 102
sulfonated polyamide 312
superoxide dismutase
see polyhemoglobin with antioxidant enzymes
surface charge and circulation time 101, 102
surface properties of artificial cell 312–314
albumin 313
antibody 313, 314
antigen 313, 314
negative charge in collodion 312
negative charge in sulfonated polyamide 312
PEG 313, 314
polysacharide 313, 314
survival of hepatectomized rats 241–243

t

T7 RNA polymerase and templates in artificial cell 329
terminologies for artificial cells 299
tetrameric hemoglobin and ECG 40–42
tetrameric hemoglobin and vasoactivity 40–42
tetrameric hemoglobin <2% 43

theophylline overdose and hemoperfusion 273–275
time line of ideas 7–11
transcription/translocation in artificial cells 328–331
transient global cerebral ischemia 90
treatment of poisoned patients 255, 256, 266–277
tumour 211, 212

u

ultrafiltration- hemoperfusion 286–288
urea converted into essential amino acids 324, 325
urease, glutamate dehydrogenase and glucose-6-phosphate dehydrogenase 321, 322
uremic metabolites and hemoperfusion 283
uremic symptoms removal 285

v

valinomycin 306–309
vasoactivity and tetrameric hemoglobin 40–42
vasopressor effects 39–43
vasopressor effects theory 39
viability of encapsulated cells 237–239

w

waste into useful products 324, 325